Structured Decision Making

Wildlife Management and Conservation | Paul R. Krausman, Series Editor

Structured Decision Making

Case Studies in Natural Resource Management

EDITED BY

MICHAEL C. RUNGE, SARAH J. CONVERSE, JAMES E. LYONS, AND DAVID R. SMITH

Published in Association with *THE WILDLIFE SOCIETY*

JOHNS HOPKINS UNIVERSITY PRESS | BALTIMORE

Printed in the United States of America on acid-free paper
9 8 7 6 5 4 3 2 1

Johns Hopkins University Press
2715 North Charles Street
Baltimore, Maryland 21218-4363
www.press.jhu.edu

Library of Congress Cataloging-in-Publication Data

Names: Runge, Michael Carl, 1967– editor.
Title: Structured decision making : case studies in natural resource management / edited by Michael C. Runge, Sarah J. Converse, James E. Lyons, and David R. Smith.
Description: Baltimore : Johns Hopkins University Press, 2020. | Series: Wildlife management and conservation | Includes bibliographical references and index.
Identifiers: LCCN 2019026094 | ISBN 9781421437569 (hardcover) | ISBN 9781421437576 (ebook)
Subjects: LCSH: Conservation of natural resources—Decision making. | Wildlife management—Decision making.
Classification: LCC S944.5.D42 S77 2020 | DDC 639.9—dc23
LC record available at https://lccn.loc.gov/2019026094

A catalog record for this book is available from the British Library.

Special discounts are available for bulk purchases of this book. For more information, please contact Special Sales at specialsales@press.jhu.edu.

Johns Hopkins University Press uses environmentally friendly book materials, including recycled text paper that is composed of at least 30 percent post-consumer waste, whenever possible.

To the staff at the National Conservation Training Center,
especially Donna C. Brewer, who passionately and tirelessly nurtured the
community of practice that undertook these case studies

Contents

Contributors

Taber D. Allison
American Wind Wildlife Institute
1990 K Street NW, Suite 620
Washington, DC 20006

Larissa L. Bailey
Colorado State University
Department of Fish, Wildlife, and Conservation Biology
1474 Campus Delivery
Fort Collins, CO 80523

Ellen A. Bean
Raleigh, NC

Clint W. Boal
US Geological Survey
Texas Cooperative Fish and Wildlife Research Unit
Texas Tech University
Lubbock, TX 79409

Gregory Breese
US Fish and Wildlife Service
2610 Whitehall Neck Road
Smyrna, DE 19977

Stefano Canessa
Department of Pathology, Bacteriology and Avian Diseases
Faculty of Veterinary Medicine
Ghent University
Merelbeke 9820
Belgium

Jean Fitts Cochrane
US Fish and Wildlife Service, retired
29 Wood Mountain Road
Grand Marais, MN 55604

Sarah J. Converse
US Geological Survey
Washington Cooperative Fish and Wildlife Research Unit
University of Washington
Seattle, WA 98195

Cami S. Dixon
US Fish and Wildlife Service
National Wildlife Refuge System
5924 19th Street SE
Woodworth, ND 58496

John G. Ewen
Zoological Society of London
Regents Park
London NW1 4RY
United Kingdom

Christelle Ferrière
Mauritian Wildlife Foundation
Grannum Road
Vacoas
Mauritius

Jill J. Gannon
US Fish and Wildlife Service
National Wildlife Refuge System
1201 Oakridge Drive, Suite 320
Fort Collins, CO 80525

Beth Gardner
School of Environmental and Forest Sciences
University of Washington
Seattle, WA 98195

Adam W. Green
Bird Conservancy of the Rockies
230 Cherry Street, Suite 150
Fort Collins, CO 80521

Justin A. Gude
Montana Fish, Wildlife, and Parks
1420 East 6th Avenue
Helena, MT 59620

Victoria M. Hunt
Chicago Botanic Garden
Glencoe, IL 60022

Kevin S. Kalasz
US Fish and Wildlife Service
28950 Watson Boulevard
Big Pine Key, FL 33043

Melinda G. Knutson
US Fish and Wildlife Service, retired
Refuges, Midwest Region
2630 Fanta Reed Road
La Crosse, WI 54650

Jim Kraus
US Fish and Wildlife Service, retired
Hakalau Forest National Wildlife Refuge
60 Nowelo Street, Suite 100
Hilo, HI 96720

Graham E. Long
Compass Resource Management
302-788 Beatty Street
Vancouver, BC V6B 2M1
Canada

Eric V. Lonsdorf
Institute on the Environment
University of Minnesota
1954 Buford Avenue
St. Paul, MN 55108

James E. Lyons
US Geological Survey Patuxent Wildlife Research Center
12100 Beech Forest Road
Laurel, MD 20708

Conor P. McGowan
US Geological Survey
Alabama Cooperative Fish and Wildlife Research Unit
Auburn University
Auburn, AL 36849

Sarah E. McRae
US Fish and Wildlife Service
PO Box 33726
Raleigh, NC 27636

Michael S. Mitchell
US Geological Survey
Montana Cooperative Wildlife Research Unit
University of Montana
Missoula, MT 59812

Clinton T. Moore
US Geological Survey
Georgia Cooperative Fish and Wildlife Research Unit
180 East Green Street
Athens, GA 30602

Joslin L. Moore
School of Biological Sciences
Monash University
Clayton Campus VIC 3800
Australia

Steven Morey
US Fish and Wildlife Service
911 NE 11th Avenue
Portland, OR 97232

Dan W. Ohlson
Compass Resource Management
302-788 Beatty Street
Vancouver, BC V6B 2M1
Canada

Charlie Pascoe
Parks Victoria
62-68 Ovens Street
PO Box 1084
Wangaratta, VIC 3677
Australia

Andrew J. Paul
Alberta Ministry of Environment and Parks
213 1 Street West
Cochrane, AB T4C 1A5
Canada

Eben H. Paxton
US Geological Survey Pacific Islands Ecosystems Research Center
Crater Rim Drive, Building 344
PO Box 44
Hawaii National Park, HI 96718

Lori B. Pruitt
US Fish and Wildlife Service
620 S. Walker Street
Bloomington, IN 47403

Michael C. Runge
US Geological Survey Patuxent Wildlife Research Center
12100 Beech Forest Road
Laurel, MD 20708

Sarah N. Sells
University of Montana
Montana Cooperative Wildlife Research Unit
Missoula, MT 59812

Terry L. Shaffer
US Geological Survey Northern Prairie Wildlife Research Center
8711 37th Street SE
Jamestown, ND 58401

Stephanie Slade
Department of Environment and Science
GPO Box 2454
Brisbane QLD 4001
Australia

David R. Smith
US Geological Survey Leetown Science Center
11649 Leetown Road
Kearneysville, WV 25430

Jennifer A. Szymanski
US Fish and Wildlife Service
2630 Fanta Reed Road
La Crosse, WI 54603

Terry Walshe
School of Biosciences
University of Melbourne
Parkville VIC 3010
Australia

Nicolas Zuël
Ebony Forest Ltd.
Seven Coloured Earth Road
Chamarel 90409
Mauritius

Foreword

Public and private institutions across the globe are searching for more effective ways to manage natural resources in the face of increasing demands for the benefits the natural world provides. Out of this need, there seems to have arisen an infinity of paradigms for managing natural resources. It is no wonder so many natural resource managers either stick to a narrow suite of tools or approaches or reject the entire field as too complex and quantitative to be useful in the real world. This book demystifies the process of applying decision analysis thinking to a wide range of natural resource management problems by way of well-thought-out examples.

Natural resource management is just about deciding how much to do of what action, or bundle of actions, in what place, and when. What shall I do, where, when, and why? All the different approaches and paradigms boil down to just 30 1-syllable words:

- What do we want? (objectives)
- What can we do? (actions)
- How does what we do change what we want? (models)
- Pick the best things to do to get what we want (algorithms)
- Do
- Learn

So, it is remarkable we make the process seem so complex and inscrutable. The most common way I see people making a mess of things is when the action becomes the objective. This often happens with invasive species management—people start thinking that killing the invasive species is the objective, rather than just an intermediate outcome that leads to more fundamental objectives, like securing species. But like all good processes, there are many ways to fail and only a few ways to get it right. This book will help you get it right.

One approach to improving natural resource management is to train people in more abstract theory, and that works for a few people. But most people learn abstract concepts from studying examples. This is what this book does—it provides a wide variety of well-worked case studies that illustrate the fundamental power of structured decision making in all kinds of natural resource management settings.

Through the case studies, the authors not only tackle a wide range of problems—from invasive species and disease management to single-species, water, and habitat management—they also address some of the more conceptually troubling issues. Risk and uncertainty are just 2 of these issues. Many a time I have been told by a manager that decision analysis won't work because there are too many random factors or there is too much uncertainty about our understanding of the system. The book contains several case studies around each of the issues of risk and uncertainty, hopefully enough to satisfy the skeptics. The case studies on uncertainty lead to discussions about my favorite topic, value of information analysis. How much time and effort should we spend gathering more information before we make a decision? Usually not much, it seems.

The authors and editors are especially well equipped to pull together the case studies. All of them have a plethora of experience dealing with, and adeptly solving, real-world problems. They know the pitfalls, and they have been able to explain some of the more arcane concepts, such as trade-offs, in plain English. It is the experience the authors have in teaching and applying their craft that makes this a very practical and readable book. This is a profoundly important book that should form part of every program in natural resource management.

Hugh P. Possingham
Chief Scientist
The Nature Conservancy
Australian Research Council Laureate Fellow
The University of Queensland

INTRODUCTION

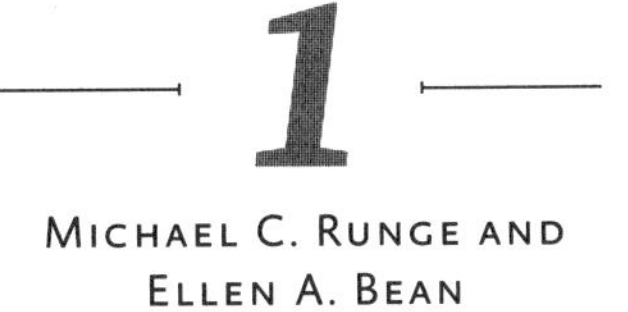

Michael C. Runge and
Ellen A. Bean

Decision Analysis for Managing Public Natural Resources

The Grand Canyon is as profound and emblematic as any natural feature in the world. It has been revered for millennia by many Native American tribes. Tourists visit by the millions to experience the vast landscape, hike the dramatic canyons, raft the famous river, and fish the tailwaters below the Glen Canyon Dam. The Colorado River, which carved the Grand Canyon, carries snowmelt from the Rocky Mountains to the Gulf of California and, along the way, provides water for 40 million people. Glen Canyon Dam, the fourth-tallest dam in the United States, annually provides over 4 billion kilowatt-hours of renewable hydroelectric power to around 5 million users in the southwestern states (US Department of the Interior 2016a). Unique plants and animals fill the buttes, canyons, river, and tributaries of the Grand Canyon. This is a remarkable place.

The management of the Grand Canyon and the Glen Canyon Dam affects all of these resources, in ways we both do and do not understand. The construction of the dam itself fundamentally changed the Colorado River ecosystem, altering the distribution and abundance of plants and animals, even bringing some species to the brink of extinction, significantly changing the flow of the river, and greatly reducing the movement of sediment downstream (Schmidt et al. 1998). Management of the dam, along with other related actions, can mitigate some of the effects on the ecosystem but often simultaneously exacerbates other effects. Like any other natural resource management question, the decisions can be contentious. How can decision makers approach such significant issues, weighing the effects of their choices on so many important outcomes, grappling with the limits of knowledge, all while transparently engaging with a large audience of stakeholders?

In December 2016, the secretary of the interior signed a new Record of Decision regarding management of the Glen Canyon Dam, based on an environmental impact statement (EIS) developed over the preceding 3 years (US Department of the Interior 2016b). To navigate the complexity of the decision, the 2 lead agencies, the Bureau of Reclamation and the National Park Service, used a number of tools from the field of decision analysis to evaluate the effects of potential actions on an array of outcomes and to understand the consequences of scientific uncertainty (Runge et al. 2015). These tools could not remove the contentiousness from the decision, but they made transparent the nature of the trade-offs and the effects of uncertainty, allowing focused and well-informed deliberations among the decision makers and stakeholders.

Purpose of This Book

Examples like the Grand Canyon story are becoming more common—agencies and stakeholders are increasingly using decision tools to help navigate complex decisions regarding natural resources. Several books and many papers have made the case for the use of decision analysis in natural resource management settings and have described the tools available (Conroy and Peterson 2013; Gregory et al. 2012; Williams et al. 2002), but only a few applications have been described. Thus, many agencies understand why decision analysis is valuable but wonder how to implement it. The primary purpose of this book is to show what decision analysis looks like in practice for natural resource management, to guide decision makers in adopting these practices to improve the achievement of their objectives. To this end, the collected case studies in this book, all real natural resource management decisions, were chosen with 3 objectives in mind: (1) to demonstrate how structuring decisions can render them more tractable, (2) to illustrate the diversity of decision analysis tools useful at various stages of the structuring process, and (3) to provide guidance on how decision analysis processes can be carried out in public environmental institutions.

The primary intended audience for this book is decision makers, their support staff, and their collaborators working in public natural resource management. Thus, we envision this book will be valuable to federal, tribal, state, provincial, and municipal agencies around the world whose jurisdiction concerns environmental resources. In addition, the book will be valuable to scientists and consultants who work, or wish to work, with such agencies. The case studies in this book come from the United States, Canada, Australia, and Mauritius, and we hope the lessons that emerge resonate even more broadly.

We envision at least 3 ways this book of case studies can be used. First, decision makers who are thinking about using decision analysis, their staff members, and others who advise them (like resource economists, ecologists, industry consultants, and academics) can use the book as inspiration, to envision a wide range of ways to apply decision analysis methods. Second, decision analysts, facilitators, and negotiation analysts working with natural resource management agencies can use the book as a reference text, to provide an introductory catalog of decision structures typical to such agencies and demonstration of corresponding methods. Third, academic faculty can use this book as a supplementary text in a course on decision making for natural resource management, to provide real-world case studies that illustrate various techniques.

Why Structured Decision Making?

Decision makers in institutions charged with management of public natural resources face a daunting task—they are asked to manage complex coupled social-ecological systems for multiple purposes, in the face of both known and unknown uncertainties, while engaging stakeholders and remaining transparent and accountable to the public. The decisions require an integration of policy and science: the legal, regulatory, and value-based aspects of the decisions are often nuanced and rarely fully resolved; the underlying science is typically complicated and uncertain; and even the distinction between these 2 realms is difficult to discern. Perhaps the biggest challenge for the decision makers is recognizing all of the different components of the decision and envisioning how to fit them together.

As applied ecologists and decision scientists have become aware of the challenges facing natural resource decision makers, many decision-support frameworks and tools have emerged (Bower et al. 2018; Schwartz et al. 2018). The *Open Standards for the Practice of Conservation* (CMP 2013) provide a framework for designing actions, managing projects, and tracking progress. *Systematic conservation planning* (Margules and Pressey 2000) provides a tool for finding cost-efficient portfolios of actions and is often used for spatial planning. *Strategic foresight* (Cook et al. 2014) is a set of methods for proactively identifying potential management interventions to address emerging threats. *Evidence-based practice* (Sutherland et al. 2004) provides disciplined methods for evaluating the evidence on the effectiveness of management actions. *Management strategy evaluation* (Bunnefeld et al. 2011; Smith 1994) is an interactive, simulation-based approach for the evaluation of fish and wildlife harvesting alternatives. *Adaptive management* (Walters 1986) provides an approach for making ongoing management decisions in the face of uncertainty, with feedback monitoring that allows decisions to improve as uncertainty is reduced. In addition, frameworks for complex decision making in other fields are often brought into the natural resource management setting. Several recent efforts have attempted to make sense of this emerging set of frameworks and the array of tools that accompany them (Bower et al. 2018; Schwartz et al. 2018), but the sheer volume of possible tools can be daunting and bewildering.

Decision analysis is the name given to the vast field of study of how humans can and do make decisions, and it spans many other fields, including economics, operations research, cognitive psychology, mathematics, computer science, behavioral ecology and evolution, philosophy, and organizational behavior. With roots that stretch back at least to Descartes, Bernoulli, Bayes, and Laplace, the modern field of decision analysis begins with von Neumann and Morgenstern (1944) and Howard (1966) and now has applications in many areas of human endeavor. In the realm of natural resource management, *structured decision making* (SDM; Gregory et al. 2012) is the term most commonly applied to the use of decision analysis. Unlike a number of the other decision-support frameworks, which were motivated by and developed for natural resource management, SDM harnesses insights and tools from the whole discipline of decision analysis, translating them into methods relevant for natural resource decision makers. But rather than view decision analysis as a competitor to all the other decision-support frameworks, there are opportunities to explore how decision analysis can enhance them by adding analytical depth to some and insights about facilitating deliberative processes to others.

Outline of Structured Decision Making

Two tenets lie at the heart of structured decision making: problem decomposition and values-focused thinking. The idea behind problem decomposition is that decisions consist of both value-based and science-based elements and, by breaking a decision problem into its constituent elements, we can distinguish separate roles for policy makers and scientists. Some of the tasks we face as decision makers, even in our personal lives, are scientific, in the sense that they ought to involve a dispassionate examination of evidence and an effort to predict what might happen as a consequence of actions we take. Other tasks are value-based, in that they involve understanding what outcomes we are trying to achieve and what means we will consider to achieve them. We need to do both types of tasks, but we do not use the same tools to do them. In complex natural resource management problems, it is even more valuable to distinguish these types of tasks because the expertise needed to undertake them is often held by different people, and conflating the tasks can lead us to a biased interpretation of evidence or hidden values driving a decision. Further, decomposition helps to turn an extraordinarily complicated problem into a set of smaller, more tractable pieces. Some detractors of decision analysis worry that the act of decomposition can break holistic concepts into meaningless parts, but decision analysis is only meant to provide insight, not to make the decision. If the insight that arises from a decision analysis does not resonate with the decision maker when the parts are reassembled, the dissonance may suggest ways to develop a more nuanced analysis of the decision.

Value-focused thinking (Keeney 1992) recognizes that all decisions are, ultimately, the expression of values the decision maker aims to achieve. This simple insight explains why natural resource management can be so contentious—because different stakeholders value different outcomes. The articulation and justification of the values being pursued in a decision can provide profound clarity, both for the deliberations of the decision maker and for communication to stakeholders. Further, early identification of these values, and consistent attention to them, can counter cognitive tendencies to focus narrowly on alternatives. Research by cognitive psychologists has shown that, when faced with difficult decisions, people tend to look at the alternative actions that are readily apparent and then make quick evaluations to discern among them. But by first examining the outcomes we value and hope to achieve, we can step back to think more creatively about how to design effective actions. Unlike in politics, advocacy, and marketing, the aim of decision analysis is not to change someone's values but rather to understand them and to use that understanding in a search for a solution.

These 2 tenets—problem decomposition and value-focused thinking—provide the underlying philosophy of decision analysis. Taken together, they provide a structured template for any decision. There are 5 constituent elements of all decisions (fig. 1.1): a context that gives the decision maker power to act (the problem framing); one or more objectives that form the desired outcomes (the objectives); a set of alternative actions to choose from (the alternatives); predictions that link the alternatives to the objectives (the consequences); and an evaluation of the trade-offs among the alternatives that leads to selection of a preferred alternative (the trade-offs). This PrOACT sequence serves both as a reminder of these elements and a process for decision analysis (Hammond et al. 1999). The PrOACT decomposition also helps to distinguish the value-based elements of the decision (defining the problem context, articulating objectives, identifying what alternatives are allowable, and navigating trade-offs) from the scientific elements (assessing which alternatives are feasible and evaluating the consequences), allowing the decision maker to bring appropriate expertise to help on each element.

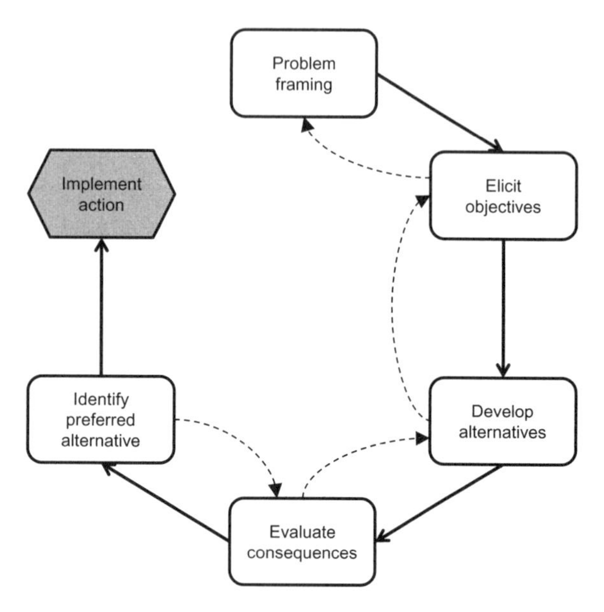

Figure 1.1. Diagram of the PrOACT sequence (Runge et al. 2013). The dashed lines indicate that the development of the elements of a decision is not necessarily linear, and insights at later stages often require revisiting earlier elements.

"Problem framing" describes the context of the decision: most important, this includes the identification of the decision maker and the authority under which they act. In decisions made by government agencies, the authority is typically statutory, and an understanding of the relevant statutes will often help in structuring other elements of the decision. The problem context also describes the spatial and temporal extent of the decision, whether a single decision or repeated decisions will be made, and often some of the relevant background that motivates the decision.

The "objectives" are the desired future conditions the decision maker is aiming to achieve. These should be fundamental objectives, not means objectives, that is, they should be outcomes that are pursued for themselves, not because they lead to something else desirable (Keeney 2007). In some fields, like finance, there may only be one objective (maximize net present value of the assets), but in natural resource management, given the nature of public trust responsibilities, there are often many objectives, including: to conserve vulnerable species; to maintain critical habitats; to provide hunting, subsistence, and recreational opportunities; to maintain ecosystem services; to uphold ethical and spiritual principles; and to minimize the costs of management (Runge et al. 2013). For decisions made by public institutions (like federal and state natural resource management agencies), the objectives are an expression of the values of the public, as transmitted through statutes, regulations, and judicial opinions and as articulated in public forums; they are not the personal values of the decision makers (Gregory et al. 2012). Choosing a desirable action depends entirely on recognizing the objectives that matter, emphasizing that any decision is an expression of values.

The "alternatives" are the options from which a decision maker can choose. The alternatives might be a simple, discrete set of options, a continuous range of values, a set of elements permuted into many portfolios, or a complex combination of all of these (Runge and Walshe 2014). The creative development of alternatives often requires expertise from many perspectives and fields and typically brings together experts from both the policy and science realms in an effort to find options that are both palatable from a values viewpoint and feasible from a technical viewpoint.

The central scientific task in any decision is the evaluation of the range of alternatives against the set of objectives, a task that requires prediction. The generation of these predictions relies on knowledge of the systems being managed, evidence from past studies, and insights into future dynamics. In the last decade or two, there has been a greater awareness that management of natural resources is often mediated through the responses of humans, so the predictions may need to recognize the connections in a coupled socioecological system (Wilson et al. 2016). Although the evaluation of the consequences is a scientific endeavor, the questions being asked are driven by value-based elements of the decision.

Provided with an evaluation of the consequences of the different alternatives in terms of achieving the various objectives, the decision maker then must choose an alternative that best balances the "trade-offs." We propose that these trade-offs can be separated into several categories: trade-offs among objectives; trade-offs among certain and uncertain outcomes; trade-offs among short- and long-term rewards; and trade-offs among short-term performance and long-term gain through learning. These trade-offs are all value judgments and reflect nuanced aspects of the relevant objectives.

In the end, a decision maker needs to take the analysis of the elements of the decision and integrate the pieces to identify an action to implement. This recomposition step can be challenging. As we describe later in this chapter, different types of decisions have particular challenges, and specific tools are available to help the decision maker through them, whether the impediment is weighing multiple objectives, understanding and managing risk, acknowledging and reducing uncertainty, or managing adaptively. In all cases, however, the same elements compose the structure of the decision (fig. 1.1), and understanding that structure is one of the profound contributions of decision analysis.

Developing Applications of Decision Analysis in Hierarchical Institutions

Many natural resource management decisions are made by public institutions—local, state, tribal, federal, or international agencies acting on behalf of their constituents—and, as the editors and authors in this book have discovered, decisions in such institutions often have unique challenges. First, public institutions often have a hierarchical structure, with multiple levels of supervision and authority. This often means it is difficult to identify the decision maker. For instance, in the United States, under the Endangered Species Act, the secretary of the interior (or the secretary of commerce for some species) has the statutory authority to make decisions. But the secretary of the interior delegates authority to the director of the Fish and Wildlife Service, who in turn delegates some of that authority to the regional directors, who in turn delegate authority to field office supervisors. Thus, while it may appear in a particular context that the decision maker is the field office supervisor, if the decision is controversial, it is possible that the authority will be pulled back up the chain of command. Second, decisions made by public institutions often draw considerable stakeholder interest, and agencies are often obliged by statute or practice to engage stakeholders in a meaningful way. Third, decisions made by public institutions are subject to public scrutiny; in some jurisdictions, transparency of the decision-making process is required by law. Fourth, public institutions are governed by statutes, regulations, and formal guidance, and these guidelines are critical to substantive aspects of the decision (especially framing, objectives, and alternatives) as well as to procedural aspects of the decision.

Deploying decision analysis tools in hierarchical public institutions, then, requires attention to the process by which the decision is made. The authors of the chapters in this book have found, collectively, that *iterative prototyping* is a helpful framework for implementing decision analysis in such settings (Garrard et al. 2017). This process (also known as *decision sketching*; Gregory et al. 2012) begins with the rapid development of an initial prototype for the decision analysis, working with the decision maker and key advisors for between half a day and 1 week to sketch the elements of the decision and to examine the impediments to finding a solution. From here, successful prototypes of increasing detail can be developed as needed to examine the elements of the decision in greater detail. Sometimes those subsequent prototypes bring in larger and larger groups of stakeholders; sometimes they require more technical analysis; and they almost always involve reflection between rounds. The iterations are complete when the decision analysis sheds enough light for the decision maker to identify and explain the choice of a preferred action.

How These Case Studies Arose

In 2007, the US Fish and Wildlife Service's National Conservation Training Center (NCTC), in Shepherdstown, West Virginia, began sponsoring structured decision-making workshops, bringing together decision makers and decision analysts to unravel difficult decisions with the tenets and tools of SDM. Between 2007 and 2017, NCTC held more than 20 workshops, covering over 80 case studies. The format and design of these workshops consisted of 2 rounds of rapid prototyping over 4.5 days with time for reflection among groups. This format provided enough time for the groups to undertake a substantive initial decision analysis; served NCTC's pedagogical purpose in allowing training opportunities for apprentice decision analysts; and permitted convenient logistical arrangements. In broader practice, a great many different forms are used for SDM processes, stretching in length from a few hours to many years, customized to the needs of the particular decision.

More than half of the case studies in this book arose initially from NCTC workshops or workshops designed in a similar manner. Some of the chapters describe the decision analysis conducted at the workshop and thus are initial prototypes that capture the essence of the decision but perhaps not all the nuance. Other chapters describe the decision analysis that resulted after more investment of time (many prototypes later). All the chapters give some indication of how the decision analysis was ultimately used.

In selecting the individual case studies for this book, we looked for several features: a well described decision structure; an analysis that led to valuable insights for the decision maker, preferably with an action being implemented; and demonstration of some useful decision analytical methods, whether quantitative or not. In choosing the set of case studies, we looked for breadth in the type of decision analysis tools used as well as diversity in the geographical and ecological settings.

Structure of This Book

Although the circumstances and details of every decision are unique, an analyst begins to recognize certain common forms with experience. Each form comes with its own common structure and challenges. Decision analysts have developed tools to help decision makers navigate each of these common types of decision. We have organized the case studies in this book around 6 typical classes of decision problems: (1) decomposing complex decisions into tractable pieces, by identifying the decision components and structure; (2) resolving trade-offs among competing objectives; (3) allocating limited resources; (4) coping with uncertainty and risk; (5) tying research to decisions; and (6) linking multiple decisions through time or space.

Each of the 6 parts of the book begins with an introduction that provides some background on the type of decision problem found in that part, discusses the impediments for the decision maker, presents an overview of the set of tools available to help, and outlines how the case studies in each section fit together. The case studies were assigned to the parts of the book after the fact, that is, after the decision analysis was complete and the structure of the decision was evident.

Part 1. Structuring Decisions

Quite often, the primary challenge the decision maker faces in wrestling with a decision is an inability to see the structure of the decision (fig. 1.1). Environmental decisions are incredibly complex, with many moving parts; identifying even what the choices are, or the desired outcomes, can be difficult. In a thoughtful and enlightening article, Keeney (2004) posits that of 10,000 decisions faced by a decision maker, perhaps only 250 require any in-depth analysis; the rest can be solved by a clear understanding of the structure of the problem. Indeed, one of the central tasks for a decision analyst is helping the decision maker identify the structure of the decision. In part 1, we discuss the insights gained by structuring decisions and share 2 case studies that exemplify these insights.

Part 2. Addressing Trade-Offs

Of the decisions that require a more detailed layer of analysis in natural resource management settings, we have found that the challenge in many of them is to identify and weigh the trade-offs among a set of competing objectives. The techniques of multicriteria decision analysis offer a valuable set of tools for navigating this type of decision. In part 2, we discuss the methods, benefits, and challenges of multiple-objective decision analysis and share 3 case studies that demonstrate several approaches.

Part 3. Addressing Resource Allocation

Sometimes, the biggest lever a natural resource management agency can pull is the amount of money it spends on any number of conservation endeavors. Budget allocation problems, or, more broadly, resource allocation problems, are challenging because the number of ways to put together a budget is nearly infinite, and discerning how all those investment categories contribute to the overall goals of the agency is complicated. As it turns out, the budget allocation problem has the same structure as reserve design problems (spatial allocation), portfolio design problems, research portfolio problems, and many other prioritization problems. By recognizing the common structure of these apparently different problems, a valuable set of framing and optimization tools becomes available. In part 3, we present the concepts of resource allocation, discuss a few of the methods, and share 3 case studies that exemplify the range of application.

Part 4. Addressing Risk

Natural systems are highly variable, and our knowledge about them is incomplete. Inevitably, some decisions have to be made in the face of this uncertainty, forcing decision makers to consider and respond to the risk or opportunity the uncertainty imposes. A structured approach helps managers make decisions involving risk carefully and transparently, clarifying the interlinked roles of scientific information and values. Risk assessment describes the potential outcomes of management alternatives and quantifies their likelihood, while risk management considers tolerance for the risk of uncertain outcomes. The field of risk analysis provides valuable methods for articulating, evaluating, and managing risk. In part 4, we discuss the principles and methods of risk analysis and share 3 case studies that demonstrate its implementation.

Part 5. Addressing Knowledge Gaps

In some decisions, we face uncertainty, but there is the possibility of resolving uncertainty before having to commit to a management action. Should we? Will the information we acquire help us make a better decision? This question is at the heart of decisions about monitoring programs, research design, and adaptive management. There are formal decision analytical tools for calculating the value of new information to the decision maker, and these tools offer valuable insights for natural resource managers. In part 5, we outline the central concepts and methods for evaluating the importance of new information and share 3 case studies that capture a range of settings in which this question is important.

Part 6. Addressing Linked and Dynamic Decisions

In the last part of this book, we address complicated decisions that have multiple decision points linked through time or space. Linked decisions provide an interesting opportunity: actions taken earlier in the sequence can possibly provide an opportunity for learning that can affect decisions made later in the sequence; this is the hallmark of adaptive management (Walters 1986). Many natural resource management decisions are iterated, quite often on an annual basis—think of hunting regulations, habitat acquisition, and habitat management, to name a few. Methods from optimal control theory provide useful tools for analyzing these kinds of decisions. In part 6, we present an overview of the analytical methods for linked and dynamic decisions and share 3 case studies that demonstrate the power of this approach.

Each of the case study chapters in this book serves as an interesting application of decision analysis in a natural resource management setting. Collectively, the case studies in a section should provide a compelling understanding of the particular class of decisions they represent. We hope the entire collection provides a vibrant and textured picture of the ways in which decision analysis can enhance our management of the natural world.

LITERATURE CITED

Bower SD, Brownscombe JW, Birnie-Gauvin K, Ford MI, Moraga AD, Pusiak RJP, Turenne ED, Zolderdo AJ, Cooke SJ, Bennett JR. 2018. Making tough choices: picking the appropriate conservation decision-making tool. *Conservation Letters* 11:e12418.

Bunnefeld N, Hoshino E, Milner-Gulland EJ. 2011. Management strategy evaluation: a powerful tool for conservation? *Trends in Ecology & Evolution* 26:441–447.

Conroy MJ, Peterson JT. 2013. *Decision Making in Natural Resource Management: A Structured, Adaptive Approach*. West Sussex, UK: John Wiley & Sons.

Conservation Measures Partnership [CMP]. 2013. *The Open Standards for the Practice of Conservation, Ver. 3.0.* The Conservation Measures Partnership.

Cook CN, Inayatullah S, Burgman MA, Sutherland WJ, Wintle BA. 2014. Strategic foresight: how planning for the unpredictable can improve environmental decision-making. *Trends in Ecology & Evolution* 29:531–541.

Garrard GE, Rumpff L, Runge MC, Converse SJ. 2017. Rapid prototyping for decision structuring: an efficient approach to conservation decision analysis. Pages 46–64 in Bunnefeld N, Nicholson E, Milner-Gulland EJ, eds. *Decision-Making in Conservation and Natural Resource Management: Models for Interdisciplinary Approaches*. Cambridge, UK: Cambridge University Press.

Gregory R, Failing L, Harstone M, Long G, McDaniels T, Ohlson D. 2012. *Structured Decision Making: A Practical Guide for Environmental Management Choices*. West Sussex, UK: Wiley-Blackwell.

Hammond JS, Keeney RL, Raiffa H. 1999. *Smart Choices: A Practical Guide to Making Better Life Decisions*. New York: Broadway Books.
Howard RA. 1966. Decision analysis: applied decision theory. Pages 55–71 in Hertz DB, Melese J, eds. *Proceedings of the Fourth International Conference on Operational Research*. New York: Wiley.
Keeney RL. 1992. *Value-Focused Thinking: A Path to Creative Decisionmaking*. Cambridge, MA: Harvard University Press.
———. 2004. Making better decision makers. *Decision Analysis* 1:193–204.
———. 2007. Developing objectives and attributes. Pages 104–128 in Edwards W, Miles RFJ, Von Winterfeldt D, eds. *Advances in Decision Analysis: From Foundations to Applications*. Cambridge, UK: Cambridge University Press.
Margules CR, Pressey RL. 2000. Systematic conservation planning. *Nature* 405:243–253.
Runge MC, Grand JB, Mitchell MS. 2013. Structured decision making. Pages 51–72 in Krausman PR, Cain JW, III, eds. *Wildlife Management: Contemporary Principles and Practices*. Baltimore, MD: Johns Hopkins University Press.
Runge MC, LaGory KE, Russell K, Balsom JR, Butler RA, Coggins LG, Jr., Grantz KA, Hayse J, Hlohowskyj I, Korman J, May JE, O'Rourke DJ, Poch LA, Prairie JR, VanKuiken JC, Van Lonkhuyzen RA, Varyu DR, Verhaaren BT, Veselka TD, Williams NT, Wuthrich KK, Yackulic CB, Billerbeck RP, Knowles GW. 2015. Decision analysis to support development of the Glen Canyon Dam Long-Term Experimental and Management Plan. *US Geological Survey Scientific Investigations Report* 2015-5176.
Runge MC, Walshe T. 2014. Identifying objectives and alternative actions to frame a decision problem. Pages 29–44 in Guntenspergen GR, ed. *Application of Threshold Concepts in Natural Resource Decision Making*. New York: Springer-Verlag.
Schmidt JC, Webb RH, Valdez RA, Marzolf GR, Stevens LE. 1998. Science and values in river restoration in the Grand Canyon. *BioScience* 48:735–747.
Schwartz MW, Cook CN, Pressey RL, Pullin AS, Runge MC, Salafsky N, Sutherland WJ, Williamson MA. 2018. Decision support frameworks and tools for conservation. *Conservation Letters* 11:e12385.
Smith ADM. 1994. Management strategy evaluation: the light on the hill. Pages 249–253 in Hancock DA, ed. *Population Dynamics for Fisheries Management, Australian Society for Fish Biology Proceedings, Perth, 24–25 August 1993*. North Beach, Western Australia: Australian Society for Fish Biology.
Sutherland WJ, Pullin AS, Dolman PM, Knight TM. 2004. The need for evidence-based conservation. *Trends in Ecology & Evolution* 19:305–308.
US Department of the Interior. 2016a. *Glen Canyon Dam Long-Term Experimental and Management Plan Environmental Impact Statement*. Washington, DC: US Department of the Interior.
———. 2016b. *Record of Decision for the Glen Canyon Dam Long-Term Experimental and Management Plan Final Environmental Impact Statement*. Washington, DC: US Department of the Interior.
von Neumann J, Morgenstern O. 1944. *Theory of Games and Economic Behavior*. Princeton, NJ: Princeton University Press.
Walters CJ. 1986. *Adaptive Management of Renewable Resources*. New York: Macmillan.
Williams BK, Nichols JD, Conroy MJ. 2002. *Analysis and Management of Animal Populations: Modeling, Estimation, and Decision Making*. San Diego, CA: Academic Press.
Wilson RS, Hardisty DJ, Epanchin-Niell RS, Runge MC, Cottingham KL, Urban DL, Maguire LA, Hastings A, Mumby PJ, Peters DP. 2016. A typology of time-scale mismatches and behavioral interventions to diagnose and solve conservation problems. *Conservation Biology* 30:42–49.

PART I STRUCTURING DECISIONS

2 Introduction to Structuring Decisions

David R. Smith

Decision structuring, also known as decision framing, provides the foundation and roadmap for analyzing a decision. For decisions that warrant a systematic approach, structuring begins with identifying the problem for analysis. This sounds simple but can be deceptively difficult because decision problems are often ill-formed at the start. Many have worked on a problem, alone or with others, only to realize down the road that it's the wrong problem, something Ron Howard—a founder of the decision analysis discipline—calls an "error of the third kind." How a decision is framed can have a profound effect on subsequent analysis and solution. Tools and templates are available to get started, but perhaps no technique is more essential than simply taking the time to ponder what the problem is all about. All of the case studies in this book have gone through decision structuring, and most followed an iterative prototyping process. In particular, the case studies in part 1 highlight the value of decision structuring to uncover hidden assumptions hindering a good solution and identify the scientific information needed.

Introduction

Decision analysis, in the words of Ron Howard (1988), is a systematic process for "transforming opaque decision problems" into "transparent decision problems by a sequence of transparent steps." By opaque, he meant "hard to understand, solve, or explain; not simple, clear, or lucid." The key feature of this definition is that important decisions are often ill-formed as presented and that the initial and perhaps the central challenge is to clarify and identify the problem for analysis. This first-order challenge is so important that Howard Raiffa (2002), when reflecting over his extraordinarily productive career, mused that he "completely missed the boat" early on by ignoring "the nonmathematical underpinnings [of decision analysis]," namely, "how to identify a problem . . . to be analyzed." To meet this challenge, a decision analyst turns to decision framing (Keeney 2004a) or decision structuring (Gregory et al. 2012), which is the topic of this section. It is only through structuring the decision that an error of the "third kind; namely, working on the wrong problem" can be avoided (Howard 1988). It is not uncommon for a collaboration among decision makers and stakeholders to spend considerable time and effort working on the wrong problem, as Howard warned, or on decisions implicitly defined differently by the participants.

What is decision structuring? Simply speaking, decision structuring is identifying the problem to be analyzed, which von Winterfeldt and Edwards

(2007, 84) likened to "hunting for the decision." To structure a decision is to define the scope of and scale for its basic elements (table 2.1). The essence of decision analysis is decomposing a problem into its elemental parts, reflecting both value-based and technical aspects of a decision. Howard (1988) separates the generic problem into the 3-part "decision basis": the choice of alternatives, the information available to evaluate the choice, and the preferences in values, time, and risk to guide the choice—sort of a 3-legged stool. Hammond et al. (1999), in contrast, split the decision problem into 5 components: problem definition, objectives, alternatives, consequences, and trade-off analysis—the PrOACT process embedded into structured decision making (Gregory et al. 2012; Runge and Bean, chapter 1, this volume, fig. 1.1). In addition to those basic components, the decision structure identifies key actors (decision makers and stakeholders), uncertainties, and constraints. Structuring roughs out those decision parts before starting the analysis, as an artist might sketch the vision of a sculpture before starting to carve. While there is not a single "correct" structure (Gregory et al. 2012), a "good solution to a well-posed decision problem is almost always a smarter choice than an excellent solution to a poorly posed one" (Hammond et al. 1999).

The structuring of the decision problem influences both the range of possible solutions and how decision makers and stakeholders feel about the solutions. While there is no single "correct" structure (Gregory et al. 2012), some framings will be more useful than others. Solving the decision problem is the end point, but all problems begin with decision framing. Insufficient time spent on framing often leads to wasted time at the analysis stage. Taking a shortcut by jumping over objectives and going straight to alternatives often results in a misplaced focus on the "means" rather than the "ends" (Keeney 1996). Depending on where the frame is placed to define the problem, the potential solutions will focus on strategic, fundamental, or means objectives (Keeney 1996, 44–47). For example, a focus on habitat improvement limits analysis to achieving a habitat goal; if instead species conservation is the target, factors other than habitat can determine viability. The framing effect can operate at a cognitive level, as in the well-known framing trap (Hammond et al. 1999; Tversky and Kahneman 1986), where whether a choice is framed in terms of gains or losses affects which alternative is preferred. In decision structuring, the framing effect can determine whether a problem is viewed narrowly or broadly.

Table 2.1. Basic elements of a decision structure

Elements
1. Identify the decision maker(s)
2. Identify other key actors
• Decision implementers and stakeholders
• Technical experts and facilitators
3. Consider the legal and regulatory context
4. Consider the decision structure
• Timing and frequency of the decision
• Temporal and spatial scope and scale
• Initial set of objectives
• Possible actions
• Constraints (perceived and real)
• Key uncertainties
5. Consider the type of analysis required
6. Revise as needed

To illustrate how framing can profoundly affect decision context, consider management of horseshoe crabs (*Limulus polyphemus*) and red knots (*Calidris canutus rufa*), which is the topic of chapter 24 (McGowan et al., this volume) and described by McGowan et al. (2015b). Two dramatically different scales can apply to this problem depending on whether we consider management at a single migratory stopover site or management across the range of a migratory species. Each spring, horseshoe crabs spawn on Delaware Bay beaches in densities so high that the crabs disturb each other's nests, bringing safely buried eggs up to the sandy surface where they are accessible to foraging shorebirds. At roughly the same time, red knots migrate from wintering areas to the breeding areas in the Arctic. Along the migratory route, the shorebird stops over at multiple places to refuel, including Delaware Bay with its abundance of fat-rich horseshoe crab eggs. Harvest of horseshoe crabs for bait and biomedical products sets up a trade-off between harvest of crabs and conservation of red knots. A horseshoe crab population decline due to overharvesting in the 1990s contributed significantly to a drop in the red knot population. The US Fish and Wildlife Service (USFWS) listed the red knot's status as threatened under the Endangered Species Act, citing that the primary future threats are "habitat loss and degradation due to sea level rise, shoreline hardening, and Arctic warming," which includes threats outside of Delaware Bay (USFWS 2014). The *broad frame* is how to recover the red knot so that it is no longer a threatened species. This broad framing would include actions throughout the red knot's wintering, stopover, and nesting habitats. The *narrow frame* is how best to manage the harvest of horseshoe crabs in Delaware Bay while providing forage for migratory shorebirds.

These framings are not mutually exclusive, and for management to be ultimately successful, both problems need to be solved. The 2 complementary frames provide insight into the nature of the trade-offs at each scale. The harvest management problem nests within the red knot recovery problem and is necessary to understand the importance of egg availability to overall recovery.

As it happened, harvest management proceeded first through the support of the Atlantic States Marine Fisheries Commission (ASMFC), which sets maximum harvest regulations for horseshoe crabs. The objective statement for the horseshoe crab harvest problem was qualitatively phrased as, "Manage harvest of horseshoe crabs in the Delaware Bay not only to maximize harvest but also to maintain ecosystem integrity and provide adequate stopover habitat for migrating shorebirds" (McGowan et al. 2015b). The USFWS is currently developing a red knot recovery plan, a product of the broad frame, with consideration for other narrowly framed conservation problems in wintering areas, additional stopover habitats, and the Arctic breeding grounds.

Who Makes the Decision?

Perhaps the most important question to ask when framing a decision is, "Who is the decision maker?" Although this seems like a simple question, the answer can be surprisingly complex or fiercely contested. Natural resource management agencies are often hierarchical; authority to make a decision is provisionally delegated, but that delegation can be rescinded if the decision garners enough attention. Natural resource management agencies also often pursue their aims through elaborate partnerships with other actors, each of which has its own authority to make decisions. Further, the authority for making a decision is often contested among independent agencies and stakeholders. Thus, identifying the decision maker may be the central impediment to decision making.

Classical decision analysis, which is the topic of this book, assumes there is a single decision maker with authority to act. Other related fields of study focus on situations where authority is shared, negotiated, or contested. *Game theory* analyzes strategies when multiple decision makers are competing and the outcomes of the decision made by one decision maker are affected by the decisions made by the other decision makers (von Neumann and Morgenstern 1944). Game theory's relevance to natural resource management has only begun to be explored (Colyvan et al. 2011). *Negotiation analysis* is a hybrid between decision analysis and game theory, in that it uses a decision-analytical understanding of the position of each actor to motivate and identify opportunities for collaboration (Sebenius 2007). Again, there is tremendous opportunity to use negotiation analysis in natural resource management, but there are so far few, if any, applications. Game theory and negotiation analysis are not explored further in this book; instead, the attention is on the fruitful use of decision analysis.

Although classical decision analysis implicitly assumes an individual decision maker acting with sole authority (Howard 1988), a full, open, and truthful exchange (FOTE) among collaborators can help them operate like a unitary decision maker and thoroughly apply the decision-analysis framework (PrOACT) promoted in this book (Gregory et al. 2012; Raiffa et al. 2002). At the start, collaborators might not possess the required level of trust, but the initial structured decision making (SDM) steps can help shift the level of trust toward a common understanding of the problem and open communication (Keeney 1996). When trust is insufficient for FOTE, then collaborators or decision-making groups can seek resolution through alternative processes (i.e., negotiation, mediation, or litigation), resulting in a potentially durable, but not necessarily optimal, solution.

Decision makers willing to find resolution collaboratively can use the decision analysis approaches described in this book. For example, the ASMFC was the decision maker for horseshoe crab harvest management with multispecies constraints (see above). The ASMFC, as a collaborative decision-making commission, receives advice from technical and stakeholder committees (McGowan et al. 2015b). When the ASMFC initiated the SDM process, significant conflict among stakeholder groups hindered efforts to reach a durable, let alone optimal, solution. However, the engagement of all stakeholders in decision structuring helped to break down barriers of communication and create the level of trust necessary to proceed with decision analysis and to implement an adaptive management framework (McGowan et al. 2015a, 2015b).

Tools for Structuring Decisions

Several useful tools and approaches are available to aid in structuring a decision. The essential technique is merely to take the time to deliberate on what the decision problem is all about (Gregory et al. 2012). Keeney (2004a) imagined that among the many decisions a person might face, the vast majority are either of small consequence or quickly resolved, but an important subset (say, 10% of all decisions) deserve careful thought, beginning with questions about problem framing. Among the questions to ponder, start with the following. What makes this particular decision challenging to resolve? Who are the key actors: who has the authority to make the decision, who can affect the decision, and who will be affected by the decision? Then move on to drafting a *decision statement*, which sums up the problem in a format that can serve as a ready reference or roadmap to follow as the decision analysis proceeds. This decision statement should be communicated to the key actors in order to ensure a shared vision and common understanding of the decision problem and the analytical approaches taken to find a solution. A shared vision of the decision problem is essential for collaborative and public policy decisions (Keeney 2004b; Raiffa et al. 2002, part V). Without a shared vision, collaboration is stymied, and conflict ensues.

A decision statement can appear in various formats. One format is a brief statement crafted from responses to a series of questions. A minimal set of topical questions includes

- Decision maker and stakeholders: Who will make the decision? Under what authority do they act? Who are the relevant stakeholders? What are their goals, aspirations, and concerns? What are they trying to achieve through the decision?
- Trigger: Why does a decision need to be made? Why does it matter?
- Alternatives: What type of options are available? What action could be taken?
- Constraints: What are the legal, financial, and political constraints? Are these constraints real or perceived?
- Frequency and timing: How often will the decision be made? Are other decisions linked to this one?
- Scope: How broad or complicated is the decision? Are there important bounds in space or time?
- Type of decision: What is the primary challenge inherent to the decision? Will decision structur-

ing illuminate the solution? Will the solution require addressing trade-offs or resource allocation? Is risk or uncertainty at the heart of the decision problem? How is the decision related to other decisions that have been or will be made? What are the likely technical analyses given the type of decision?

Regarding the last question in the list, the contents of this book are organized to match common decision types: (1) decomposing complex decisions into tractable pieces (which applies to all decisions worthy of systematic thought); (2) resolving trade-offs among competing objectives; (3) allocating limited resources; (4) coping with uncertainty and risk; (5) tying research to decisions; and (6) linking multiple decisions through time or space (Runge and Bean, chapter 1, this volume). Preliminary identification of a decision type, along with the other parts of the decision statement, provides insight into what challenges lie ahead on the way to solving the problem. In other words, forewarned is forearmed (Hammond et al. 1998).

In some situations, a template for decision statements can be convenient. For example, a template used in SDM courses offered at the USFWS National Conservation Training Center in Shepherdstown, West Virginia, results in a succinct decision statement (Romito et al. 2015):

> Decision Maker (*D*) is trying to do *X* to achieve *Y* over time *Z* and in place *W* considering *B* where *D* = the Decision maker(s), *X* = the type(s) of action that needs to be taken, *Y* = the ultimate goal(s) to be achieved by implementing *X*, *Z* = the temporal extent of the decision problem, *W* = the spatial extent of the decision problem, and *B* = potential constraints (legal, financial, and political) and important uncertainties (scientific or other).

Templates, while a bit constraining, can be highly useful for a quick distillation of the problem at hand. Also, when developing a common understanding of the problem in a group decision setting, the template's well-defined terms can be valuable.

In some collaborative efforts, a formal statement with an agreement between the decision makers and key stakeholders is useful. Gregory et al. (2012, 67) recommended a decision charter with the following components to formalize a decision statement:

1) The pending decision, its relationship to other decisions, and confirmed scope constraints (what's in and what's out)
2) The scope of alternatives under consideration (not their details, just the general scope)
3) Preliminary objectives and performance measures
4) Uncertainties and trade-offs that are expected to be central to the decision
5) The expected approach to analysis and consultation
6) Roles and responsibilities including who is/are the decision maker(s)
7) Milestones for decision maker input
8) An implementation plan, budget and timeline

While Gregory et al. (2012) suggested this list as a template for a formal decision charter, it is generally applicable for group decisions to help a collaborative effort move beyond the initial scoping phase. An initial deliberation guided by this template can help identify appropriate stakeholder involvement, types of expertise required, decision class, and likely analysis needed to solve the problem.

The critical nature of involving the decision maker, as Gregory et al. (2012) emphasize in the decision charter list, cannot be overstated. A decision analysis without decision-maker input from the start and throughout the process is unlikely to be successful. Recall Howard's 3-legged decision basis: what is to be achieved, what can be done to achieve that, and what is known about the outcome of possible actions. Decision makers have relevant input for all 3 legs of the stool, but they have the final say on what is to be achieved, which is the ultimate target of values-focused thinking (Keeney 1996). As Keeney (2007, 108) said, "Without knowing what you [the decision maker] want to achieve, there doesn't seem to be any reason to care or think about a decision." In practice, if the decision analysis is not based on the framing from the decision maker's perspective, then the solution will almost certainly miss the mark. This can happen, for example, when there is a mismatch in decision framing between a recommendation team and the ultimate decision maker. Differences can arise from any part of the frame, such as relevance and importance of the objectives, acceptability of alternatives, or level of trust and transparency among stakeholders.

A decision statement or structure will not be a finished product on the first try. Several revisions, with testing and learning between each revision, might be required before a decision structure is provisionally final (von Winterfeldt and Edwards 2007). The term "provisional" is used here because the decision analyst should view a decision structure as continually subject to revision until a decision is made, although after a few iterations, we would expect there to be less need for change. What we are describing here is the design and engineering practice of prototyping, which fits well to structured decision making (Blomquist et al. 2010; Garrard et al. 2017). An engineered product does not start off in final form. Rather, the design progresses in phases or prototypes with the initial phase merely a sketch of the product. Gregory et al. (2012) use the term "decision sketch" to refer to the initial prototype, which roughly describes the basic elements (table 2.1) resulting in the first draft of a decision structure.

Test, revise, and repeat is the prototyping mantra. By "test," we mean solve the prototype decision problem with what you have available at the time to gain insights into whether all the relevant perspectives or objectives were considered, whether the constraints were real or perceived, whether the alternatives were sufficient and comprehensive, and whether the decision choice appears to be sensitive to underlying assumptions or key uncertainties. The aim is to converge on a durable decision structure as quickly as possible. By definition, a prototype is a simplification, but much can be learned about the essential ingredients for solving a decision problem by starting simple and adding complexity only as needed. Prototyping is also a "fail fast" approach. The idea is to fail when the investment is low, learn from that failure, and revise the approach to address the shortcomings. A large investment into a polished product makes it very hard to acknowledge shortcomings and make needed revisions. Blomquist et al. (2010) provide an excellent example of a decision problem worked in the form of 2 prototypes with different frames (structures). Garrard et al. (2017) provide an excellent synopsis of rapid prototyping with sage advice on how to proceed.

Case Studies

Decision structuring is an iterative and clarifying process. Decision problems worthy of systematic thought are often complex and difficult to wrap one's arms around. Environmental decision problems are characterized by competing objectives, alternatives with multiple steps or multiple components, and an overlay of constraints (some real and some perceived), among other complexities. Environmental managers, when faced with such complexity, intuitively rely on the status quo option or other heuristics for solutions, which are often demonstrably inferior choices. Thinking systematically about the decision problem can help alleviate the frustration and confusion caused by overwhelming complexity. The 2 case studies in part 1 illustrate how decision structuring alone can clarify the problem and provide key insights leading to a solution or to critical next steps. The same process of structuring, often through iterative prototyping, is also evident in all the other case studies in this book.

In chapter 3, Michael Runge describes an analysis of how best to allocate funding under the National Fish Habitat Action Plan. The previous allocation approach had resulted in inequities and questionable effectiveness. The decision structure that resulted from a prototyping workshop identified fundamental, means, process, and strategic objectives, developed a set of alternative allocation approaches that span and contrast the range of possibilities, and built a simple but serviceable predictive model that could be used with available information. The prototype confirmed that the underlying structure for the decision was to choose an allocation strategy that balanced multiple objectives, articulated a comprehensive set of fundamental objectives, and exposed hidden assumptions regarding the roles of leverage and efficiency, which had been sources of past disagreements and once exposed allowed for transparent steps to reach a solution to the problem. The insights from the prototype allowed the agency to focus on the critical tasks needed to develop a final allocation approach, which was then implemented.

In chapter 4, Eben Paxton and Jim Kraus analyze options for conservation of Hawai'ian birds in the face of climate change and consequent environmental changes. Because climate change elevates the risk of mosquito-vectored disease, there is an urgent need to identify effective actions. The decision analysis started at a prototyping workshop where objectives and alternatives were structured. In this case study, the predictive model matters a great deal. The models that are needed to predict the effectiveness of actions, are not yet available, but the framing and prototyping identified the inputs and outputs for such a model. The inputs are related to the actions, and outputs are understood in terms of the objectives and measurable attributes. By prototyping and identifying objectives and alternatives, the model building can be much more definite and targeted to the requirements of the decision analysis.

LITERATURE CITED

Blomquist SM, Johnson TD, Smith DR, Call GP, Miller BN, Thurman WM, McFadden JE, Parkin MJ, Boomer GS. 2010. Structured decision-making and rapid prototyping to plan a management response to an invasive species. *Journal of Fish and Wildlife Management* 1:19–32.

Colyvan M, Justus J, Regan HM. 2011. The conservation game. *Biological Conservation* 144:1246–1253.

Garrard GE, Rumpff L, Runge MC, Converse SJ. 2017. Rapid prototyping for decision structuring: an efficient approach to conservation decision analysis. Pages 46–64 in Bunnefeld N, Nicholson E, Milner-Gulland EJ, eds. *Decision-Making in Conservation and Natural Resource Management: Models for Interdisciplinary Approaches*. Cambridge, UK: Cambridge University Press.

Gregory R, Failing L, Harstone M, Long G, McDaniels T, Olson D. 2012. *Structured Decision Making: A Practical Guide to Environmental Management Choices*. West Sussex, UK: Wiley-Blackwell.

Hammond JS, Keeney RL, Raiffa H. 1998. The hidden traps in decision making. *Harvard Business Review* 76(5):47–58.

———. 1999. *Smart Choices: A Practical Guide to Making Better Life Decisions*. New York: Broadway Books.

Howard RA. 1988. Decision analysis: practice and promise. *Management Science* 34:679–695.

Keeney RL. 1996. *Value-Focused Thinking*. Cambridge, MA: Harvard University Press.

———. 2004a. Making better decision makers. *Decision Analysis* 1:193–204.

———. 2004b. Framing public policy decisions. *International Journal of Technology, Policy and Management* 4(2):95–115.

———. 2007. Developing objectives and attributes. Pages 104–128 in Edwards W, Miles RFJ, von Winterfeldt D, eds. *Advances in Decision Analysis: From Foundations to Applications*. Cambridge, UK: Cambridge University Press.

McGowan CP, Lyons JE, and Smith DR. 2015a. Developing objectives with multiple stakeholders: adaptive management of horseshoe crabs and red knots in Delaware Bay. *Environmental Management* 55:972–982.

McGowan CP, Smith DR, Nichols JD, Lyons JE, Sweka JA, Kalasz K, Niles LJ, Wong R, Brust J, Davis M, Spear B. 2015b. Implementation of a framework for multi-species, multi-objective adaptive management in Delaware Bay. *Biological Conservation* 191:759–769.

Raiffa H. 2002. Decision analysis: a personal account of how it got started and evolved. *Operations Research* 50:179–185.

Raiffa H, Richardson J, Metcalfe D. 2002. *Negotiation Analysis: The Science and Art of Collaborative Decision Making*. Cambridge, MA: Harvard University Press.

Romito AM, Cochrane JF, Eaton MJ, Runge MC. 2015. Problem definition. Module 04 in Runge MC, Romito AM, Breese G, Cochrane JF, Converse SJ, Eaton MJ, Larson MA, Lyons JE, Smith DR, Isham AF, eds. *Introduction to Structured Decision Making*. Shepherdstown, WV: US Fish and Wildlife Service, National Conservation Training Center.

Sebenius JK. 2007. Negotiation analysis: between decisions and games. Pages 469–488 in Edwards W, Miles RFJ, von Winterfeldt D, eds. *Advances in Decision Analysis: From Foundations to Applications*. Cambridge, UK: Cambridge University Press.

Tversky A, Kahneman D. 1986. Rational choice and the framing of decisions. *The Journal of Business* 59: S251–S278.

US Fish and Wildlife Service. 2014. Endangered and threatened wildlife and plants; threatened species status for the rufa red knot. *Federal Register* 79:73705–73748.

von Neumann J, Morgenstern O. 1944. *Theory of Games and Economic Behavior*. Princeton, NJ: Princeton University Press.

von Winterfeldt D, Edwards W. 2007. Defining a decision analytic structure. Pages 81–103 in Edwards W, Miles RFJ, von Winterfeldt D, eds. *Advances in Decision Analysis: From Foundations to Applications*. Cambridge, UK: Cambridge University Press.

Michael C. Runge

Allocating Funds under the National Fish Habitat Action Plan

Each year, the director of the US Fish and Wildlife Service (Service), with advice from a Fisheries Management Team, allocates funding to support the National Fish Habitat Action Plan. The Service distributes the funds to Fish Habitat Partnerships (FHPs), who, in turn, undertake projects that "protect, restore, or enhance fish and aquatic habitats or otherwise directly support habitat-related priorities of Fish Habitat Partnerships." Initially, this allocation was made based on a simple formula: larger FHPs received twice the allocation of smaller FHPs. But as the number of partnerships grew, and as funding grew at a slower rate, inequities developed among the FHPs. In 2012, the Service convened a structured decision-making process to develop a more equitable, transparent, and strategic formula for annual funding allocation. The initial decision analysis, which focused on strategic aspects of the allocation, is described in this chapter. Deliberate consideration of decision analysis concepts brought about 2 advances: a focus on the fundamental long-term objective of maximizing the sustainability of aquatic species populations and recognition that the benefits of the relatively small investment by the Service occur through leveraging contributions from management partners and increasing the efficiency of on-the-ground projects. Four allocation strategies were evaluated, using formal expert judgment methods, against an array of ecological and administrative objectives. The resulting consequence table was presented to Service managers to illustrate the considerations that underlie an allocation strategy. The insights of this initial decision analysis led to further internal discussions within the Service and development of a fully articulated allocation method. In December 2013, the director of the Service approved this new, competitive, performance-based method for allocating funds to FHPs, and it has been used since then to guide decision making. This case study illustrates the power of problem framing, the importance of articulating fundamental objectives, and the value of making transparent the hidden predictions at the heart of any decision.

Problem Background

The decisions at the heart of natural resource management are often decisions about investment of scarce resources. Given a limited budget or finite staff time, how should we distribute those resources across worthy activities? Such decisions, which are familiar to economists as cost-benefit analyses and to mathematicians as constrained optimization problems, can be challenging in the natural resources setting, because the objectives may be obscure and multifaceted, the full range of options elusive, and the system dynamics complex; indeed, the form of the decision problem may not even be recognizable. The first step, as with any decision problem, is to see the crux of the issue, to recognize the nature of the decision, to understand the framework of the problem. In this case study, involving the annual distribution of federal funding to management partners for fish habitat restoration and management, the value of structured decision making for problem framing is demonstrated.

Decision Maker and Their Authority

Each year, the director of the US Fish and Wildlife Service allocates funding to support the activities of a set of partnerships established under the National Fish Habitat Action Plan. The Action Plan was developed by a diverse group convened by the Association of Fish and Wildlife Agencies (AFWA) and was signed by leaders of the AFWA, the Department of the Interior, and the Department of Commerce in 2006 (AFWA 2006). The Action Plan established a voluntary framework for public and private partners to improve conservation of fish and aquatic communities by protecting, restoring, and enhancing their habitats. The primary work units under the Action Plan are Fish Habitat Partnerships (FHPs), geographically based collections of partner agencies who undertake projects in important aquatic habitats. The number of FHPs grew quickly in the initial years, from 5 pilot FHPs in 2006 to 18 FHPs in early 2012. Projects undertaken by the FHPs are funded collectively by the member partners and through grants.

The Action Plan has enjoyed strong support of state agencies, major nongovernmental organizations, federal agencies, and Congress. Congress provided funds to the Service in 2006 for support of the Action Plan, and the administration has requested funds each year since (AFWA 2006, 2012). The Service distributes approximately $3 million annually to address strategic priorities of FHPs and supports FHP operations through its Regional Fisheries Programs. A central tenet of this distribution is that Service funds will be used to motivate and leverage funding from other partners, enhancing the benefits for aquatic species.

Ecological Background

Fish populations in the United States have benefited from habitat protection through regulatory approaches, such as under the Clean Water Act (33 USC §§1251–1387) and the Endangered Species Act of 1973 (16 USC §1531 et seq.). These measures, however, have not kept pace with human population growth and changes in land use. Fish habitats continue to decline in many places across the United States, and as of August 2019 only 2 fish species listed under the Endangered Species Act have been recovered (the Modoc sucker, *Catostomus microps*, and the Oregon chub, *Oregonichthys crameri*; https://ecos.fws.gov/ecp0/reports/delisting-report). The Action Plan provides a framework to enlist landowners, local communities, businesses, and other diverse partners to address root causes of fish habitat decline, through voluntary nonregulatory action. Protection of healthy habitat and populations from future degradation is understood to be the most cost-effective approach to achieve the goals of the Action Plan, but restoration and enhancement of degraded habitat are also valuable tools.

Decision Structure

Each year, the director of the Service allocates funding to address priorities of the FHPs, and the Service then distributes the funds to projects that "protect, restore, or enhance fish and aquatic habitats or otherwise directly support habitat-related priorities of Fish Habitat Partnerships" (US Fish and Wildlife Service Manual 717 FW 1). Initially, this allocation was made by a simple formula: larger FHPs received twice the allocation of smaller FHPs. As the number of partnerships grew (to 18 as of January 2012), funding grew at a slower rate. Older partnerships maintained the status quo while additional funds were dispersed equally among additional FHPs as they were recognized, resulting in an ad hoc distribution of project funds that was neither equal nor consciously designed to achieve the aims of the program. When the number of FHPs began to stabilize in early 2012, the Service recognized a need to develop a more fair, transparent, and strategic formula for annual funding allocation.

In November 2011, the Service's Fisheries Management Team (FMT) agreed to a conceptual allocation framework, structured around 3 functional classes of funding: a base allocation, equal across FHPs, to support operations; a proportion of the remaining budget allocated through a competitive process for science and monitoring activities; and a variable allocation to FHPs for conservation projects, taking into account past successes, strategic planning, and complexity of operations. In December 2011, the FMT tasked a team of regional coordinators to use a structured decision-making process to develop the allocation method.

The leadership of the Service desired that the allocation formula help achieve the long-term vision and goals of the Action Plan to promote effective fish habitat conservation, using sound science and principles of accountability. The FHPs vary widely in geographic extent, in aquatic habitat quality, in the number of trust aquatic species, in the complexity of their operations, and in their fish habitat conservation priorities. Changes in allocation over time could result from performance of the FHPs, new information gathered from habitat assessments developed by each FHP, or changes in aquatic and landscape conditions, management opportunities, and fish habitat priorities in the FHPs. Given the dynamic nature of the National Fish Habitat Partnership, the Service wanted the allocation method to accommodate increases or decreases in the number of FHPs as well as changes to the total budget allocated for this work.

Thus, the decision problem is how to allocate an annual budget to the existing number of FHPs, following the 3-tiered framework described above. The *objectives* are multifaceted and reflect what is necessary to reach the desired future aquatic habitat condition nationally. The *alternatives* represent a spectrum of possible approaches to allocating a fixed budget to eligible FHPs. The allocation method is founded, at least conceptually, on a *predictive model* that links each alternative to the long-term outcomes of the program. Ideally, the solution would be expressed as a formula that provides an *optimal allocation* to achieve the objectives. The formula would take as input the total available budget, the number of FHPs, and proxy descriptors of the FHPs (such as area, aquatic habitat condition, number of trust species, stressors, and partnership performance).

Decision Analysis

This chapter describes the prototype decision analysis developed during a 5-day workshop held in February 2012 at the National Conservation Training Center in Shepherdstown, West Virginia. The participants in the workshop were staff members from the Service's regional fisheries management programs (all regions except 1 were represented; see Acknowledgments). Most of the participants reported directly to a member of the FMT, which is the body that carries recommendations to the director of the Service. Two decision analysts (one from the US Geological Survey and one from a different program within the Service) facilitated the workshop. Several conference calls were held before the workshop to orient the participants and develop an initial problem statement. Several conference calls were also held after the workshop to coordinate a report that was transmitted to the FMT. The charge to the workshop participants was to develop and recommend an allocation method for deliberation by the FMT. This chapter describes the product of the workshop, emphasizing the structure of the decision problem, with only tangential mention of the details of the process as needed.

Objectives and Performance Metrics

The mission of the National Fish Habitat Action Plan (NFHAP, "to protect, restore and enhance the nation's fish and aquatic communities through partnerships that foster fish habitat conservation and improve the quality of life for the American people") is bold, multifaceted, and motivational for the Service and its partners. The primary fundamental objective identified for the allocation decision, to maximize the sustainability of aquatic species populations, reflected the overarching goal of the Action Plan. This objective seeks the long-term persistence of self-sustaining populations of fish and aquatic invertebrates, as well as the habitats and ecological communities on which they rely. Some of the species of interest are managed by the Service in the public trust: threatened, endangered, and candidate species under the Endangered Species Act; interjurisdictional species; species of management concern; Tribal trust species; anadromous and diadromous species; other native, non-game species; and fish on Service lands. Some of the species of interest are fished for recreational, subsistence, or commercial purposes. On a national scale, however, the allocation of somewhere around $3 million in a given year can have little direct effect on sustainability of aquatic populations. Instead, the Service sees the benefit of this funding acting through the creation, maintenance, and motivation of the voluntary landscape partnerships. The performance metric for this objective reflected the Service's role in supporting these partnerships, as measured by the effective annual support for FHP projects, taking into account direct project funding, the leveraged funding motivated by operational support, and increases in efficiency of management induced by scientific investment.

A number of other objectives were identified by the workshop participants as important considerations in the development of an allocation methodology (table 3.1). Some of these were fundamental objectives, some process objectives, and some strategic objectives (Keeney 2007), but all were treated as fundamental objectives for the purpose of this decision analysis. To develop performance metrics that reflect achievement of those objectives and that can be used to compare alternatives, while recognizing the time constraints of a week-long workshop, several types of measurement scales were used. Two of the performance metrics described in table 3.1 are on natural scales that directly reflect the corresponding objective, but most use a constructed scale that measures achievement in a clear, but abstract, manner. The purpose of all of them, however, was to translate an objective into a metric that could be used to evaluate alternative allocation strategies.

Maximize the capacity of the Service to support Fish Habitat Partnerships. In addition to direct funding, the Service supports FHPs through coordination at the regional level and through field office support and delivery of on-the-ground projects. Both of these activities require staff time. With competing demands and shrinking budgets, Service staff can be pulled away from support of FHPs. To the extent that the Service's FHP funding can be retained within the Service for staff support, the Service can maintain or increase its ability to support the FHPs. Thus, the performance metric for this objective was the fraction of FHP funding that could be retained within the Service.

Maximize the Service's flexibility to cope with short-term budget challenges. At the time of the workshop, the Service was facing some difficult budget circumstances, brought about by Congressional budget cuts and an inability to move or reduce permanent positions. One of the considerations raised was whether funding dedicated to FHP support could be used, on an emergency basis, to cover short-term budget shortfalls within the Service. This was a controversial topic, and the workshop participants discussed whether it was appropriate even to articulate this objective. Ultimately, they included it in the interests of generating full and transparent discussion. The performance metric was a constructed scale reflecting how much of the FHP allocation could be used on a short-term basis for emergency

Table 3.1. Fundamental objectives and performance metrics for evaluating allocation strategies for funds to support the Fish Habitat Partnerships

Objective and description	Performance metric
Maximize sustainability of aquatic species populations	Effective spending on FHP projects (in millions of dollars per year), where the calculation accounts for both leverage of external funds and increases in efficiency
Maximize FWS capacity to support FHPs: Maximize the amount of FHP funding that can be kept in house for supporting, building, and engaging in fish habitat partnerships	Fraction of funding easily accessible for FWS use
Maximize flexibility to cope with short-term budget challenges: Maximize the amount of FHP funding that could be flexibly used within a region for short-term purposes	Constructed scale (−1, 1), where −1: Allocation ties up a substantial amount of the budget in long-term commitments or competitions 0: Some flexibility exists, but it may not be easy +1: Regions have full, flexible access to a substantial portion of the budget
Maximize transparency: Maximize the degree to which our partners will perceive the allocation strategy to be transparent	Constructed scale (−1, 1), where: −1: Strategy cannot be explained such that people will think it is transparent 0: Acceptable, transparency is not an issue +1: Strategy will receive high praise for its transparency
Maximize cohesiveness of the National Fisheries Program: Maximize the degree to which the process inherently builds cohesiveness in the Fisheries Program	Constructed scale (−1, 1), where −1: Strategy is detrimental to a cohesive fisheries program, creates division 0: Acceptable, not an issue +1: Strategy will lead the way toward a harmonious, cohesive program
Maintain NFHAP structure and maximize dynamic partnerships: Encourage stronger, but not more, FHPs, with expanded participation, capabilities, and strategic focus over the long term	Constructed scale (−1, 1), where −1: Some FHPs would cease to exist due to lack of participation 0: Status quo +1: Greater participation, more leveraging, more projects, achieving strategic goals
Motivate performance-based feedback: Maximize the degree to which the funding strategy financially rewards or penalizes partnerships in future years based on past performance	Constructed grading scale (0–100), where 0: No mechanism exists for feedback <75: Failing, even if a mechanism exists, it is unlikely to create an incentive to improve performance 75: Just passing >75: Excelling, encouraging feedback, FHPs are motivated to continually improve
Maximize ease of implementation: Maximize the ease of developing a new allocation system, making the transition to the new system, and implementing annual allocations from an approved congressional budget to the FHPs	Constructed grading scale (0–100), where: <75: Failing, cumbersome process, takes forever, requires a lot of resources 75: Just passing >75: Excelling, readily implemented in a timely manner

budget situations (for example, a strategy that funded a full-time coordinator for each FHP would tie up much of the funding and would not allow short-term redirection).

Maximize transparency. One of the concerns raised about FHP funding over the several years leading up to the workshop was that the allocation mechanism kept changing and the FHPs and partners did not understand how those decisions were made. Thus, one of the fundamental objectives for the decision was to have an allocation strategy that was transparent to the partners. The performance metric was a constructed scale reflecting the degree to which partners would perceive the allocation strategy as transparent.

Maximize cohesiveness of the National Fisheries Program. Fisheries management within the Service is largely delivered at the regional and field-office levels, rather than from the national headquarters, which provides at least the potential for the philosophy and approaches of the different regions to develop in different directions. The Action Plan was perceived as a way to generate some cohesiveness in how the regions and field offices approached fisheries management. The performance metric was a constructed scale reflecting the degree to which the funding strategy would help build cohesiveness within the national program.

Maintain NFHAP structure and maximize dynamic partnerships. In 2006, the Action Plan had established a vision for voluntary, partner-driven fisheries habitat management across the country, and those partnerships had flourished in the intervening years. The participants in the workshop thought that the funding allocation strategy could affect the maintenance of that structure and the engagement of partners in the FHPs. This objective seeks stronger partnerships, with expanded participation, enhanced capabilities, and a strategic focus over the long term. The performance metric was a constructed scale indicating whether the allocation strategy would improve or degrade the engagement in the FHPs.

Performance-based feedback. One of the objectives of Service leadership was to motivate improvement in the delivery of fish habitat conservation over time, and they saw performance-based funding as a mechanism for such improvement. Taking into account past performance of projects would reward successful partnerships with increased funding. The performance metric was a grading scale (0–100) that took into account whether a mechanism existed in the allocation strategy to account for performance of past projects and how well that mechanism was likely to work.

Maximize ease of implementation. Recognizing the bureaucratic challenges of change and ongoing implementation, the last objective was to seek an allocation strategy that could be developed easily, could readily replace the existing methods, and could be implemented easily on an annual basis. The performance metric was a grading scale (0–100) that reflected the ease with which the allocation strategy could be implemented.

Alternatives

The number of ways to divide an annual budget among 18 FHPs is so large as to be impractical even to list. But rather than search among all the possible allocations, the Service was looking to develop a repeatable and transparent rule for allocation, so the focus of the decision was on alternative rules. The number of possible rules is also very large. A common approach when exploring a large number of potential alternatives, especially in an early prototype stage, is to create a small set of alternatives that spans the range of possibilities, provides stark contrast among the alternatives, and is expected to generate valuable insights about the inherent trade-offs in the decision.

Alternative allocation strategies were developed to consider different ways of motivating and supporting voluntary conservation efforts by partners in the FHPs. The annual funding available ($3.3 million in 2012) is not by itself sufficient to make a large, direct difference in fish habitat conservation nationwide; all the funds could be dedicated to the conservation priorities of any one FHP to good effect. The alternatives differed in how they allocated funds based on 3 classes of investment, each class designed to increase leverage or efficiency of conservation efforts. Class I funding ("operations") is intended to provide staffing and operational support to ensure the effectiveness of FHPs. Class II funding ("science") is dedicated to biological planning, conservation design, outcome-based monitoring, and assumption-based research, to increase the effectiveness and efficiency of conservation delivery. Class III funding ("project") is for on-the-ground, cost-shared projects to protect, restore, and enhance fish habitats or otherwise directly support habitat-related priorities of FHPs. The 4 strategies all allocate the same amount of annual funding ($3.3 million) but differ in how they base funding to the FHPs on these 3 classes of investment (table 3.2).

Coordinator focus. This alternative is driven by the belief that if limited funds were used to pay the salary and operations of a skilled, energetic, driven individual to coordinate the operations, outreach, project management, and other activities of an FHP, that FHP would be more effective in getting projects completed on the ground because it would be able to leverage far more resources than in the absence of a coordinator. This philosophy also assumes that there would be an increase in the effectiveness of the partnership through strategic project selection with a dedicated, well-informed, and focused coordinator driving decision making.

Assessment focus. This alternative assumes that using scientific information, such as habitat condition assessments, allows an FHP to make more strategic and targeted decisions with respect to putting habitat projects on the ground. By focusing the use of funds to generate the necessary assessment information, efficiency is gained by choosing better-quality projects (those more likely to help the FHP meet their objectives). It also anticipates that the partnership will create additional leverage by providing assessments and other scientifically derived tools to partners, local and regional governments, and local conservation actors to make more informed decisions that could also meet FHP objectives when selecting and completing projects.

Project focus. Under this alternative, FHPs would be given funds primarily to complete on-the-ground projects, with a smaller level of funding for operations and scientific investment. The alternative assumes that direct investment in on-the-ground projects is a good way to achieve conservation, demonstrate success, and encourage other partners to participate as well.

Regional discretion. Under this alternative, a formula (not yet specified) would be developed to allocate funding to regions, but the allocation from regions to classes of funding and then to FHPs would be determined by regional discretion. This alternative is

Table 3.2. Alternative strategies considered for allocating funding to Fish Habitat Partnerships

Alternative	Class I funding (operations)	Class II funding (science)	Class III funding (project)
A Coordinator focus	Up to $140K per FHP (for a coordinator)	None (i.e., no national competition)	Remainder (~$700K) distributed to FHPs for use in Classes II & III
B Assessment focus	$90K per FHP (FHP could use for projects)	None (i.e., no national competition)	Remainder (~$1.7M) distributed to FHPs. Initially, spend on assessment tool to be completed by 2015. Discretion of FHP between Classes II & III thereafter.
C Project focus	$50K per FHP (funding goes to lead region)	20% (~$460K), for assessment & monitoring. Seek 1:1 match from LCCs.[a]	80% (~$1.84M) distributed to FHPs for use in Classes II & III
D Regional discretion	Funding is allocated across service regions based on criteria to be defined. Regions then distribute funding across classes and FHPs as they see fit.		

Note: Each alternative describes how funding would be allocated to FHPs based on 3 classes of investment: Class I funding at a fixed level per FHP to support operations; Class II funding for science and monitoring activities; and Class III funding for project support. The total annual funding under each alternative is $3.3 million.

[a] Landscape Conservation Cooperatives.

derived from the belief that FHPs may have many different needs, that they are all unique, and that the Service's regional coordinators and other local staff are in the best position to decide what and how much to fund in order to maximize effective conservation.

These 4 alternatives describe how the national funding would be allocated across the classes of investment but not how the funding would then be allocated to FHPs. The workshop participants recognized that further development of the alternatives would ultimately be needed.

Predictive Model

To evaluate the alternatives against the fundamental objective of sustained aquatic populations, a simple predictive model was developed. The ability to reach this objective through allocation of funds is a function of the availability of on-the-ground project funds and the efficiency of implementation of those projects. Funding in each of the 3 investment classes (operations, science, and projects) affects conservation of aquatic populations through 2 mechanisms: leveraging of funds and increases in efficiency of conservation delivery (fig. 3.1). For each dollar invested in Class i, a multiplier of L_i is expected to be invested in project delivery, through the addition of voluntary partner contributions. In addition, investments in Class I and II can also increase the efficiency of project investment (at rates E_1 and E_2, respectively).

In the context of this model, leverage is defined as the ability to attract additional cash and in-kind

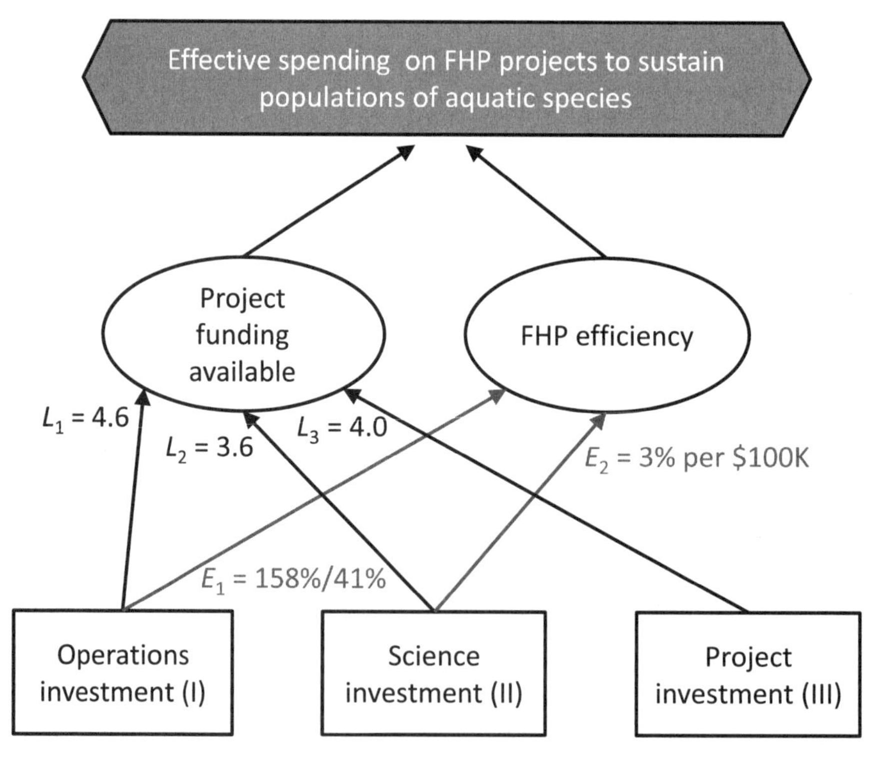

Figure 3.1. Influence diagram describing the effect of 3 classes of investment on the achievement of the fundamental objective associated with sustained populations of aquatic species. The 3 classes of investment have leverage effects (L_i), which increase the pool of funding available for conservation projects, and efficiency effects (E_i), which increase the conservation outcomes relative to the amount of money spent. The leverage effects are expressed as the ratio of the total project funding available to the amount allocated to investment class i. The efficiency effects are expressed as the percent increase in sustained aquatic populations as a result of investment in a full-time coordinator (158%), a half-time coordinator (41%), or scientific research (3% per $100,000 invested in science).

resources to advance the objectives of an FHP and could arise from investment in coordination, science, or projects. Investment in full-time, stable coordination can result in significant additional investment, assuming the coordinator writes grant applications, manages them, and actively seeks contributions from partner agencies. Investment in science and assessment could also increase resources by directing the investment of other partners. Finally, direct project funding could also leverage partner resources, particularly if projects are favored that bring substantial matching funds with them.

Efficiency is defined as the extent to which the FHP can identify the best projects in the best places and use the best practices to protect and restore habitat to sustain aquatic populations. The better informed an FHP, the more efficient it is likely to be in implementing conservation projects.

During the workshop, the leverage and efficiency parameters were estimated through a formal process of expert judgment (Hanea et al. 2017), with the workshop participants serving as the expert panel. The 3 classes of investment had similar mean leverage ratios, ranging from 3.6 for Class II investment to 4.6 for Class I investment (fig. 3.1), with wide confidence intervals reflecting considerable uncertainty about these predictions. The gains in efficiency were expected to be highest from investment in a full-time coordinator and lowest from investment in scientific assessment (fig. 3.1).

Consequence Analysis

The 4 alternative allocation strategies were evaluated against the 8 objectives, and the results were assembled into a consequence table (table 3.3). The model described in figure 3.1 was used to predict the performance of alternatives A, B, and C against the objective of sustaining populations of aquatic species. For the remaining entries in the table, formal methods of expert judgment were used (Hanea et al. 2017), with the workshop participants serving as the expert panel.

With regard to the objective of maximizing the sustainability of aquatic populations, alternative A (coordinator focus) was predicted to perform the best and alternative C (project focus) the worst. This pattern was driven by the estimates of leverage and efficiency associated with Class I investment in a dedicated, full-time coordinator to organize the operations of each

Table 3.3. Consequence analysis of 4 alternative allocation strategies against 8 objectives

		Alternatives			
Objective	Scale	A Coordinator focus	B Assessment focus	C Project focus	D Regional discretion
Sustainability of aquatic species	$	$37.92M	$21.66M	*$19.85M*	$30.00M
Capacity to support FHPs	%	62%	*58%*	62%	94%
Flexibility to cope with short-term budget challenges	−1 to +1	*−0.44*	0.06	0.39	1.00
Transparency	−1 to +1	0.72	0.44	0.50	*−0.78*
Cohesiveness of the National fisheries program	−1 to +1	0.17	0.33	0.56	*−0.44*
NFHAP structure and partnerships	−1 to +1	−0.11	0	0.72	*−0.28*
Performance-based feedback	0 to 100	73	80	85	*68*
Ease of implementation	0 to 100	81	79	77	*58*

Note: For each objective, the best-performing alternative is shaded in gray; the worst-performing alternative is in italic font.

FHP. Alternative B (assessment focus) performed nearly as poorly as alternative C, but the higher investment in operations and coordination and the efficiency gained by Class II investment provided a slight improvement for alternative B over alternative C. Alternative D was harder for the panel to evaluate, because the discretion left to the individual regions meant it was difficult to predict the investment in the 3 classes; still, the panel believed that the regions would be fairly astute in their investments, ranking alternative D second with regard to this objective.

The coordinator focus alternative (A) performed best across the alternatives for the sustainability, transparency, and ease of implementation objectives but performed worst on the flexibility objective (table 3.3). Establishing a fixed investment in each FHP to support a dedicated, full-time coordinator would provide considerable transparency in the allocation strategy and be easy to implement on an annual basis (in part because the allocations would not change much from year to year), but much of the funding would be tied up in permanent positions, leaving little flexibility to address short-term budget challenges.

The assessment focus alternative (B) did not perform best on any objective and performed worst on only 1 objective (capacity to support FHPs). It ranked second for the performance-based feedback objective, because the emphasis in establishing strong assessment frameworks means the FHPs would have the tools to evaluate and improve their practices.

The project focus alternative (C) emphasized allocating funds directly to on-the-ground projects. While this focus meant the predicted value of the effective conservation delivery (first objective) was lowest among the alternatives, it performed best among the alternatives for the cohesiveness, NFHAP structure, and performance-based feedback objectives. The emphasis on projects was seen as a visible way to motivate both the Service's fisheries programs and the NFHAP partnerships.

The regional discretion alternative (D) differed starkly from the other 3 alternatives in leaving most of the allocation decisions to the individual regions. Although this meant the effective conservation delivery was high ($30 million) and the capacity to support FHPs and the flexibility to cope with budget challenges were predicted to be highest among the alternatives, this alternative also performed worst on the remaining 5 objectives. In particular, this mechanism of allocation would not be transparent, and the implementation might be challenged on a regular basis as a result. Further, the regional focus undermines the objectives that focus on the national fisheries and NFHAP programs.

The patterns of performance of the 4 alternative allocation strategies against the 8 objectives revealed substantial trade-offs among the objectives, such that no single alternative was evidently best or worst. In fact, no alternative dominated or was dominated by the others. The choice of an allocation strategy, then, comes down to how to balance the objectives against one another. Further decision analysis along those lines, using techniques from multi-criteria decision analysis, was not undertaken for this prototype; those methods are addressed in later chapters of this book.

Three salient insights arose from this decision analysis and helped to remove the impediments to establishing a stable, satisfactory, and efficient allocation strategy. First, this prototype decision analysis established the structure of the underlying decision, namely, the search for an allocation strategy that balanced multiple objectives. Second, the fundamental objectives were articulated. Previous efforts had made the first objective clear (sustainability of aquatic populations), but the other objectives had not been expressed clearly enough to be evaluated. Third, 2 important hidden assumptions were revealed: the roles of leverage and efficiency in how investment decisions translate into conservation outcomes. This insight provided clarity that could help explain past disagreements over the relative effectiveness of different strategies and allow a more transparent evaluation of the alternatives.

Decision Implementation

After the workshop at which this prototype decision analysis was developed, the participants documented the work in an internal report (Gallagher et al. 2012) and provided briefings to the members of the Service's Fisheries Management Team. The participants, who were regional fisheries coordinators, but not decision makers themselves, recommended that the FMT proceed through the remaining steps of a multi-criteria decision analysis and select a preferred alternative. With such a strategy selected, further details of implementation could then be developed.

Over the course of the next year, from late 2012 until late 2013, the Service undertook continued internal deliberation about the NHFAP allocation strategy. The prototype decision analysis described here provided conceptual guidance and insight, but more formal methods of decision analysis were not pursued. In December 2013, the director of the Service approved a new, competitive, performance-based method for allocating funds to FHPs. The allocation method provides a fixed allocation to each FHP for operational support (Class I investment) and a performance-based competitive allocation for discretional use by the FHP for Class II and III investments; thus, it most resembles alternative A in the prototype described in this chapter. The performance-based portion is allocated based on scores assigned to each FHP that consider, among other criteria: whether projects completed in the last 3 years address Service priorities; the completion rate of projects; the amount of leveraged funds that are brought to the projects; and whether the projects include monitoring and evaluation plans. Beginning with fiscal year 2014 funding, this allocation strategy has been used annually to provide funding to the FHPs.

Discussion

This case study demonstrates the value of decision structuring. Prior to the workshop described in this chapter, the Service had undertaken several efforts to develop an allocation rule for sending funding to the FHPs, but consensus had been difficult to reach. In part, this was because the full set of objectives held by the decision makers and stakeholders had not been explicitly articulated, nor had fundamental objectives been distinguished from means objectives. As a result, it was difficult to evaluate any proposed allocation scheme. The time taken to articulate the decision structure, particularly the development of the fundamental objectives and the recognition that the choice was between alternative methods of allocation, opened up the discussion and provided new avenues for evaluation.

The decision analysis described in this chapter, however, is a prototype, not a finished product, and many of the elements could be refined. The objectives (table 3.1) are a mixture of process (the transparency objective), strategic (cohesiveness of the fisheries program and NFHAP structure objectives), and fundamental objectives. Generally, multiple-objective analyses should focus on fundamental objectives, so some additional thought could be given to whether these strategic objectives need to be treated as fundamental in this case, and whether transparency is something the decision maker really wants to consider trading off. The performance metrics (table 3.1) are mostly constructed scales, with few guideposts, that require expert scoring; they were understood by and valuable to the participants, but they are difficult for others to interpret. This is a common feature of prototypes—the performance metrics are often rough constructed scales that capture the essence of the trade-offs but also identify where more development is needed. The alternatives in this prototype (table 3.2) were an incomplete set of partially described options. Alternative D (the "regional discretion" alternative) was vague by design, as it was meant to capture a starkly different approach that did not specify the allocation method at the national scale, but it is difficult to understand the consequence analysis associated with it. All of the alternatives left out details about how the funding would be allocated from funding classes to FHPs, a necessary step in any final allocation protocol.

This case study, however, demonstrates the value of using a process of rapid prototyping to understand the decision structure (Garrard et al. 2017). The purpose of any prototype in a sequence of prototypes is not to be perfect but rather to be better than the last prototype and to generate insights that lead to further improvement. Three insights stand out in this case study.

1. The transparent articulation of 8 objectives was important; prior to this workshop, disagreements had been difficult to understand because 7 of the objectives were only implicit. Making them explicit allowed focused conversation about their suitability and importance.
2. The consequence analysis regarding the first objective (sustainability of aquatic species) produced profound insights about the nature of the allocation decision, especially that the funding available would never be enough to directly achieve the first objective but instead had its most important effects through its ability to attract other funding and to improve the efficiency of investments. These concepts (leverage and efficiency) became central to how the participants thought about the allocation question and represent important insights for the large class of funding allocation decisions in natural resource management.
3. The specific estimates from this case study for the efficiency parameters were somewhat surprising, in particular, the low estimate for the efficiency gained through investment in scientific studies (the expert panel estimated that each $100,000 investment in scientific assessment would increase the efficiency of on-the-ground projects by 3%, fig. 3.1). This estimate deserves scrutiny but reveals the sense by the experts that in this setting, there is more to be gained by the dissemination of existing information (e.g., through a coordinator) than by investment in new information.

All of these insights were important in shaping the thinking of the Service as it proceeded with subsequent development of the allocation strategy.

The type of decision problem exemplified by this case study—an allocation decision that seeks to achieve multiple objectives—is common in many settings. Often, the most direct decisions we make in natural resource management concern how we allocate funding to worthy endeavors. There are quantitative tools available for analysis of allocation decisions (Converse et al. 2011; Joseph et al. 2009), which are discussed in greater length in later chapters of this book. The initial steps of decision structuring, however, often provide critical insight that allows the decision maker to identify an effective solution.

ACKNOWLEDGMENTS

The regional fish coordinators and others who participated in the workshop in February 2012 generously shared their time, expertise, and patience to develop this decision analysis: Maureen Gallagher, Tom Busiahn, Bob Clarke, Vicki Finn, Jennifer Fowler-Propst, Jarrad Kosa, Steve Krentz, Callie McMunigal, Cecil Rich, and Cindy Williams. Maureen Gallagher was instrumental in documenting the work. Terry Doyle helped to facilitate the workshop. Insightful comments that helped in the preparation of this chapter were provided by Cecelia Lewis, Lynn Maguire, and Terry Walshe.

LITERATURE CITED

Association of Fish and Wildlife Agencies [AFWA]. 2006. *National Fish Habitat Action Plan*. Washington, DC: Association of Fish and Wildlife Agencies.

———. 2012. *National Fish Habitat Action Plan*. 2nd ed. Washington, DC: Association of Fish and Wildlife Agencies.

Converse SJ, Shelley KJ, Morey S, Chan J, LaTier A, Scafidi C, Crouse DT, Runge MC. 2011. A decision-analytic approach to the optimal allocation of resources for endangered species consultation. *Biological Conservation* 144:319–329.

Gallagher M, Busiahn T, Clark B, Doyle T, Finn V, Fowler-Propst J, Kosa J, Krentz S, McMunigal C, Rich C, Runge MC, Williams C. 2012. *Allocating FWS Funds Available for*

Projects under the National Fish Habitat Action Plan. A case study from the Structured Decision Making Workshop, February 6–10, 2012. Shepherdstown, WV: US Fish and Wildlife Service, National Conservation Training Center.

Garrard GE, Rumpff L, Runge MC, Converse SJ. 2017. Rapid prototyping for decision structuring: an efficient approach to conservation decision analysis. Pages 46–64 in Bunnefeld N, Nicholson E, Milner-Gulland EJ, eds. *Decision-Making in Conservation and Natural Resource Management: Models for Interdisciplinary Approaches*. Cambridge, UK: Cambridge University Press.

Hanea A, McBride M, Burgman M, Wintle B, Fidler F, Flander L, Twardy C, Manning B, Mascaro S. 2017. I nvestigate D iscuss E stimate A ggregate for structured expert judgement. *International Journal of Forecasting* 33:267–279.

Joseph LN, Maloney RF, Possingham HP. 2009. Optimal allocation of resources among threatened species: a project prioritization protocol. *Conservation Biology* 23:328–338.

Keeney RL. 2007. Developing objectives and attributes. Pages 104–128 in Edwards W, Miles RFJ, Von Winterfeldt D, eds. *Advances in Decision Analysis: From Foundations to Applications*. Cambridge, UK: Cambridge University Press.

4 Keeping Hawai'i's Forest Birds One Step Ahead of Disease in a Warming World

Eben H. Paxton
and Jim Kraus

Hawai'i's high-elevation forests provide a critical refuge from disease for native forest birds. However, global warming is facilitating the encroachment of mosquitoes and the diseases they transmit into increasingly higher elevations of remaining refugia, threatening the viability of the forest birds across the islands. Multiple management actions to address the threat of disease have been proposed, but there is an urgent need to identify which actions (or series of actions) should be prioritized as most effective, most cost-efficient, and most likely to produce results at a pace sufficient to stay ahead of climate change. A group of scientists, managers, and policy makers convened to evaluate a set of possible conservation strategies under a structured decision-making framework, focusing on management of Hakalau Forest National Wildlife Refuge, which was established to protect native Hawai'ian forest birds. The biological models necessary to evaluate the set of conservation actions identified are not yet available, but the process of developing the framework for the decision analysis was immensely valuable for framing the issues and identifying information needs. Lessons learned from Hakalau Forest will be applicable to many other areas in Hawai'i facing the same threat to forest birds.

Problem Background

Hakalau Forest National Wildlife Refuge (Hakalau) was created specifically to be a high-elevation refuge for Hawai'ian forest birds and their habitat, including 3 endangered species, with a mandate to protect these threatened birds as well as all the native species the refuge harbors. Hakalau provides a critical refuge for native birds from vector-borne diseases and is one of the only places in the state where native forest bird populations are stable or increasing (Camp et al. 2010, 2015). Unfortunately, global warming is predicted to facilitate the encroachment of mosquitoes and diseases into increasingly higher elevations of the refuge while intensifying disease at lower elevations (Liao et al. 2015), requiring urgent conservation strategies to mitigate the threat. However, conservation strategies are constrained by uncertainty on effective management responses, limited budgets, land availability (e.g., jurisdictional issues and habitat availability), and public perception and cooperation. As a result, a process for developing and comparing alternate management actions at Hakalau is needed to preserve the refuge's native birds from avian disease and other threats. While refuge managers have direct control over on-the-ground actions at the refuge, some actions require cooperation from surrounding land managers and the community. Actions to confront these threats to native bird populations may need to be started now, even though the actual peril could be years in the future. The 2 key uncertainties to this issue are the speed at which disease will intensify or move up in elevation and the effectiveness of various management actions in slowing or preventing the incursion of disease or increasing resiliency of existing populations to disease. Although these management strategies will apply specifically to Hakalau and to surrounding landholdings, lessons learned from Hakalau should be applicable to other areas in Hawai'i facing similar threats to their forest birds.

Decision Maker and Their Authority

The decisions on which management actions to implement at Hakalau ultimately rest with the refuge manager. Hakalau exists to protect native fauna and flora within its boundaries and surrounding areas and must conduct management actions consistent with the preservation of native bird populations as typically required by policies and rules the refuge must act under. Multiple executive orders and legislative acts apply to the management of Hakalau, including at least 11 broad federal regulatory requirements, specific rules governing National Wildlife Refuges, and guidance specific to Hakalau. These rules and regulations provide a multitiered policy framework designed to guide refuge management. However, the spatial scale at which many possible management actions should be conducted exceed the boundaries of the refuge, and therefore neighboring land managers also make decisions that could influence the ultimate outcome of the problem.

Ecological Background

The Hawai'ian Islands have evolved a highly endemic avifauna as a result of geographical isolation, diverse topography ranging from sea level to mountains exceeding 4,000 meters above sea level (m asl), and habitats ranging from tropical lowland rain forests to subalpine tundra to deserts over distances as small as 40 km. Native Hawai'ian forest birds have experienced one of the highest rates of extinction in the world (Banko and Banko 2009) because of habitat loss and the introduction of alien plants and animals. While the conservation efforts to date have been substantial and have achieved successes, the scale of action has not matched either the scale of the threats or the conservation goals. Unfortunately, we are losing, not gaining ground in the race against extinction. In the past 25 years, 10 species of endemic Hawai'ian birds have been lost to extinction, and only 24 of the 46 historically known forest bird species still survive, with 13 listed as endangered (Banko and Banko 2009) and the entire community on Kauai rapidly declining (Paxton et al. 2016).

Today, most Hawai'i forest birds are restricted to high-elevation forests (over 1,500 m asl) largely as a result of 2 major factors. First, most low-elevation native habitats have been lost or heavily degraded, leaving remaining intact lowland forests scattered and fragmented. Secondly, introduced diseases, specifically avian malaria and avian pox, along with an introduced mosquito (southern house mosquito, *Culex quinquefasciatus*) that efficiently transmits these diseases (LaPointe et al. 2005), have largely displaced susceptible native birds from low-elevation forests where disease transmission occurs throughout the year (Atkinson and LaPointe 2009; Samuel et al. 2011) and many species in midelevation forests where disease is seasonally epizootic (Atkinson and Samuel 2010). These elevational disease patterns are driven, in part, by the effects of temperature and rainfall on mosquito dynamics (Ahumada et al. 2004; Samuel et al. 2011) and sporogonic development of the avian malaria parasite (*Plasmodium relictum*) within the mosquito vector (LaPointe et al. 2010) across an elevational gradient. The threshold temperature for malarial development is 13°C, which coincides with the 1,800 m asl elevation contour (mean annual temperature); this temperature threshold creates a disease-free refuge in forests above approximately 1,800 m asl (Atkinson and LaPointe 2009; Benning et al. 2002). At elevations between 1,500 and 1,800 m asl (corresponding to 13–17°C), seasonally favorable conditions can allow disease transmission to occur, typically in the late summer and fall (LaPointe et al. 2010).

Because avian malaria and pox are primarily spread through mosquitos, distribution and abundance of mosquitoes are key to understanding the distribution of these avian diseases. In most mid- and high-elevation Hawai'ian forests, *Culex quinquefasciatus* populations are limited by the availability of aquatic larval habitat; on Hawai'i Island this is primarily rain-filled tree fern (hapu'u) cavities created by foraging feral pigs and rock pools along the margin of intermittent stream beds (Atkinson and LaPointe 2009). Unmanaged artificial water impoundments can also contribute significantly to available larval habitat (Reiter and LaPointe 2009). Approximately 95% of Hakalau lands lie at elevations above the 17°C isotherm for seasonal disease transmission, effectively creating a refugia from year-round disease, although periodic disease outbreaks may occur in the lower and mid parts of the refuge (Freed et al. 2005; VanderWerf 2001). Thus, the refuge (fig. 4.1) preserves habitat for limited but stable populations of 3 endangered forest birds, the Hawai'i 'Akepa (*Loxops coccineus*), Hawai'i Creeper (*Oreomystis mana*), and Akiapola'au (*Hemignathus*

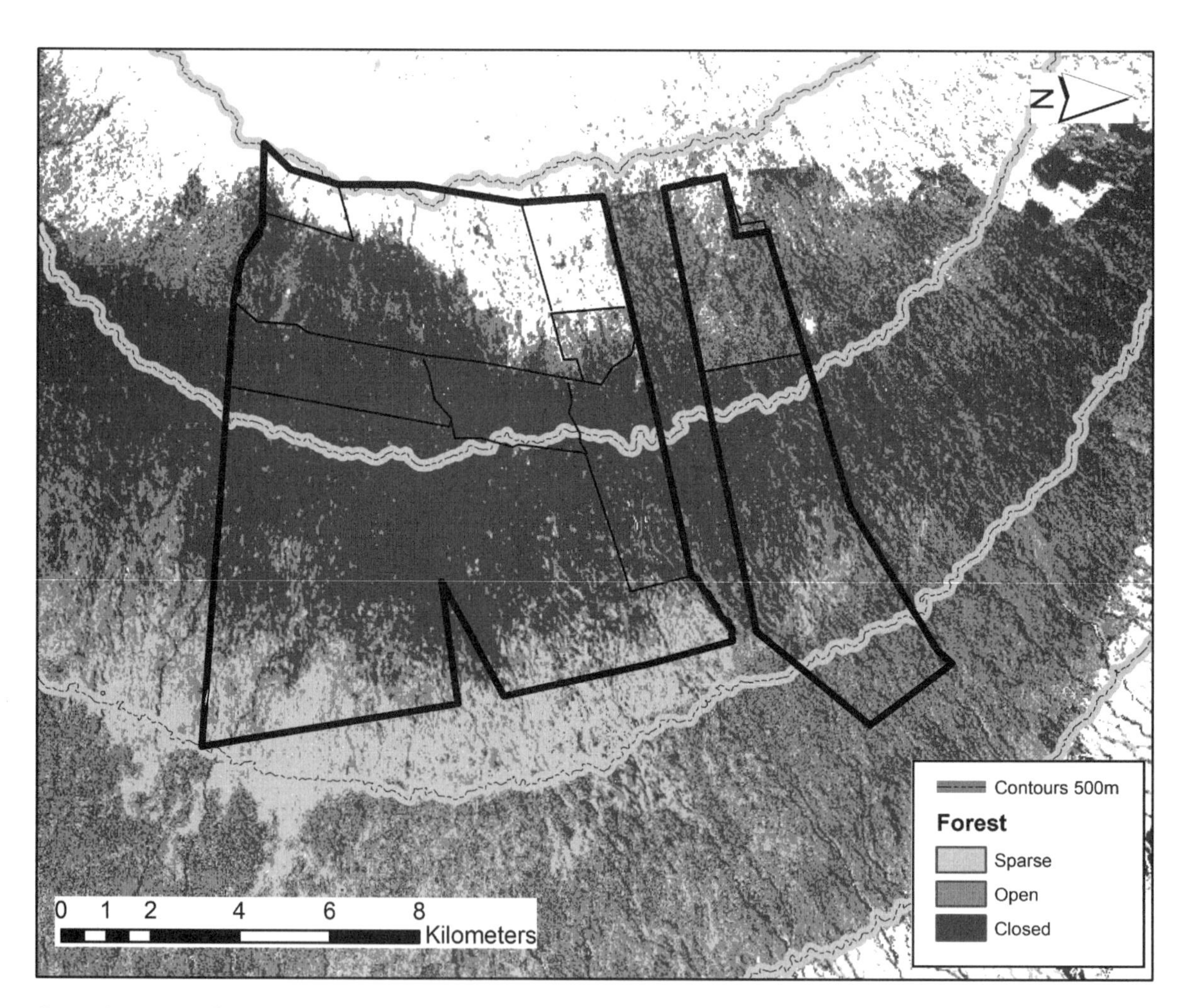

Figure 4.1. Map of Hakalau Forest NWR (thick black lines) and its management units (thin black lines). Light gray areas within the refuge are actively forested with koa trees. Active ungulate control occurs in the upper units. Area between the 2 parcels is Piha Forest Reserve. The 13,355 ha Hakalau was created in 1985 to protect rainforest that supports endangered forest birds. Located on the windward slope of Mauna Kea Volcano on Hawai'i Island, the refuge contains some of the best remaining examples of native rainforest in the state.

munroi) as well as larger populations of non-endangered Iʻiwi (*Drepanis coccinea*), ʻApapane (*Himatione sanguinea*), Hawaiʻi ʻAmakihi (*Chlorodrepanis virens*), Hawaiʻi ʻElepaio (*Chasiempsis sandwichensis*), and Omaʻo (*Myadestes obscurus*) (Camp et al. 2015).

Because temperature is a critical element in Hawaiʻi's disease-bird cycle, global warming is considered a grave threat to the disease-free sanctuaries of high-elevation forests for native forest birds. Currently, disease exposure in Hakalau forest birds is limited in the upper elevations, with a 2012 survey of avian malaria exposure (based on sampling blood from wild birds) indicating highs of 14% in Omaʻo and 9% in ʻApapane but low to zero levels in the remaining species (LaPointe et al. 2016). Surprisingly, prevalence of malaria was significantly lower in 2012 (4.9% out of all birds sampled) than 1998 (12.4%), the last period that birds were sampled for avian malaria in Hakalau. In addition, of 1,004 birds examined, 5% had evidence of active avian pox infections (LaPointe et al. 2016). However, temperatures are rising, particularly at higher elevations (Diaz et al. 2011), and recently developed downscaled climate change models predict a temperature increase of 2.5°C or more across Hawaiʻi by 2100 (Zhang et al. 2012). The increasing temperatures will effectively eliminate this high-elevation transmission-free habitat within the current boundaries of the refuge (Benning et al. 2002), with expected profound negative effects on forest bird populations in Hakalau and across the Hawaiʻian Islands (Fortini et al. 2015; Liao et al. 2015). However, recent evidence of disease resistance in native Hawaiʻian forest birds (Atkinson et al. 2014; Woodworth et al. 2005) demonstrates that at least some species, in at least some areas, are developing resistance or tolerance to disease, and there is hope for similar evolution occurring in other species and populations.

Structuring the Decision

A group of 12 researchers, managers, and policy experts (see Acknowledgments) came together in 2011 to develop strategies for managing the looming problem of climate change and disease in forest birds. The group identified multiple objectives and developed alternative management strategies that encompassed a range of identified conservation actions. The biological models necessary to evaluate the alternative management strategies do not yet exist, but the process was immensely valuable for defining the problem, identifying key information needs, and providing a framework for how to move forward on addressing the issue.

Objectives

The primary objective was maintaining the abundance and diversity of forest birds at Hakalau. However, other objectives were identified as important and needed to be addressed. The 4 objectives were defined as follows:

1) Maintain each forest bird species at or above their current abundance. Our primary objective was to maintain (or increase) abundance of each native forest bird, which also implicitly ensures that current diversity is maintained on the refuge. Management actions would be assessed by using the most recent survey results (Camp et al. 2010, 2015) as a baseline to gauge alternative management strategies (table 4.1). We chose a 200-year period as the time scale, so that strategies were long-term in nature and would consider climate change and other long-term trends.
2) Minimize cost of management. The cost of alternative actions was identified as a key objective to recognize the reality that funds are limited and, all else being equal, the set of alternative actions that cost the least amount would be favored. Different management alternatives have different up-front and long-term costs (for example, building fences has significant upfront costs but comparatively low maintenance costs). We calculated costs for all sets of actions in a 15-year framework, which is about the typical refuge conservation plan cycle; these costs are approximations but give an indication of the spread of resources needed

Table 4.1. Abundance estimates with 95% confidence interval (CI) for native Hawai'ian forest birds at Hakalau Forest National Wildlife Refuge

Species	Scientific name	Abundance (95% CI)
Akiapola'au	*Hemignathus wilsoni*	575 (306–1,044)
Hawai'i 'Akepa	*Loxops coccineus*	11,012 (7,331–15,740)
Hawai'i Creeper	*Oreomystis mana*	13,106 9,276–17,626)
I'iwi	*Drepanis coccinea*	111,917 (94,078–130,165)
Oma'o	*Myadestes obscurus*	16,396 (12,583–20,836)
Hawai'i 'Elepaio	*Chasiempis sandwichensis*	18,196 (14,738–22,039)
'Apapane	*Himatione sanguinea*	77,811 (59,668–98,538)
Hawai'i 'Amakihi	*Chlorodrepanis virens*	48,691 (39,073–60,430)

Source: Camp et al. 2015.

Note: Objective 1 was to maintain populations at or above the current abundance levels based on the most recent survey results from 2012.

to achieve each alternative (table 4.2). Units for management costs were in US dollars.

3) Maximize probability of acceptance by the public. While the management alternatives that we considered are largely focused on the refuge, which could be conducted with a focus solely on refuge priorities, many actions would benefit from general public acceptance and cooperation with neighboring landowners. The refuge is surrounded by other landowners that have forest habitats adjacent to the refuge, and a conservation strategy that is adopted by neighboring land managers is likely to increase the success on the refuge. For example, a strategy that considered mosquito control through the application of a larvacide might require public input for the authorization stage. There are multiple approaches for measuring this means objective, including elucidating expert opinion, conducting surveys, or leading active outreach programs that measure attitudes toward different conservation actions.
4) Abide by all relevant legal statutes. The group realized that some actions might not be allowed under current law (e.g., aerial broadcasting of rodenticides to control rats), while other actions might indirectly violate other laws, such as spraying insecticides to control mosquitoes, which could harm endangered insects or run contrary to water quality regulations. Therefore, to constrain any possible actions into what is legally permissible, we identified the metric for this process objective as a simple binary yes/no.

Alternatives

To accomplish the primary objective (maintain current population levels of the 8 forest birds native to Hakalau for 200 years) in the face of rising temperatures and disease risk would require the adoption of new landscape-scale management initiatives of unknown efficacy. Without such initiatives, the disease-free area of the refuge would contract as global temperatures rise, until at some future time when no areas within the refuge boundary would be disease-free, and birds sensitive to disease would suffer population declines or local extinction. To prevent extinctions and extirpations, management actions are needed, such as providing new disease-free refugia, rendering existing habitat disease-free, or reducing the mortality of disease on native forest birds.

The group identified specific management actions that could be grouped into 4 broad categories, with associated costs. The groups of management actions included disease control approaches that were "low tech" (e.g., ungulate control, ungulate-proof fencing, mosquito control) and "high tech" (e.g., sterile male mosquito release, bio-engineered mosquitoes or *Plas-*

Table 4.2. Detailed breakout of costs associated with each alternative strategy over a 15-year time frame

Status quo	
Ungulate control ($175K/yr.)	$2,625K
Fence maintenance/replacement ($175K/yr. per 20-year replacement cycle)	$2,625K
Reforest koa ($75K)	$1,125K
Habitat enrichment ($75K)	$1,125K
New fencing (18.5 mi./5.6K additional acres)	$1,954K
Weed Control ($100K/yr.)	$1,500K
	$10,954K
A. Status quo + refuge-wide mosquito control	
Status quo	$10,954K
Additional ungulate control ($175K/yr.)	$2,625K
Additional fence maintenance/replacement ($175K/yr.)	$2,625K
Hapu'u cavity removal (eliminate larval habitat; $600K/yr.)	$9,000K
Streambed spray (treat twice/yr., $51K/yr.)	$765K
Heli-time ($775/hr. twice per yr., $15.5K/yr.)	$188K
Acquire or land swap for Piha GMU land	Unknown
	$26,157K +
B. Status quo + new mauka forest habitat (4,000 acres DHHL Humuʻula lands)	
Status quo	$10,954K
New fencing (10 miles @ 100K/mi.)	$1,000K
Fence replacement (50-year cycle/3x)	$900K
Ungulate eradication (DHHL 1yr. ungulate free)	$175K
Reforest koa and native understory ($2,000/acre)	$600K
Water system development	$1,000K
Cooperative management ($100K/yr. × 15-yr. joint restoration and coordination)	$1,500K
Weed control (DHHL $50K/yr. × 15 yr.)	$750K
	$16,879K
C. Status quo + disease compensation	
Status quo	$10,954K
Predator control ($200K/yr.)	$3,000K
Enrich natural food source ($75K/yr.)	$1,125K
Artificial food source ($50K/yr.)	$750K
	$15,829K
D. Status quo +building resistance	
Status quo	$10,954K
Augment native birds ($n = 20$/yr. @ $3,000 per; $60K/yr.)	$900K
Augment w/ disease-resistant birds ($n = 100$ @ $1,000 per; $100K/yr.)	$1,500K
	$13,354K

modium parasite). We also considered habitat manipulations (e.g., removal of mosquito larval habitat, reforestation of refuge or adjacent lands, removal of invasive weeds, enhancement of food resources) and population manipulations (e.g., removal of mammalian predators, augmentation of bird populations to increase genetic diversity, and translocation of individuals from disease-tolerant or -resistant populations to Hakalau). Finally, we considered land management arrangements that would help achieve conservation goals (e.g., comanagement with adjacent landowners, purchase of key land parcels).

Combinations of these actions resulted in 12 strategies, of which we considered 4 in detail (table 4.2), as they represented distinct approaches to addressing the looming conservation problem. These 4 new strategies assume continuation of the current management strategies and activities ("status quo"), but we felt it was important to explicitly consider the status quo (i.e., baseline management actions) as a standalone alternative action to quantitatively evaluate the consequences of "doing nothing." Each strategy varied in terms of the area affected (e.g., upper refuge, entire refuge), whether management was directed at improving habitat or directly targeted at bird populations, and whether cooperation with surrounding landowners was necessary. The strategies examined were:

- Status quo. Current management activities designed to benefit native forest birds at Hakalau, which are envisioned to continue for perpetuity, represent a baseline level of management effort. However, this strategy entails no additional action in the face of new threats from climate change, and as such we included it in the decision process to evaluate the added benefit of actively mitigating for the new threats, or the risks of "doing nothing."
    - *Area*: Current 8 fenced units for ungulate control, largely disease-free, 14,000 total acres in the upper portion of the refuge.
    - *Habitat management*: Reforestation of former pasture with koa and understory plant species, maintain fences and ungulate control efforts.
    - *Population management*: None.
    - *Co-management*: None.
    - *Cost (15 yrs.)*: $10,954,000.
- Reserve-wide mosquito kill plus. Strategy A aims to provide a disease-free safe haven by completely removing larval mosquito habitat from across the entire refuge. The mosquito-free area would be less than the total refuge area due to the ability of adult mosquitoes to penetrate some distance (approximately 1 km) from bordering unmanaged lands, but the reduction of disease transmission (regardless of temperature) would allow vulnerable species to expand or at least maintain their current distribution.
    - *Area*: Full refuge extent, 32,000 acres + intervening 4,000 acres in Piha, a state forest reserve lying between the 2 units of Hakalau.
    - *Habitat management*: Status quo actions plus new ungulate fencing and pig removal from lower portions of refuge, removal of artificial habitat for mosquitoes, larvicide treatment of streambeds and hapu'u cavities.
    - *Population management*: None.
    - *Co-management*: Management of Piha State Forest Reserve consistent with refuge management.
    - *Cost (including status quo, 15 yrs.)*: $26,157,000.
- New refugia upslope. Strategy B would expand the forested area of the refuge upslope, providing more high-elevation, disease-free habitat to compensate for the encroachment of disease into lower areas of the refuge that are currently disease-free. This expansion is limited by the trade-wind inversion, which determines the tree line through precipitation and is expected to continue to confine forests below approximately 2,500 m (Cao et al. 2007).
    - *Area*: Current 8 fenced units (14,000 acres) for ungulate control in upper portion of refuge + 2,000 acres upslope of refuge.
    - *Habitat management*: Status quo actions plus new ungulate fencing, cattle and pig removal, reforestation with koa and native understory plants, removal of artificial mosquito larval habitat in new areas upslope of the refuge.
    - *Population management*: None.
    - *Co-management*: Co-management for wildlife with landowners of adjacent upslope lands.
    - *Cost (including status quo, 15 yrs.)*: $16,879,000.
- Disease compensation/facilitating evolution. Strategy C implements predator management and enhancement of food resources to increase survivorship, productivity, and carrying capacity for bird populations with the intent that the increase in demographic rates would offset disease-related mortality and decrease the risk

of population declines. One approach would be to target midelevation populations where there is moderate disease transmission, with the hope that disease-resistant or disease-tolerant genotypes may be more likely to survive and become more abundant in the population (Kilpatrick 2006). This action assumes that increasing the population size of native birds should slow disease-driven declines, providing time for disease resistance or tolerance to develop (Kilpatrick 2006).

- *Area*: Focused on area of refuge with seasonal disease transmission (midelevation), which will vary over time.
- *Habitat management*: Status quo plus enhancement of food resources through outplanting of understory and artificial feeders. Research would be needed to determine most effective approaches to enhance food resources.
- *Population management*: Intensive suppression of mammalian predators, specifically trapping for rats to decrease nest predation and possible predation of incubating females.
- *Co-management*: None.
- *Cost (including status quo, 15 yrs.)*: $15,829,000.

- Building resistant genotypes. Strategy D uses a different strategy to promote the evolution of disease resistance. Translocations to the refuge would be used to maximize genetic diversity of resident bird populations and to augment their gene pools by introducing disease-resistant individuals from other populations.
  - *Area:* Entire refuge (32,000 acres).
  - *Habitat management:* Status quo, only in current 14,000 acres.
  - *Population management:* Translocate cohorts of all 8 species into refuge to increase genetic diversity, including the translocation of disease-resistant birds of any species with populations persisting at low elevations (e.g., Hawai'i 'Amakihi from the Puna district of Hawai'i Island).
  - *Co-management:* None.
  - *Cost (including status quo, 15 yrs.):* $13,354,000.

Consequences and Trade-Offs

The alternative strategies identified here represent different approaches to maintaining bird populations in the face of almost certain failure of the status quo approach over the long term. Identifying and describing key objectives and alternative strategies to address this problem was a major accomplishment of the group, and the clarity brought from that effort has helped guide subsequent discussions on how to mitigate the threat. However, formal assessment of the consequences and trade-offs of each alternative action requires a population model that can predict the potential outcome of each management action based on the changing rates of disease transmission over time. The model would need to be spatially explicit to account for refuge boundaries, management area effects, and impacts of bird and mosquito movement across the landscape. To evaluate strategies C and D would also require incorporating evolution of disease resistance as a submodel. Because many of the actions would take decades to have their full effect, the rate of temperature (and rainfall) change due to global warming is a critical variable in modeling the potential of each strategy to achieve the long-term objectives.

One of the major benefits of this process was identifying key gaps in knowledge that contribute to our uncertainty in how forest bird will respond to future climates and in the effectiveness of management actions to mitigate any negative changes. Since this effort began, several new studies have been conducted, many as a direct result of this decision analysis process that highlighted research needs. For example, Fortini et al. (2015) used recently developed downscaled climate projections for Hawai'i to develop climatic-based species distribution models for all forest birds and projected future distributions under 2100 climate conditions showing how the ranges of forest birds would severely contract as disease spreads into high-elevation forest. Liao et al. (2015) projected disease dynamics in future years on 3 species of Hawai'i honeycreepers by applying a detailed epidemiological model of the Hawai'ian forest bird disease system (Samuel et al. 2011) to downscaled climate models, highlighting that responses to changing disease distribution will vary among species. In addition, behavior differences among species can alter their risk to changing conditions, as Guillaumet et al. (2017) demonstrated by modeling landscape-level movements of the nectarivorous 'I'iwi in search of flowering trees that can take it to lower-elevation forests where disease occurs. Importantly, both the Liao et al. (2015) and Guillaumet et al. (2017) efforts developed models that link changing climate conditions to forest bird demographics, a class of models needed to fully evaluate the decision analysis presented in this chapter. Additionally, information on survival, recruitment (Guillaumet et al. 2015), and productivity (Cummins et al. 2014) provide the basic demographic rates needed to estimate population processes. Studies of forest bird populations in restoration forests (Paxton et al. 2017) provide important information on rates of colonization for proposed reforestation efforts.

Once the models are available to estimate the effects of alternative actions on our primary objective, we can move to the next step of choosing preferred alternatives in light of all objectives. The main approach we identified for conducting trade-offs is the SMART table (simple multi-attribute rating technique), where alternative actions could be assessed by how well they addressed the identified objectives (Gregory et al. 2012). Alternatively, a Pareto optimal set or efficiency frontier could be explored to choose alternatives that achieve the best ecological results for least cost (Keeney and Raiffa 1993). We also considered several approaches to incorporating the considerable uncertainties in this complex system. One approach would be to use a sensitivity analysis within a trade-off analysis, where predicted consequences for each alternative would be varied to see whether changes in expected outcomes would alter the overall decisions. If changes in the predicted consequences of an alternative strategy do not change the ultimate decision, then uncertainty does not matter for the purposes of making decisions. Likewise, the trade-off analysis could be conducted under a reasonable range of different climate change scenarios, each with their own predicted environmental responses. If the decision outcome were the same under the different climate scenarios, then climate change uncertainty could be removed from the decision process. However, if the sensitivity analysis indicated that different climate scenarios would affect the decision process, then climate change uncertainty would need to be considered when deciding which alternatives to adopt.

Discussion

Developing long-term conservation strategies to maintain the viability of Hawai'i's forest birds requires addressing the substantial issues of avian disease, habitat degradation, and other negative effects from non-native plants and animals (Paxton et al. 2018). The conservation challenges are many (Pratt et al. 2009), and the window of opportunity to effectively act on behalf of Hawai'i's forest birds is rapidly declining (Paxton et al. 2016). A primary limiting factor in addressing these issues is the lack of funds to carry out many of the basic conservation needs (Leonard 2008), let alone implement new strategies. However, the identification of key management strategies, as identified here, will help prioritize scarce resources to maximize the conservation benefits.

While recent studies have increased our understanding of Hawai'i's forest bird dynamics in present and future climates, there is still tremendous uncertainty in what future conditions will be, how they will manifest themselves in Hawai'i, and what the ecological response will be (Fortini et al. 2015). We expect that temperature increases in Hawai'i will produce significant increases in malaria transmission throughout mid- and high-elevation forests. However, climate models are uncertain on changes in precipitation, which can result in unpredictable consequences for avian malaria (Atkinson et al. 2014). For example, in a drier future with fewer high-precipitation events, permanent streams on Mauna Kea may become intermittent and thereby provide more larval mosquito habitat, thus requiring more direct management. These climate uncertainties add to the uncertainties associated with the efficacy and feasibility of proposed management actions. A structured decision-making approach allows mangers to develop fully articulate and transparent strategies, with input from experts and stakeholders, which will be easier to explain and justify and can be easily modified as conditions change and new information is gained.

Value of Decision Structuring

Effective conservation of Hawai'i's forest birds will require close collaboration between researchers and managers to jointly identify threats and management responses, implemented at ecologically meaningful scales, and continue to work at reducing key uncertainties about the impacts of threats and the efficacy of management actions. A formal decision-analysis approach can help managers justify actions taken in the face of uncertainty and provide the foundation for adjusting strategies as new information becomes available in an adaptive management framework (Runge 2011). Ideally, future model developments and improvements can be embedded within an adaptive management process combining both research and management programs to continually improve our knowledge about system dynamics and simultaneously evaluate different conservation strategies. Therefore, an adaptive management approach may be the best long-term approach to facilitate the decision process given the many uncertainties (Nichols et al. 2011). At the end of a predetermined time period (e.g., 15 years), the models would be revisited to determine whether (1) the predicted conditions have changed and (2) the models should be updated with new and informative information. If so, then a revised set of models would be run for the next management period (e.g., 16–30 years).

What Advances Did Decision Analysis Provide for This Problem?

For the initial evaluation of this problem, a diverse group of scientists and stakeholders was convened, including an equal representation of decision makers with resource management backgrounds and scientists who develop new knowledge to aid decision makers' management decisions. The strategies identified by the group represent different approaches to maintaining healthy bird populations in the face of the almost-certain failure of the status quo approach over the long term. Formal assessment of these conservation strategies to accomplish the objectives still requires a biological model that can predict the potential outcome of each management action based on the changing distribution and intensity of disease over time. At the close of the workshop, the group was pleased with its progress but clearly recognized this was only the first step toward a more detailed process involving stakeholder meetings, acquiring new information to inform existing models, and developing new models for assessing evolution of disease resistance under different scenarios.

Resource managers have recognized the importance of addressing this issue to prevent further avian extinctions, but an overall sense of exasperation over the complexity and scale of the problem, coupled with limited financial resources for integrative research and adaptive management, has hindered progress in developing practical solutions. The effort described in this chapter represents the first time that resource managers and scientists sat at the same table to discuss and prioritize potential strategies to solve this problem—an important development in itself. Although this effort did not result in the immediate implementation of specific actions, it did help prioritize which management actions are likely to have the most beneficial impact on bird populations and where there needed to be more research to fill in knowledge and technology gaps. Having a plan to move forward, even if exact steps were still uncertain, provided a sense of hope and purpose.

ACKNOWLEDGMENTS

This chapter summarizes the results of a workshop sponsored by the National Conservation and Training Center held in Volcano, Hawai'i. The group was co-led by Jeff Burgett (Pacific Islands Climate Change Cooperative, Honolulu, HI) under the guidance of Eve McDonald-Fadden (School of Biological Sciences, University of Queensland, St. Lucia, Australia) and Ellen Bean (US Geological Survey Patuxent Wildlife Research Center, Laurel, MD). Workshop participants included Carter T. Atkinson and Dennis A. LaPointe (US Geological Survey Pacific Island Ecosystems Research Center, Hawai'i National Park, HI), Donna Ball (USFWS Partners Program, Hilo, HI), Colleen Cole (Three Mountain Alliance, Hawai'i National Park, HI), Lisa H. Crampton (Kauai Forest Bird Recovery Project, Waimea, HI), Loyal Mehrhoff (USFWS Ecological Services, Honolulu, HI), and Michael D. Samuel (US Geological Survey Wisconsin Cooperative Wildlife Research Unit, University of Wisconsin, Madison, WI). The workshop also received important help and guidance from Donna C. Brewer (USFWS National Conservation Training Center, Shepherdstown, WV), Sarah J. Converse (US Geological Survey Patuxent Wildlife Research Center, Laurel, MD), and Steve Morey (USFWS Region 1, Portland, OR).

LITERATURE CITED

Ahumada JA, Lapointe D, Samuel MD. 2004. Modeling the population dynamics of Culex quinquefasciatus (Diptera: Culicidae), along an elevational gradient in Hawai'i. *Journal of Medical Entomology* 41:1157–1170.

Atkinson CT, Lapointe DA. 2009. Introduced avian diseases, climate change, and the future of Hawai'ian honeycreepers. *Journal of Avian Medicine and Surgery* 23:53–63.

Atkinson CT, Samuel MD. 2010. Avian malaria *Plasmodium relictum* in native Hawai'ian forest birds: epizootiology and demographic impacts on 'apapane Himatione sanguinea. *Journal of Avian Biology* 41:357–366.

Atkinson CT, Utzurrum RB, Lapointe DA, Camp RJ, Crampton LH, Foster JT, Giambelluca TW. 2014. Changing climate and the altitudinal range of avian malaria in the Hawai'ian Islands: an ongoing conservation crisis on the island of Kaua'i. *Global Change Biology* 20:2426–2436.

Banko WE, Banko PC. 2009. Historic decline and extinction. Pages 25–58 in Pratt TK, Atkinson CT, Banko PC, Jacobi JD Woodworth BL, eds. *Conservation Biology of Hawai'ian Forest Birds: Implications for Island Avifauna.* New Haven, CT: Yale University Press.

Benning TL, Lapointe D, Atkinson CT, Vitousek PM. 2002. Interactions of climate change with biological invasions and land use in the Hawai'ian Islands: modeling the fate of endemic birds using a geographic information system. *Procceedings of the National Academy of Science* 99:14246–14249.

Camp RJ, Brinck KW, Gorresen PM, Paxton EH. 2015. Evaluating abundance and trends in a Hawai'ian avian community using state-space analysis. *Bird Conservation International* 1:1–18.

Camp RJ, Pratt TK, Gorresen PM, Jeffrey JJ, Woodworth BL. 2010. Population trends of forest birds at Hakalau Forest National Wildlife Refuge, Hawai'i. *The Condor* 112:196–212.

Cao G, Giambelluca TW, Stevens DE, Schroeder TA. 2007. Inversion variability in the Hawai'ian trade wind regime. *Journal of Climate* 20:1145–1160.

Cummins GC, Kendall SJ, Paxton EH. 2014. Productivity of forest birds at Hakalau Forest NWR. *Hawai'i Cooperative Studies Unit Technical Report HCSU-056*, University of Hawai'i at Hilo. https://dspace.lib.Hawaii.edu/bitstream/10790/2608/1/PaxtonHakalau110514.pdf.

Diaz HF, Giambelluca TW, Eischeid JK. 2011. Changes in the vertical profiles of mean temperature and humidity in the Hawai'ian Islands. *Global and Planetary Change* 77:21–25.

Fortini LB, Vorsino AE, Amidon FA, Paxton EH, Jacobi JD. 2015. Large-scale range collapse of Hawai'ian forest birds under climate change and the need for 21st century conservation options. *PLOS ONE* 10:e0140389.

Freed LA, Cann RL, Goff ML, Kuntz WA, Bodner GR. 2005. Increase in avian malaria at upper elevation in Hawai'i. *The Condor* 107:753–764.

Gregory R, Failing L, Harstone M, Long G, Mcdaniels T, Ohlson D. 2012. *Structured Decision Making*. Oxford, UK: Wiley-Blackwell.

Guillaumet A, Kuntz WA, Samuel MD, Paxton EH. 2017. Altitudinal migration and the future of an iconic

Hawai'ian honeycreeper in response to disease, climate change and management. *Ecological Monographs* 87:410–428.

Guillaumet A, Woodworth BL, Camp RJ, Paxton EH. 2015. Comparative demographics of a Hawai'ian forest bird community. *Journal of Avian Biology* 47:185–196.

Keeney RL, Raiffa H. 1993. *Decisions with Multiple Objectives: Preference and Value Tradeoffs*. Cambridge, UK: Cambridge University Press.

Kilpatrick AM. 2006. Facilitating the evolution of resistance to avian malaria in Hawai'ian birds. *Biological Conservation* 128:475–485.

Lapointe D, Gaudioso-Levita JM, Atkinson CT, Egan AN, Hayes K. 2016. Changes in the prevalence of avian disease and mosquito vectors at Hakalau Forest National Wildlife Refuge: a 14-year perspective and assessment of future risk. *Hawai'i Cooperative Studies Unit Technical Report HCSU-073*, University of Hawai'i at Hilo. https://dspace.lib.Hawaii.edu/bitstream/10790/2670/6/TR073LaPointeAvDis.pdf.

Lapointe DA, Goff ML, Atkinson CT. 2005. Comparative susceptibility of introduced forest-dwelling mosquitoes in Hawai'i to avian malaria, *Plasmodium relictum*. *Journal of Parasitology* 91:843–849.

———. 2010. Thermal constraints to the sporogonic development and altitudinal distribution of avian malaria Plasmodium relictum in Hawai'i. *Journal of Parasitology* 96:318–324.

Leonard DL. 2008. Recovery expenditures for birds listed under the US Endangered Species Act: the disparity between mainland and Hawai'ian taxa. *Biological Conservation* 141:2054–2061.

Liao W, Elison Timm O, Zhang C, Atkinson CT, Lapointe DA, Samuel MD. 2015. Will a warmer and wetter future cause extinction of native Hawai'ian forest birds? *Global Change Biology* 21:4342–4352.

Nichols JD, Koneff MD, Heglund PJ, Knutson MG, Seamans ME, Lyons JE, Morton JM, Jones MT, Boomer GS, Williams BK. 2011. Climate change, uncertainty, and natural resource management. *Journal of Wildlife Management* 75:6–18.

Paxton EH, Camp RJ, Gorresen PM, Crampton LH, Leonard DL, VanderWerf EA. 2016. Collapsing avian community on a Hawai'ian island. *Science Advances* 2:e1600029.

Paxton EH, Laut M, Vetter JP, Kendall SJ. 2018. Research and management priorities for Hawaiian forest birds. *The Condor* 120:557–565.

Paxton EH, Yelenik SG, Borneman TE, Rose ET, Camp RJ, Kendall SJ. 2017. Rapid colonization of a Hawai'ian restoration forest by a diverse avian community. *Restoration Ecology* 26:165–173.

Pratt TK, Atkinson CT, Banko PC, Jacobi JD, Woodworth BL. 2009. *Hawai'ian Forest Birds: Their Biology and Conservation*. New Haven, CT: Yale University Press.

Reiter ME, Lapointe DA. 2009. Larval habitat for the avian malaria vector, *Culex quinquefaciatus* (Diptera: Culicidae), in altered mid-elevation mesic-dry forests in Hawai'i. *Journal of Vector Ecology* 34:208–216.

Runge MC. 2011. An introduction to adaptive management for threatened and endangered species. *Journal of Fish and Wildlife Management* 2:220–233.

Samuel MD, Hobbelen PHF, Decastro F, Ahumada JA, Lapointe DA, Atkinson CT, Woodworth BL, Hart PJ, Duffy DC. 2011. The dynamics, transmission, and population impacts of avian malaria in native Hawai'ian birds: a modeling approach. *Ecological Applications* 21:2960–2973.

VanderWerf EA. 2001. Distribution and potential impacts of avian poxlike lesions in 'Elepaio at Hakalau Forest National Wildlife Refuge. *Studies in Avian Biology* 22:247–253.

Woodworth BL, Atkinson CT, Lapointe DA, Hart PJ, Spiegel CS, Tweed EJ, Henneman C, Lebrun J, Denette T, Demots R, Kozar KL, Triglia D, Lease D, Gregor A, Smith T, Duffy D. 2005. Host population persistence in the face of introduced vector-borne diseases: Hawai'i amakihi and avian malaria. *Proceedings of the National Academy of Sciences* 102:1531–1536.

Zhang C, Wang Y, Lauer A, Hamilton K. 2012. Configuration and evaluation of the WRF model for the study of Hawai'ian regional climate. *Monthly Weather Review* 140:3259–3277.

PART II ADDRESSING TRADE-OFFS

5 Introduction to Multi-criteria Decision Analysis

Sarah J. Converse

Multiple-objective problems are ubiquitous in natural resource management. Often, solving these problems requires dealing with trade-offs between multiple potentially competing objectives. In this chapter, I provide an overview of approaches to solving multiple-objective problems. I begin by contrasting 2 general classes of solution methods within multiple-objective decision making: multi-objective programming approaches and multi-criteria decision analysis. Multi-objective programming focuses on technical solutions to identifying preferred alternatives from large sets of implicitly defined candidate alternatives. Multi-criteria decision analysis is typically used to evaluate smaller numbers of explicitly defined alternatives, with a unifying theme of evaluating trade-offs between objectives in decision-making processes. I introduce 3 primary classes of methods within multi-criteria decision analysis: methods based on multi-attribute value or utility, the analytic hierarchy process, and outranking methods. I also briefly consider deliberative, discussion-based approaches to dealing with trade-offs and then introduce 3 multiple-objective case studies from the United States, Australia, and Canada.

Introduction

Arguably, nearly all decision problems are multiple-objective problems. Indeed, multiple-objective decision problems are so central to decision making that some experts define structured decision making explicitly as a process designed to address multiple-objective problems (Gregory et al. 2012). Dealing with trade-offs between different values is the key challenge in multiple-objective problems, and certainly this is the case in many natural resource management applications. Decision makers need to understand and utilize the methods available to make these trade-offs—economic development versus environmental protection, recreation access versus wilderness values, persistence of species *X* versus persistence of species *Y*, and more—to make transparent decisions that are most likely to meet their objectives.

When decisions are highly visible and complex, when the values at stake are difficult to quantify, and when stakeholder viewpoints about the proper course of action are entrenched, it is hard for decision makers to deal with trade-offs transparently and deliberatively. Natural resource management problems exemplify this challenge (e.g., Retief et al. 2013). Against this backdrop, how can we design multiple-objective decision-making processes that will move us forward?

Several points deserve mention prior to a discussion about dealing with trade-offs. First, the critical factors determining the success of a multiple-objective decision process are most often the organization and facilitation of the process, rather than the technical method chosen for dealing with trade-offs (Saarikoski et al. 2016). Organization and facilitation encompass considerations about who should be engaged, how the decision can best be framed, how work on the process will be undertaken, and how the process and results will be communicated. A capable decision analyst who can guide the process is valuable, and in high-stakes and high-visibility multiple-objective decisions, an experienced decision analyst is indispensable.

Second, decision-making processes that focus on dealing with difficult trade-offs must be collaborative to receive wide acceptance (Gregory et al. 2012; Williams et al. 2007). Identifying acceptable solutions to these problems is contingent on understanding the values of stakeholders and the trade-offs they are willing to make. Carefully considering who the stakeholders are and how to engage them in a process is likely to pay benefits when it comes to implementation of the resulting decision (Gregory et al. 2012).

Third, there will often need to be substantial time investment in the articulation of objectives. Clarity about objectives is central to good decision making (Keeney 1992). Furthermore, formalizing objectives permits stakeholders to see that their concerns are included in the analysis of a problem. Explicitly identifying objectives takes vague assurances about commitment to engaging stakeholders and gives them content and meaning. However, identifying objectives can be a key challenge for a variety of reasons. Some values relevant to natural resource management are inherently difficult to express. The values people place on wild places, clean water, clean air, and healthy populations of native species don't necessarily lend themselves to easy articulation or quantification. Also, frequently what are originally identified by stakeholders or decision makers as important objectives are in fact means to achieve some higher-level (fundamental) objective, and analysis based on means rather than fundamental objectives can lead to poor decisions (Keeney 2002). For instance, increasing survival rate in a threatened species may appear to be an objective, but it is more likely a means to a higher-level objective, such as long-term persistence of the species. Developing objective hierarchies to clearly understand the various objectives and their relationship to each other can be a powerful approach to clarifying multiple-objective problems (Gregory et al. 2012; Keeney and Raiffa 1976).

Fourth, for multiple-objective problems, time spent generating alternatives can be particularly important. Time invested in developing creative new alternatives might allow decision makers and stakeholders to identify "win-win" alternatives, under which everyone is happier compared to the status quo (Gregory et al. 2012). In this way, some of the more unpalatable trade-offs can be avoided. Ultimately, the best alternative that can be chosen is only as good as the best alternative that is considered.

Once we've set up the problem, identifying a thoughtful and sensible approach to dealing with trade-offs will be key to solving multiple-objective problems. The case studies in this part emphasize different approaches to dealing with trade-offs, including quantitative approaches and deliberative approaches that focus on structuring discussions around trade-offs. I provide here an introduction to quantitative approaches based in multi-criteria decision analysis, briefly discuss deliberative approaches to negotiating trade-offs, and then introduce 3 case studies from the United States, Australia, and Canada.

Multi-criteria Decision Analysis

Multi-criteria decision analysis (MCDA) arose as an alternative to single-criterion approaches in decision analysis. Single-criterion approaches stem from the assumption that a single relevant criterion can be developed *a priori* to evaluate any of the alternative courses of action under consideration (Roy 2005). MCDA rests on the recognition that there are often multiple relevant criteria and that they exist on different scales, such that no single criterion can be easily developed to evaluate alternatives. The field of MCDA is enormous, and it is not possible to provide a comprehensive review of even major classes of methods in this short review. Instead, I introduce some major classes of MCDA methods and focus on those that are used most in the case studies in this part (and elsewhere in this book): methods based in multi-attribute value theory.

To develop an overview of MCDA (table 5.1), it is useful to consider first the wider field of multi-objective decision making, which includes both multi-objective programming (MOP) and MCDA. In MOP problems, there is typically not a defined set of alternatives (Ehrgott 2005). Instead, alternatives are numerous and only identified implicitly. MOP problems are characterized by having 1 or more decision variables that are continuous or take a large set of discrete values. Solution approaches tend to be computationally intensive. A relevant application is the selection of land parcels for protection (Ball et al. 2009; Golovin et al. 2011; Williams 1998; see also Converse et al., chapter 11, this volume). For a set of available land parcels, we want to select a subset that, for example, maximizes the number of species protected while minimizing acquisition cost. The set of alternatives increases as an exponential function $(2x-1)$ of the number (x) of possible parcels. There are a large variety of approaches to solving MOP problems, only a few of which I mention here; Ehrgott (2005) provides a comprehensive overview. One is the ε-constraint method, wherein all but 1 of the objectives are converted to constraints and the optimal solution is the one that maximizes attainment of the remaining objective while meeting all established constraints. Another is the weighted-sum approach, wherein the decision maker specifies weights indicating the relative importance of the different objectives, and through use of these weights a single metric is developed on which to evaluate the alternatives (Coello Coello 1999). Both the ε-constraint method and the weighted-sum method have clear analogs in MCDA. Finally, another approach worth

Table 5.1. Major classes of approaches used in multi-objective decision making and introduced in this chapter, along with key characteristics

Approaches	Characteristics
Pareto-based approaches	Relevant to multi-objective programming or multi-criteria decision analysis problems; focus is on identifying a (possibly complete) set of alternatives that are nondominated, i.e., a gain in performance on 1 objective cannot be obtained without a loss on some other objective
Multi-objective programming	Decision variables are continuous or composed of a large set of discrete values; distinct alternatives are not specifically identified
ε-constraint method	All but 1 of the objectives are converted to constraints, while performance on the remaining objective is maximized
Weighted-sum method	Relevant weights are assigned to objectives and used to develop a single criterion
Goal programming	Objectives are expressed in terms of desired levels of performance
Multi-criteria decision analysis	Alternatives are discrete and specifically identified
Outranking methods	Outranking relationships are determined given indifference, weak preference, and strong preference on each of a set of criteria for pairs of alternatives (e.g., ELECTRE, PROMETHEE)
Multi-attribute value theory	Predicted performance on objectives is ideally quantitative, and weights are placed on objectives; uncertainty is relatively unimportant or can be addressed through sensitivity analysis (e.g., SMARTS)
Multi-attribute utility theory	Predicted performance on objectives is quantitative but presented in terms of utility, a transformation of value to account for risk tolerance in the face of uncertainty
Analytic hierarchy process	Pairwise comparisons of criteria are based on their influence on the decision, and these are combined with pairwise comparisons of alternatives on each criterion to develop alternative scores

Note: See main text for sources.

mentioning because of its relative familiarity is goal programming (Tamiz et al. 1998), wherein a desirable level of performance is specified for each objective. The optimal alternative is the one that minimizes deviance from the desirable level across the objectives. In any case, solving an MOP problem requires dealing, either implicitly or explicitly, with the relative importance of the objectives to the decision maker. In this sense MOP problems are like MCDA problems: to solve any multiple-objective problem, we face trade-offs.

In MOP problems, there is often substantial attention paid to whether selected alternatives are Pareto efficient, and identification of Pareto-efficient alternatives can be challenging. Pareto efficiency is a state wherein improvement in performance on 1 objective cannot be achieved without a loss on some other objective (Williams and Kendall 2017). The Pareto frontier is the set of Pareto-efficient alternatives in a decision problem. Pareto efficiency is also relevant to MCDA problems.

In MCDA, solution methods begin with the assumption that we have a well-defined set of discrete alternatives (unlike MOP problems) and multiple objectives relevant to the selection of a course of action. A sensible initial step involves simplification of the problem to as great a degree as possible before directly wrestling with trade-offs. There are several ways to do this. One is through identification of the set of Pareto-efficient alternatives. Sometimes this can be done via simple inspection and tends to be much simpler than in MOP problems. Another way to simplify is through identification and elimination of irrelevant objectives. An irrelevant objective is one that is important to the decision maker on the surface yet does not help to distinguish among alternatives. For example, a decision maker may identify the cost of an alternative as an important consideration, but if all identified alternatives cost the same, cost becomes irrelevant in making the decision. Finally, the even swaps method can be used to simplify MCDA problems (Gregory et al. 2012; Hammond et al. 2000). When 2 objectives can be expressed in the same currency with relative ease, we can cancel out the differences between alternatives on 1 objective, thereby creating an irrelevant objective, and reflect those differences in the other objective (Hammond et al. 2000). We then delete the irrelevant objective from the analysis.

Once the problem has been simplified to the greatest degree possible, the analysis should have exposed the key trade-offs with which decision makers will ultimately need to grapple. The goal should be to identify a method for evaluating trade-offs that is explicit, transparent, and can be communicated to stakeholders (Gregory et al. 2012).

One simple approach for evaluating trade-offs is the MCDA equivalent of the ε-constraint method: convert all objectives to constraints save 1 and choose the alternative that maximizes that objective while satisfying all constraints. The method requires identifying what minimum level of performance is satisfactory on the objectives that will be converted to constraints, while also deciding which objective should be retained for maximization. However, it does not require explicitly considering how much of a loss on 1 objective is made up for by a gain on another. While the simplicity of this approach is attractive, information is lost by converting objectives into constraints, and the resulting solution may be less satisfactory than one arising from a process in which trade-offs are evaluated explicitly. For formal evaluation of trade-offs, we consider 3 classes of MCDA approaches: methods based on multi-attribute value theory / multi-attribute utility theory, the analytic hierarchy process, and outranking methods. The last 2 of these are covered in relatively sparse detail.

Multi-attribute value theory (MAVT) and multi-attribute utility theory (MAUT) build quantitative functions to represent decision-maker preferences for alternatives. The value or utility function can be used to score and evaluate the alternatives, given performance predictions on the individual criteria. Construction of this value or utility function is recognized as a key challenge. The distinction between

MAVT and MAUT is that MAUT is concerned with utility rather than value. Utility is a transformation of value designed to account for attitudes toward risk in cases of decision making under uncertainty (Keeney and Raiffa 1976). A utility function can be treated as a value function for the purpose of scoring alternatives (see Runge and Converse, chapter 13, this volume and the case studies in part 4 for more on utility).

The basic case in MAVT/MAUT is an additive value function, V, for alternative j, constructed as

$$V_j = \sum_{i=1}^{n} w_i x_{i,j} \tag{1}$$

where $x_{i,j}$ are performance predictions on each of a set of $i = 1 \ldots n$ criteria (attributes), evaluated for each of the alternatives $j = 1 \ldots K$. These criteria are used to distinguish between alternatives, that is, they represent the decision maker's objectives (Keeney and Raiffa 1976). The w_i represent weights associated with each attribute and capture the relative importance of the attributes to the decision maker. The w are chosen such that $\sum_{i=1}^{n} w_i = 1$. If we have, for example, $w = (0.2, 0.8)$ for 2 attributes $i = 1, 2$, which typically are measured on different scales, the decision maker is indifferent to trading off a loss of 4 units of attribute 1 for a gain of 1 unit of attribute 2. For a criterion $i = I$, if we place each of the $x_{i=I,j}$ on a standardized scale (e.g., where 0 represents the poorest performance on that attribute and 1 represents the best performance; see also Martin and Mazzotta 2018a) and repeat that process for each of the $i = 1 \ldots n$ criteria, we can calculate the value of any alternative, given a set of weights, using eq. 1. This is the process behind, for example, the simple multi-attribute rating technique with swings (SMARTS; Edwards 1977; Edwards and Barron 1994). SMARTS assumes that value scales linearly across the range of performance values for any $x_{i=I,j}$, and it makes use of *swing weighting* for determining objective weights (von Winterfeldt and Edwards 1986). Swing weighting arises from the idea that the weight of an objective should be a function of not just its general importance to the decision maker but also the degree to which the alternatives vary on that objective. A decision maker might claim that cost is important, but if the alternatives range in cost between, say, $20,000 and $20,500, the cost may not carry a lot of weight in the decision. SMARTS is demonstrated in this part by both Smith and McRae (chapter 6) and Walshe and Slade (chapter 7). Other methods for eliciting weights include those based on ordinal scales of importance—that is, ranks—of objectives (e.g., rank order centroid weights; Edwards and Barron 1994) as well as additional methods based—as swing weights are—on cardinal scales, including direct elicitation of weights. Srivastava et al. (1995) describe and evaluate 7 weighting methods based on experimental data and tentatively conclude that rank order centroid weights, derived from an interim step in the swing-weighting procedure, offer a favorable combination of predictive performance and ease of elicitation. The process of specifying weights can provide insights and guide the creation of additional creative alternatives, in that the best alternatives will help to achieve the most important and therefore highest-weighted objectives.

Closely related to, but distinct from, MAVT methods is the analytic hierarchy process (AHP; Saaty 1980; Saaty and Vargas 2012). In AHP, the decision maker performs pairwise comparisons of objectives on ratio scales (e.g., 9 = reference objective is 9 times as important as the comparator, 1 = reference objective is exactly as important as the comparator, 1/9 = reference objective is 1/9 as important as the comparator). This step is followed by pairwise comparisons of alternatives in terms of their performance on each objective. Pairwise comparisons of objectives can be used to calculate objective weights from the eigenvectors of the matrix of pairwise comparisons, and in combination with the output of comparisons of alternatives they can be used to develop quantitative scores for alternatives.

Outranking methods do not assume that a decision maker has a value or utility function that is known, nor do they assume that judgments about preferences are certain. Instead, the focus is on producing outranking relationships, or degrees of preference, for the alternatives under consideration. Outranking methods are an outgrowth of voting theory and particularly suit processes in which at least some of the relevant criteria are qualitative in nature (Kangas et al. 2001). Key concepts include indifference and preference thresholds. A difference in performance between 2 alternatives on a given criterion (objective) that is less than an indifference threshold indicates that the decision maker is indifferent between the alternatives on that criterion. A difference in performance that is greater than a preference threshold indicates that the decision maker has strong, or strict, preference for the alternative with the more favorable value on that criterion. Between the indifference and preference thresholds, a decision maker will have weak, or fuzzy, preference for the alternative with the more favorable value. Within this general framework, the various outranking methods differ in the approach used to ascertain and quantify indifference, weak preference, and strong preference and on how overall outranking relationships are assessed given the set of criteria under consideration. Major classes of outranking methods include the ELECTRE family (Figueira et al. 2005; Roy 1991) and the PROMETHEE family (Brans and Mareschal 2005). Kangas et al. (2001) describe the use of outranking approaches in forest management strategy selection, and Martin and Mazzotta (2018b) describe an application to wetland restoration.

Deliberative Approaches to Negotiating Trade-Offs

Not all problems need to be solved through quantitative analysis of trade-offs. In some cases, just by going through the preceding steps—clearly defining the decision, articulating objectives, identifying alternatives, making predictions, and simplifying the problem to eliminate dominated alternatives—a decision maker will find that a clear winner emerges. Perhaps, through emphasizing creative, and often iterative, generation of alternatives, an alternative will be identified with which all stakeholders are happier compared to the status quo. If a clear winner does not emerge, the decision maker and stakeholders should still be much better prepared to understand and discuss trade-offs because the process of framing and simplifying the problem should make the key trade-offs explicit. At this point, some groups may be prepared to undertake approaches to trade-off analysis that focus on structured discussions rather than quantitative techniques. This may involve direct negotiations around a reduced set of alternatives. If only 2 or 3 objectives are important to the decision, a plot of the alternatives on the Pareto frontier can be useful in facilitating this discussion. Such deliberative approaches may also be used in conjunction with quantitative approaches. For example, Gregory et al. (2012) advocate use of multiple methods in conjunction, including (1) comparative discussions of pairs of alternatives to identify practically dominated alternatives that can be eliminated without controversy, (2) direct ranking of alternatives, and (3) quantitative analyses. Inconsistency in the outcomes from direct ranking of alternatives and quantitative analyses can expose weaknesses in the framing, such as missing objectives. Discussion itself can provide insights to decision makers that might be missed if there were too strong an emphasis on the results of quantitative analyses. Ultimately, the goal should be a decision that is most likely to satisfy the decision maker and stakeholders, and several pathways may lead to that goal.

Case Studies

In chapter 6, David Smith and Sarah McRae describe a process undertaken to support the development of a conservation strategy for an endangered mussel endemic to North Carolina, in the southeastern United States. Management objectives included maximizing probability of persistence and the genetic diversity of the mussel, maximizing public support, and minimizing management costs. Thus, the problem exemplifies a fundamental challenge in species conservation problems: How do we trade off conservation against monetary costs? Alternatively, how do we efficiently allocate our precious monetary resources? The persistence objective contained further complexity: Managers wanted to conserve the species in each of 2 watersheds where it is endemic. Therefore, this problem illustrates another common aspect of species conservation problems: where conservation happens matters. Frequently managers want to conserve things in many places, but some spatial allocations of resources may perform worse than others in terms of overall species conservation. We then must consider how to trade off local benefits for global benefits. The analysts used SMART to deal with trade-offs. The decision analysis contributed to a management plan for the species that can be used to determine how to allocate relatively scarce monetary resources when they become available. The plan emphasized actions in the watershed with a relatively intact population and deemphasized actions in the watershed with greater threats. Many conservation spending decisions tend to be made without careful attention to their impacts on fundamental objectives, resulting in an inefficient use of scarce resources. This case study emphasizes the value of a structured approach so that efficient decisions can be identified.

In chapter 7, Terry Walshe and Stephanie Slade describe a process for deciding among alternatives for fisheries closures to protect finfish on the Great Barrier Reef off the coast of Queensland, Australia. The problem was made challenging by the strongly differing values of stakeholders and the tight timeline that was set for completing the process. In addition to protection of finfish populations, objectives included maximizing access by fishers in commercial, recreational, and charter sectors and realizing ecosystem benefits, among others. The analysts used a novel approach to make the problem tractable for the time allowed. They began with eliciting *global weights* on a wide suite of objectives. Global weights are based on a global scale of values derived from the decision maker's experience or imagination, while *local weights*, such as swing weights, are based on a local scale derived from the specific alternatives available (Monat 2009). That is, the bounds of a global scale are determined by the decision maker's experience or imagination about the extreme values that a given criterion might take, while the bounds of a local scale are based on the best- and worst-performing alternatives on that criterion in the alternative set. The weakness of global weights is that they don't account for the range of performance on the objectives. Walshe and Slade used global weights to reduce the problem to a manageable scope. They then used a SMART analysis with context-specific (local) weights to evaluate alternatives. The analysis was completed individually for a range of stakeholders with strongly varying views. The analysts demonstrated how to identify alternatives that are robust to uncertainty, using sensitivity analysis and a minimax optimization approach. They also identified alternatives that represented reasonable compromises for a variety of stakeholders, as opposed to alternatives that were polarizing. The information gained directly informed a decision by the minister responsible for fisheries. Walshe and Slade emphasize the value of a technically simple approach, like SMART, in the context of a contentious decision and limited time.

In chapter 8, Dan Ohlson and colleagues describe a process for determining allowable water withdrawals from the Lower Athabasca River for oil sands mining in Alberta, Canada. Water withdrawals have potential effects on a multitude of values. The decision analysts, in this case, used a negotiation-oriented approach to trade-offs between values represented by different interest groups (oil sands mining industry, First Nations, and nonprofit environmental organizations). Through a modeling platform that allowed for creation and evaluation of water withdrawal alternatives in real time, stakeholders collaborated to build alternatives that struck an acceptable balance between different stakeholders' values. In other words, trade-offs were not evaluated formally, but instead, the process was designed to facilitate collaboration and negotiation to identify mutually acceptable trade-offs. The results of the process were consensus recommendations on water withdrawals under most flow conditions and articulation of exactly where disagreements lay under extreme low-flow (drought) conditions. That information directly informed government regulators' decisions. While the analysts initially undertook the problem with a plan to use quantitative trade-off methods, their realization was that, given that values trade-offs were quite apparent, a quantitative trade-off approach (i.e., weighting objectives) risked increasing polarization. An informal approach preserved their collaborative environment and led to a search for creative and widely palatable alternatives.

Open Questions and Sticky Issues in Multi-objective Decision Making

Perhaps more than any other, a key challenge in multiple-objective decision problems is our ability to fully capture important values. The set of criteria identified in a multiple-objective process should be exhaustive, cohesive, and nonredundant (Roy 2005). Unfortunately, attaining an exhaustive set of objectives is not simple. Stakeholders and decision makers often find it challenging to identify all their objectives. In a series of experiments, Bond et al. (2008) found that decision makers, when making a decision that was of substantial personal consequence, missed nearly half of the objectives that they later identified as important. Furthermore, they identified the objectives that they initially missed as being as important as the objectives they did identify. This is startling: People are only able to identify about half of the things that are (or ought to be) influencing their decisions! Failing to recognize all important objectives can seriously undermine a decision-making process; unrecognized objectives cannot be used to help generate new alternatives, and they cannot be accounted for in making trade-offs. Bond et al. (2010) and Keeney (2013) provide suggestions for improving the generation of objectives; the most effective methods they identified experimentally included categorizing objectives and challenging participants to do better by warning them that important objectives are (or might be) missing.

One class of objectives is particularly challenging to work with: There may be a tendency to ignore or downplay values that do not easily lend themselves to quantification or even monetization, such as indigenous or other cultural values (Chan et al. 2012; Turner et al. 2008). The values that people place on ecosystems, for example, often include cultural components, including spiritual, aesthetic, artistic, and educational values derived from experiences in nature (Runge et al. 2011; Satterfield et al. 2013). While cultural values cannot easily be quantified, they are often our most personal and closely held values. Furthermore, certain cultural values are sacred to those who hold them, and trade-offs between sacred and secular values may be considered taboo, such that stakeholders asked to consider these trade-offs find the question not just impossible to answer but offensive to consider (Daw et al. 2015; Fiske and Tetlock 1997; Tetlock 2003; Tetlock et al. 2000), although evidence indicates that the degree to which certain types of trade-offs are considered taboo may be context-specific (Baron and Leshner 2000). The difficulty in measuring cultural values and the existence of taboo trade-offs suggest that substantial care must be taken in considering and characterizing cultural values in decision-making processes. Satz et al. (2013) identified challenges associated with inclusion of cultural values in environmental decision making, including interconnected benefits, which make it difficult to develop independent measures of cultural and noncultural benefits; fundamental noncomparability with noncultural values; and others.

MCDA approaches have been identified as an alternative to conventional cost-benefit approaches that rely on monetization of values in environmental decision making (e.g., de Groot et al. 2010; Langemeyer et al. 2016). An important benefit of MCDA in this context is its ability to account for nonmonetary values. Saarikoski et al. (2016) contrast economic valuation and MCDA approaches in environmental decision making and conclude that MCDA offers benefits in terms of its ability to capture nonmonetary values. However, MCDA does not alleviate the challenge of eliciting and quantifying such values; it simply provides a framework for analysis if this elicitation can be achieved. Gould et al. (2014) describe an interview-based approach to assist with elicitation and description of cultural values provided by ecosystems and note that, while not providing quantitative measures, their protocol could guide the design of surveys to obtain quantitative metrics. There are 3 basic types of quantitative measurable attributes (units in which performance on an objective is measured; Gregory et al. 2012): natural measures, proxy measures, and constructed measures. Natural measures reflect achievement on an objective in natural units (e.g., dollars to measure cost); proxy measures are correlated with, but do not directly measure, performance on an objective (e.g., number of fish caught to measure fisher satisfaction); constructed measures are based on scales constructed for the decision at hand, typically to measure things that are otherwise difficult to measure (e.g., scale of 1–7 to measure recreational opportunity). Presumably, cultural values will often be quantified through use of constructed scales. In one of the few MCDA efforts in natural resource management to delve deeply into elicitation and quantification of cultural values, Runge et al. (2011) conducted a process in which spiritual values of Native American tribes were included explicitly in a decision process, using constructed measures. Further approaches for, and examples of, formalizing cultural values are needed.

There is no silver bullet to address the challenges of identifying and appropriately describing objectives in multi-objective decision problems. Perhaps the best practice is developing in decision makers and stakeholders a willingness to see the process of decision analysis as iterative. Once it is recognized that an objective is missing or characterized inappropriately—and this may not happen until a nominally optimal alternative is identified—it is important to go back and consider how this needs to be integrated into the process. Each successive attempt at solving the decision problem is a prototype (Garrard et al. 2017), and the need for further prototyping should be evaluated based on the decision maker's needs and preferences.

LITERATURE CITED

Ball IR, Possingham HP, Watts M. 2009. Marxan and relatives: software for spatial conservation prioritisation. Pages 185–195 in Moilanen A, Wilson KA, Possingham HP, eds. *Spatial Conservation Prioritisation: Quantitative Methods and Computational Tools.* Oxford, UK: Oxford University Press.

Baron J, Leshner S. 2000. How serious are expressions of protected values? *Journal of Experimental Psychology: Applied* 6:183–194.

Bond SD, Carlson KA, Keeney RL. 2008. Generating objectives: can decision makers articulate what they want? *Management Science* 54:56–70.

———. 2010. Improving the generation of decision objectives. *Decision Analysis* 7:238–255.

Brans J-P, Mareschal B. 2005. PROMETHEE methods. Pages 163–195 in Figueira J, Greco S, Ehrgott M, eds. *Multiple Criteria Decision Analysis: State of the Art Surveys.* Boston, MA: Springer.

Chan KMA, Guerry AD, Balvanera P, Klain S, Satterfield T, Basurto X, Bostrom A, Chuenpagdee R, Gould R, Halpern BS, Hannahs N, Levine J, Norton B, Ruckelshaus M, Russell R, Tam J, Woodside U. 2012. Where are *cultural* and *social* in ecosystem services? A framework for constructive engagement. *Bioscience* 62(8):744–756.

Coello Coello CA. 1999. A comprehensive survey of evolutionary-based multiobjective optimization techniques. *Knowledge and Information Systems* 1:269–308.

Daw TM, Coulthard S, Cheung WWL, Brown K, Abunge C, Galafassi D, Peterson GD, McClanahan TR, Omukoto JO, Munyi L. 2015. Evaluating taboo trade-offs in ecosystems services and human well-being. *Proceedings of the National Academy of Sciences* 112:6949–6954.

de Groot RS, Alkemade R, Braat L, Hein L, Willemen L. 2010. Challenges in integrating the concept of ecosystem services and values in landscape planning, management and decision making. *Ecological Complexity* 7:260–272.

Edwards W. 1977. How to use multiattribute utility measurement for social decisionmaking. *IEEE Transactions on Systems, Man and Cybernetics* 7:326–340.

Edwards W, Barron FH. 1994. SMARTS and SMARTER: improved simple methods for multiattribute utility measurement. *Organizational Behavior and Human Decision Processes* 60:306–325.

Ehrgott M. 2005. Multiobjective programming. Pages 667–722 in Figueira J, Greco S, Ehrgott M, eds. *Multiple Criteria Decision Analysis: State of the Art Surveys*. Boston, MA: Springer.

Figueira J, Mousseau V, Roy B. 2005. ELECTRE methods. Pages 133–162 in Figueira J, Greco S, Ehrgott M, eds. *Multiple Criteria Decision Analysis: State of the Art Surveys*. Boston, MA: Springer.

Fiske AP, Tetlock PE. 1997. Taboo trade-offs: reactions to transactions that transgress the spheres of justice. *Political Psychology* 18:255–297.

Garrard GE, Rumpff L, Runge MC, Converse SJ. 2017. Rapid prototyping for decision structuring: an efficient approach to conservation decision analysis. Pages 46–64 in Bunnefeld N, Nicholson E, Milner-Gulland E, eds. *Decision-Making in Conservation and Natural Resource Management: Models for Interdisciplinary Approaches*. Cambridge, UK: Cambridge University Press.

Golovin D, Krause A, Gardner B, Converse SJ, Morey S. 2011. Dynamic resource allocation in conservation planning. *Proceedings of the AAAI Conference on Artificial Intelligence* 25:1331–1336.

Gould RK, Klain SC, Ardoin NM, Satterfield T, Woodside U, Hannahs N, Daily GC, Chan KMA. 2014. A protocol for eliciting nonmaterial values through a cultural ecosystem services frame. *Conservation Biology* 29:575–586.

Gregory R, Failing L, Harstone M, Long G, McDaniels T, Ohlson D. 2012. *Structured Decision Making: A Practical Guide to Environmental Management Choices*. Oxford, UK: Wiley-Blackwell.

Hammond JS, Keeney RL, Raiffa H. 2000. The even-swap method for multiple objective decisions. Pages 1–14 in Haimes YY, Steuer RE, eds. *Research and Practice in Multiple Criteria Decision Making*. Berlin: Springer.

Kangas A, Kangas J, Pykäläinen J. 2001. Outranking methods as tools in strategic natural resources planning. *Silva Fennica* 35:215–227.

Keeney RL. 1992. *Value-Focused Thinking: A Path to Creative Decisionmaking*. Cambridge, MA: Harvard University Press.

———. 2002. Common mistakes in making value trade-offs. *Operations Research* 50:935–945.

———. 2013. Identifying, prioritizing, and using multiple objectives. *EURO Journal on Decision Processes* 1:45–67.

Keeney RL, Raiffa H. 1976. *Decisions with Multiple Objectives: Preferences and Value Trade-Offs*. New York: John Wiley & Sons.

Langemeyer J, Gomez-Baggethun E, Haase D, Scheuer S, Elmqvist T. 2016. Bridging the gap between ecosystem service assessments and land-use planning through Multi-Criteria Decision Analysis (MCDA). *Environmental Science & Policy* 62:45–56.

Martin DM, Mazzotta M. 2018a. Non-monetary valuation using Multi-Criteria Decision Analysis: sensitivity of additive aggregation methods to scaling and compensation assumptions. *Ecosystem Services* 29:13–22.

———. 2018b. Non-monetary valuation using Multi-Criteria Decision Analysis: using a strength-of-evidence approach to inform choices among alternatives. *Ecosystem Services* 33:124–133.

Monat JP. 2009. The benefits of global scalinig in multi-criteria decision analysis. *Judgment and Decision Making* 4:492–508.

Retief F, Morrison-Saunders A, Geneletti D, Pope J. 2013. Exploring the psychology of trade-off decision-making in environmental impact assessment. *Impact Assessment and Project Appraisal* 31:13–23.

Roy B. 1991. The outranking approach and the foundations of ELECTRE methods. *Theory and Decision* 31:49–73.

———. 2005. Paradigms and challenges. Pages 3–24 in Figueira J, Greco S, Ehrgott M, eds. *Multiple Criteria Decision Analysis: State of the Art Surveys*. Boston, MA: Springer.

Runge MC, Bean E, Smith DR, Kokos S. 2011. Non-native fish control below Glen Canyon Dam: report from a structured decision-making project. US Geological Survey Open-File Report 2011-1012. http://pubs.usgs.gov/of/2011/1012/.

Saarikoski H, Mustajoki J, Barton DN, Geneletti D, Langemeyer J, Gomez-Baggethun E, Marttunen M, Antunes P, Keune H, Santos R. 2016. Multi-criteria decision analysis and cost-benefit analysis: comparing alternative frameworks for integrated valuation of ecosystem services. *Ecosystem Services* 22:238–249.

Saaty TL. 1980. *The Analytic Hierarchy Process*. New York: McGraw-Hill.

Saaty TL, Vargas LG. 2012. *Models, Methods, Concepts and Applications of the Analytic Hierarchy Process*. International Series in Operations Research and Management Science, edited by Hillier FS. Vol. 175. New York: Springer.

Satterfield T, Gregory R, Klain S, Roberts M, Chan KM. 2013. Culture, intangibles and metrics in environmental management. *Journal of Environmental Management* 117:103–114.

Satz D, Gould RK, Chan KMA, Guerry A, Norton B, Satterfield T, Halpern BS, Levine J, Woodside U, Hannahs N, Basurto X, Klain S. 2013. The challenges of incorporating cultural ecosystem services into environmental assessment. *Ambio* 42:675–684.

Srivastava J, Connolly T, Beach LR. 1995. Do ranks suffice? A comparison of alternative weighting approaches in value elicitation. *Organizational Behavior and Human Decision Processes* 63:112–116.

Tamiz M, Jones D, Romero C. 1998. Goal programming for decision making: an overview of the current state-of-the-art. *European Journal of Operational Research* 111:569–581.

Tetlock PE. 2003. Thinking the unthinkable: sacred values and taboo cognitions. *TRENDS in Cognitive Sciences* 7:320–324.

Tetlock PE, Kristel OV, Elson SB, Green MC, Lerner JS. 2000. The psychology of the unthinkable: taboo trade-offs, forbidden base rates, and heretical counterfactuals. *Journal of Personality and Social Psychology* 78:853–870.

Turner NJ, Gregory R, Brooks C, Failing L, Satterfield T. 2008. From invisibility to transparency: identifying the implications. *Ecology and Society* 13:7.

von Winterfeldt D, Edwards W. 1986. *Decision Analysis and Behavioral Research*. Cambridge, UK: Cambridge University Press.

Williams BK, Szaro RC, Shapiro CD. 2007. *Adaptive Management: The US Department of the Interior Technical Guide*. Washington, DC: Adaptive Management Working Group, US Department of the Interior.

Williams JC. 1998. Delineating protected wildlife corridors with multi-objective programming. *Environmental Modeling and Assessment* 3:77–86.

Williams PJ, Kendall WL. 2017. A guide to multi-objective optimization for ecological problems with an application to cackling goose management. *Ecological Modelling* 343:54–67.

6

David R. Smith and
Sarah E. McRae

Strategic Conservation of an Imperiled Freshwater Mussel, the Dwarf Wedgemussel, in North Carolina

To be effective, managers of imperiled species must resolve the unavoidable trade-off between conservation benefits and constrained budgets and must not be paralyzed by scientific uncertainty. Decision analysis can help meet these challenges when used to develop cost-effective strategies to recover or improve the status of species. The US Fish and Wildlife Service, along with state partners, developed a structured decision analysis to guide conservation of Dwarf Wedgemussel (*Alasmidonta heterodon*) in North Carolina. The Dwarf Wedgemussel is federally listed as endangered, and North Carolina is the southernmost extent of its range, where small and vulnerable populations occur in the Tar and Neuse River Basins. The main threat in the Neuse River Basin is habitat loss due to anthropogenic land use changes. In contrast, the Tar River Basin primarily has been affected by recent drought and stream habitat loss due to beaver impoundments. A collaborative team used multiple-objective decision analysis to compare the ability of conservation strategies to maximize species persistence while accounting for uncertainty in management effectiveness and variation in the importance of different management objectives. The decision analysis helped managers evaluate trade-offs regarding Dwarf Wedgemussel distribution within the Neuse River and Tar River Basins. The most cost-effective and robust strategies traded off some opportunity for persistence in the Neuse River for protection of populations in the Tar River Basin. The decision analysis is being used to guide efforts to conserve Dwarf Wedgemussel in North Carolina, although challenges continue due to constrained budgets, workload management, and limited regulatory tools.

Problem Background

Decision analysis can be used to help figure out how best to recover or improve the status of endangered species (Gregory et al. 2013). Conservation plans should be focused first on what managers want to achieve before settling on specific conservation actions—that is, managers should sort out what they want and then figure out how to get it (Dunn 2004; Linke et al. 2011). Managers of imperiled species typically strive to minimize species' risk of extinction or, conversely, maximize species' viability in an efficient and effective way (Gregory et al. 2012). They do so by determining what specific actions could be taken and then allocating effort (staff and funding) among management actions to form a strategy. Because there is more than one possible allocation of effort, multiple alternative strategies can be formed. So, an important decision revolves around which alternative strategy to select and implement. Ideally, managers allocate effort so that the conservation benefits (e.g., maximizing viability) can be achieved at minimal public costs (Bottrill et al. 2008; Joseph et al. 2009; Linke et al. 2011). Also, conservation plans should account for uncertainty in management effectiveness and trade-offs arising from stakeholder disagreements regarding the importance of different objectives (Gregory and Long 2009; Runge et al. 2011). Finally, managers should consider species' risk of extinction separate from their tolerance for that risk (Doremus and Tarlock 2005; Smith et al. 2018; Vucetich et al. 2006). A species' risk is a scientific prediction (e.g., 5% risk of extinction in 50 years), whereas risk tolerance reflects values revealed in law or policy (e.g., whether 5% risk of extinction is acceptable to policy makers). Decision analysis can help managers face all of these challenges because it focuses on values and deconstructs a decision problem to allow for the analysis of trade-offs, exploration of uncertainty, and clarification of the roles of science and policy (Conroy and Peterson 2013).

The Dwarf Wedgemussel (*Alasmidonta heterodon*) is a federally listed endangered species and a freshwater mussel (Unionidae), which is among the faunal groups with the highest extinction rates in North America (Ricciardi and Rasmussen 1999). The species' range extends from New England to North Carolina (Strayer et al. 1996; USFWS 2013). Dwarf Wedgemussel populations are stable in some locations, but in North Carolina, populations are small and restricted to the Tar and Neuse River Basins (fig. 6.1). The species was listed as endangered in 1990, and a recovery plan was completed in 1993 (USFWS 1993). At the outset of the work described here, the plan had not been updated, leaving the species without a formalized conservation strategy based on new information and a better understanding of the different challenges in the northern and southern portions of its range (USFWS 2013). In the absence of an up-to-date conservation strategy, management actions tended to be taken in reaction to pending impacts rather than based on effectiveness and efficiency considerations. A strategic and prospective plan for conservation of *A. heterodon* in North Carolina was needed to replace reactive and *ad hoc* management.

In 2011, the USFWS convened a team of biologists and managers from USFWS, the North Carolina Wildlife Resources Commission, and North Carolina State University, along with other experts on freshwater mussel ecology, to develop and evaluate alternative strategies for conservation of Dwarf Wedgemussel. This expert team identified the factors limiting species recovery, which included an Allee effect, unsuitable habitat (physical, flow, or water quality), and beaver alteration of habitat. An Allee effect is a reduced population growth rate at low abundance (i.e., depensation; Courchamp et al. 1999). In freshwater mussels, the hypothesized mechanism for depensation is reduced reproduction, particularly reduced proportion of gravid females, at low densities (McLain and Ross 2005). Workshop participants ranked an Allee effect as one of the top limiting factors in all stream reaches and especially in the Neuse River Basin.

The team followed a structured decision-making (SDM) process during 2 workshops held in 2011 and 2012. The conservation strategy that resulted from the decision analysis was adopted and has guided conservation for the species since. In this chapter, we summarize the SDM process and the resulting conservation plan. Greater detail on the analyses can be found in Smith et al. (2015). We also review the current implementation of the strategy and discuss lessons learned.

Decision Analysis

The team framed the decision problem as the development of a cost-effective strategy to guide where and how to conserve Dwarf Wedgemussel in North Carolina over the next 20 years.

Objectives

The overall goal of the strategy was to maximize persistence of Dwarf Wedgemussel within its historical

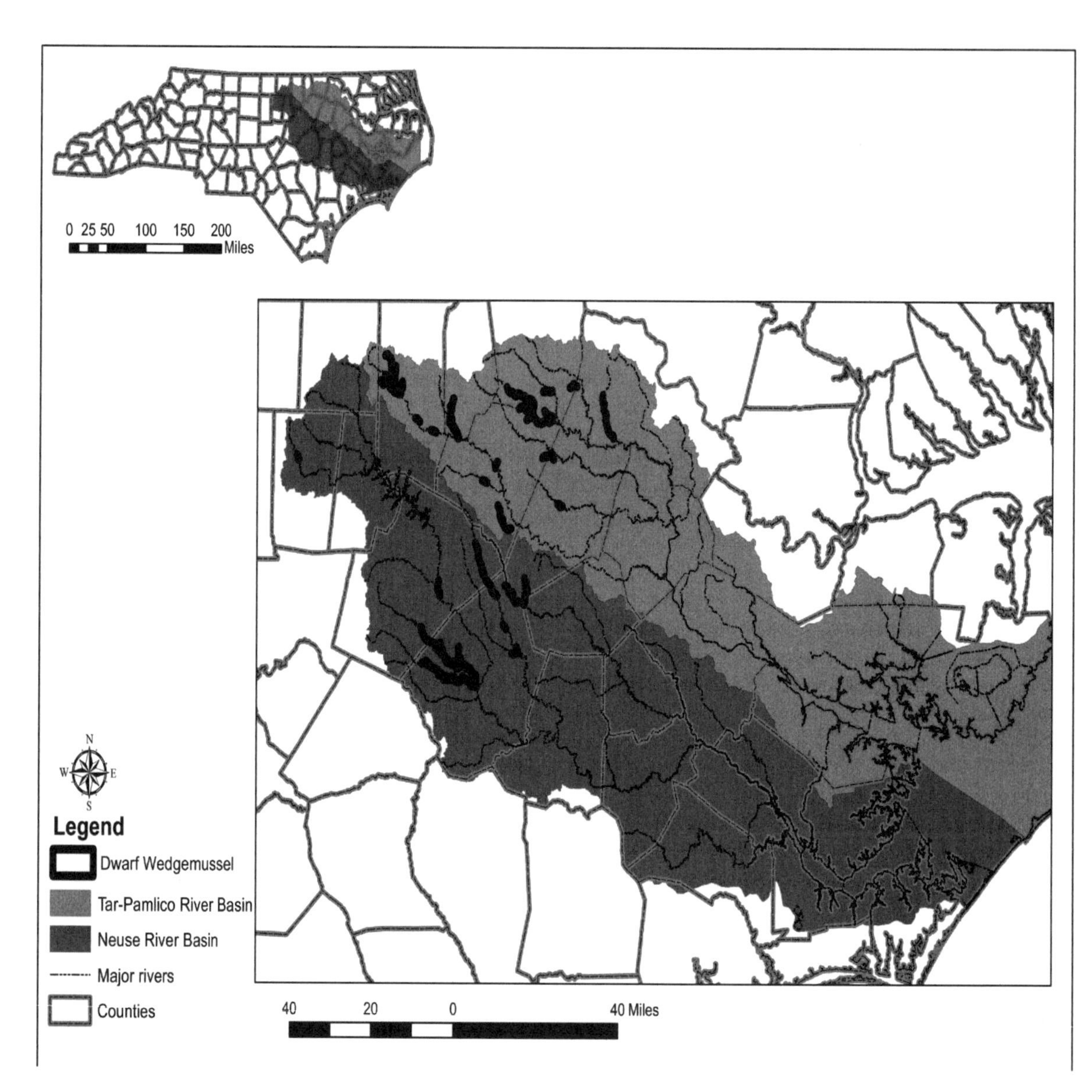

Figure 6.1. Locations of stream reaches with Dwarf Wedgemussel within the upper Tar and Neuse River Basins in North Carolina.

range while considering costs; this goal of maximizing persistence while considering costs is nearly ubiquitous for imperiled species conservation (Joseph et al. 2009; Naidoo et al. 2006). Stepping down from the overall goal, the team identified the following fundamental objectives: maximize the probability of persistence of Dwarf Wedgemussel within North Carolina, maintain genetic diversity (separate from persistence), maximize public support, and minimize management costs (Smith et al. 2015). The team discussed whether to treat genetic diversity as a means to persistence. However, for this analysis, the team chose to consider genetic diversity as a separate objective to explore trade-offs between persistence within the planning horizon of 20 years and maintenance of genetic diversity, which could contribute to adaptive capacity well beyond the planning horizon (Lankau et al. 2011). In other words, a conservation plan might maximize probability of persistence in the short term but do so in a way that compromised long-term viability by reducing genetic diversity. Further, although public support can be viewed as a means of achieving conservation benefits, the team found it helpful to keep track of the effect of alternative strategies on public support and, all else being equal, would choose a strategy with high public support. The Tar and Neuse River Basins comprise the species' historical range in North Carolina, so maximizing persistence within each basin constituted 2 sub-objectives within the persistence objective (fig. 6.2). This allowed for the exploration of trade-offs in persistence between the river basins within North Carolina.

The performance measures for persistence at the reach scale were abundance (categorical based on individuals detected per person-hour of search), distribution (stream reach occupied or not), and recruitment over the most recent 3 years (yes or no). Measures for persistence were predicted at the reach scale and then aggregated to the basin level. Although continuous measures allow for distinguishing finer-scale differences between alternatives, the team used categorical measures because of limitations in the available data and knowledge. For genetic diversity, the performance measure was the probability of no loss of genetic diversity. Public support was measured on a constructed scale to capture the likelihood and strength of opposition or support for a conservation strategy. Active opposition can prevent an action; on the other hand, active support can expedite an action. Thus, the performance measure was a categorical scale from −2 to +2 to represent active and passive opposition, neutrality, or passive and active support, respectively. Costs were in dollars; agency costs for staff and operations were measured separately from nonagency costs for land acquisitions, grants, and conservation banks.

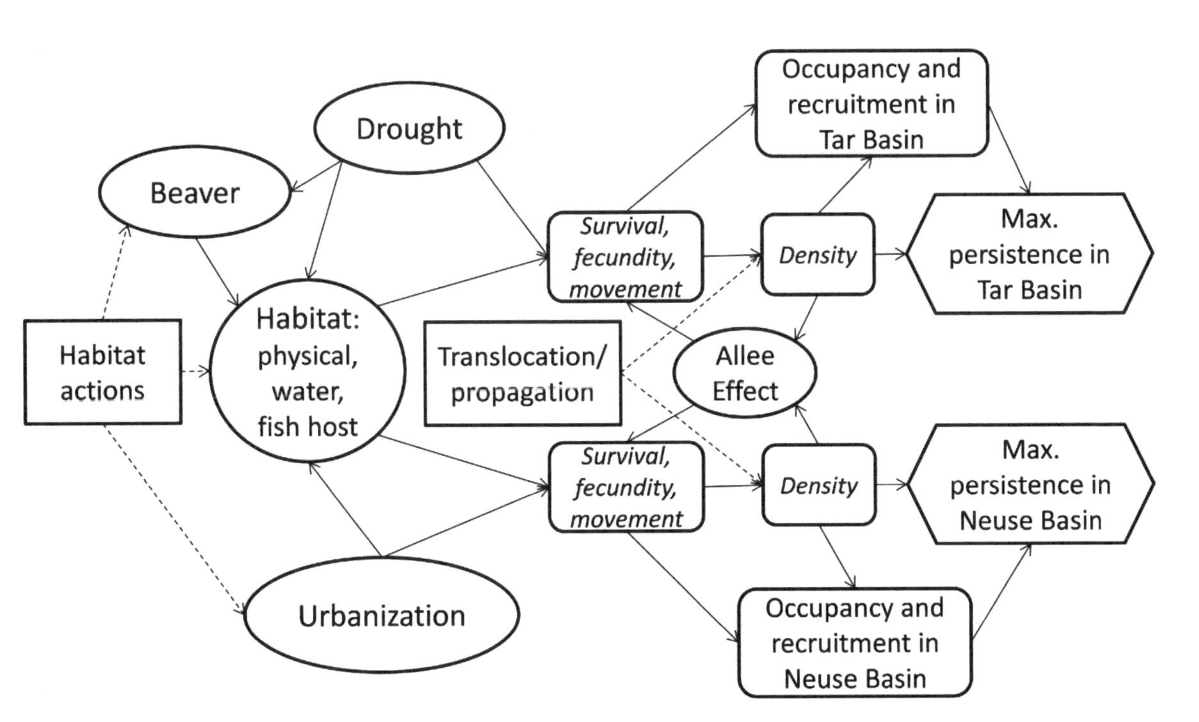

Figure 6.2. Conceptual model for population factors (round-cornered rectangles), ecological and environmental processes (circles and ellipses), and management actions (rectangles) that influence the fundamental objectives (hexagons) of maximizing persistence within the Tar and Neuse Basins. Dashed lines indicate uncertainty in management effectiveness.

Alternatives

The team developed 4 alternative strategies, each differing in how funding was allocated among types of management actions (table 6.1). Alternative 1 was the status quo allocation of effort in reaction to project-specific threats constrained by available funding. Beyond the status quo, the team wanted to explore the consequences of strategies that either focused on protecting the best remaining populations or focused on expanding the species' distribution in North Carolina. In effect, these alternative strategies contrasted allocations of effort between the Neuse or Tar River Basins. Protecting the best remaining populations put the focus on the Tar River Basin, where the best populations were located. And expanding the distribution put the focus on the Neuse River Basin, where reintroduction opportunities existed. In a way, these alternative strategies were bookends representing the extremes of spatial allocation of conservation effort within North Carolina. A fourth strategy bridged the extremes by creating a combination or hybrid strategy.

The types of management actions considered under the alternative strategies included population management, habitat management, research and monitoring, and partnership building or outreach (table 6.1). While research and monitoring do not directly influence the biological objectives, budgets often include these actions, and including them in the strategies allowed for an explicit analysis of their contribution to meeting objectives. Allocation of effort was at the level of the management actions. For example, population management included the following actions: implementation and enforcement of the Endangered Species Act (Sections 7 and 10) regulations, supplementation (release of propagated individuals into existing populations), reintroduction (release of individuals to establish new populations), and management of captive populations. Habitat management included development and enforcement of best practices for habitat management, land or easement acquisitions, restoration of instream or riparian habitat, and maintenance or restoration of connectivity. The team found it practical to identify the allocation under status quo and then make adjustments to create alternative strategies. Most allocations were in terms of the number of stream reaches where actions were applied or the level of effort associated with a particular action (table 6.1). For example, under status quo, no propagated animals had been released in North Carolina, but alternative strategies specified the number of juveniles that would be released per year at various locations to supplement existing populations or establish new populations. Also, restoration of instream and riparian habitats would increase, decrease, or maintain the status-quo levels of effort.

Consequences

A population model was developed for Dwarf Wedgemussel to predict the effectiveness and cost of propagation and release (details in Smith et al. 2015). The model, which included depensation, was used to determine ages and levels of propagation and release that would be expected to establish populations or overcome an Allee effect in extant populations. The parameters of the model were literature-based. Expert judgment was used to parameterize components of the model for which no information was otherwise available, including consequences of actions other than releases. Abundance and occupancy were predicted at the stream reach level and summarized at the basin level.

A simple multiple-attribute rating technique (SMART) trade-off analysis (Converse, chapter 5, this volume; Goodwin and Wright 2007) was conducted to compare the performance of the conservation strategies (see Smith et al. 2015 for detailed results). The SMART analysis results in a standardized and weighted average (score) for each alternative with the weights reflective of objective importance, which can vary among stakeholders. An additional set of analyses was undertaken to determine the sensitivity of the comparison to uncertainty

Table 6.1. Strategy table showing level of implementation for management actions among 4 strategies for management of Dwarf Wedgemussel in North Carolina

Management type	Management action	Unit	Conservation strategy			
			Status quo	Protect the best	Expand the distribution	Hybrid
Population management	Implement ESA section 7 and 10 regulations and influence agency enforcement	Number of stream reaches	20	18.5	23	19.5
	Increase extant populations	Number of juveniles released per year and location	0	3,000 in Shocco Creek (Tar)	3,000 in Little River and Swift Creek (Neuse)	1,000 in Shocco Creek (Tar)
	Establish new populations	Number of juveniles released per year and location	0	2,000 in Fishing Creek (Tar)	2,000 in Deep Creek (Neuse)	4,000 in Little River (Neuse)
	Use available means to protect or establish populations	Number of stream reaches	5	9	13	5
	Manage captive populations	Yes/no	No	Yes	No	No
Habitat management	Regulatory protection of stream and riparian habitats including state riparian buffers	Number of stream reaches	31	31.5	31.5	34.5
	Land acquisition or easements	Number of stream reaches	13	16.5	14.5	15
	Restoration of in-stream and riparian habitat	Number of stream reaches	10	8	10	9
	Maintenance or restoration of connectivity	Effort relative to status quo and dam removal location	2	1 and Oxford dam removal	2 and no dam removals	2
Monitoring and research	Genetics monitoring and research	Number of projects	0	4	3	0
	Assessment surveys of freshwater mussel communities	Number of stream reaches	21	21	21	21
	Targeted monitoring of extant populations	Number of stream reaches	12	17	14.5	12
	Population viability analyses	Location of population	Swift Creek	Swift Creek	Swift Creek	Swift Creek
	Life history	Yes/no	Yes	Yes	Yes	Yes
	Propagation and stocking protocols	Status	In revision	Finalize	In revision	In revision
	Captive management research	Yes/no	No	Yes	No	No
	Assess existing and potential habitat	Average level of assessment per species	2.7	2.7	2.7	2.7
	Habitat monitoring	Number of stream reaches	3	11	11	3
	Water quality research	Yes/no	Yes	No	Yes	Yes
Partnership building/ outreach	Work with industry	Number of stream reaches	2	2	7	2
	Work with partners	Number of stream reaches	15	20	16.5	15

Source: Reproduced from Smith et al. 2015.

Note: "Status quo" strategy reflects current management. The "protect the best" strategy focuses on management and protection of the extant populations in the Tar River Basin. The "expand the distribution" strategy focuses on management of extant populations and establishment of new populations in the Neuse River Basin. The "hybrid" strategy combines elements from each strategy to balance conservation between the Tar and Neuse River Basins. Units describe how the level of implementation was measured. ESA = Endangered Species Act.

about how effective management might be and to variation in how the objectives were weighted. Our approach was to explore each strategy's performance across a gradient in stakeholder values and management effectiveness and look for robust performance (i.e., strategies that performed well regardless of uncertainty level or variation in objective weighting). Management effectiveness was reduced by decreasing the difference in performance of an action relative to the status quo.

Overall, we found that different strategies performed best depending on the relative importance of the objectives, that is, how the objectives were weighted (table 6.2). If species' persistence regardless of river basin was most important, then the strategy that protected the best remaining populations came out on top. However, if the species' distribution across basins and genetic diversity were most important, then the hybrid strategy rose to the top. The relative strategy performance was not sensitive, and therefore robust, to management effectiveness, in the sense that the top-performing strategy did not switch across a wide range of uncertainty in management effectiveness. See Smith et al. (2015) for details on the outcomes of the SMART analysis.

Decision Implementation

The decision analysis developed by Smith et al. 2015 has been available to help guide and communicate to stakeholders the best ways to conserve Dwarf Wedgemussel in North Carolina. Funding for conservation remains tied primarily to the Endangered Species Act Section 7 consultation or mitigation funds; however, as funding becomes available, the strategic plan can be referenced to determine which type of management actions to support and where to apply those actions. The decision analysis calls for more conservation to focus on the Tar River Basin than does the status quo approach. This may create a perception of "walking away" from the populations in the Neuse River Basin. In reality, however, regulatory obligations prevent the habitat and populations in the Neuse River from being disregarded. Nevertheless, given staff time and limits in operating funds, decisions must be made about where to best focus conservation (Bottrill et al. 2008). The actionable insight from this decision analysis was that it is best, from the standpoint of species persistence, to implement more conservation activities in the Tar River Basin compared to the Neuse River Basin. The existence of an accepted conservation strategy

Table 6.2. Trade-off analysis and sensitivity of the optimal strategy for conservation of Dwarf Wedgemussel in North Carolina to variation in objective weighting

	Objective weights					Strategies			
Objective weighting emphasis	Persistence in the Neuse	Persistence in the Tar	Genetic diversity	Public support	Management costs	Status quo	Protect the best	Expand the distribution	Hybrid
Equal	20	20	20	20	20	0.25	**0.47**	0.30	0.38
Conservation	25	25	25	13	13	0.17	0.38	0.38	**0.43**
Persistence statewide	29	29	14	14	14	0.19	**0.43**	0.36	0.35
Persistence in the Neuse	33	17	17	17	17	0.21	0.39	**0.42**	0.36
Persistence in the Tar	17	33	17	17	17	0.22	**0.50**	0.25	0.36
Genetic diversity	17	17	33	17	17	0.21	0.39	0.33	**0.48**
Cost	17	17	17	17	33	0.38	**0.50**	0.25	0.37
Cost (extreme)	11	11	11	11	56	**0.59**	0.56	0.17	0.36

Source: Reproduced from Smith et al. 2015.

Note: The simple multi-attribute rating technique (SMART) was used to rate each strategy based on a trade-off analysis. Ratings under the strategy columns are the weighted average standardized scores from the trade-off analysis using the objective weights in this table divided by 100. The rating for the optimal strategy is highlighted in bold for each objective weighting.

has helped to justify the allocation of resources to areas outside of project boundaries.

Although the strategy is available for guidance, there are impediments to its influence. First, workload decisions must be in sync with the allocations recommended in the strategy (Converse et al. 2011); otherwise, staff time will remain consistent with the status quo, and change will not occur. Second, new regulatory tools (e.g., mitigation or conservation banks) could create the flexibility needed to accumulate and redirect conservation funding and effort from several small opportunities to contribute to the larger strategic effort. Third, the single-species framing of the strategy does not take into account that there are other species of conservation interest. In retrospect, it might have been useful to include co-occurring species and frame the problem as a multispecies conservation strategy (Smith et al. 2017) to evaluate allocation of effort among multiple species and explore opportunities to conserve multiple species.

Some areas of progress can be traced to this conservation strategy. Mitigation funds, which became available through a coal ash spill settlement (www.justice.gov/opa/pr/duke-energy-subsidiaries-plead-guilty-and-sentenced-pay-102-million-clean-water-act-crimes), were used for research and development of Dwarf Wedgemussel propagation, which was identified as important to support population management. Further, beaver management in the Tar River Basin, which was identified in the strategy as important for habitat restoration, is achieving some success. And although outreach and partnership building to create public support requires considerable staff time, these efforts are continuing. Overall, the SDM process has been shown to be beneficial because the conservation strategy provides a rational and defensible basis upon which to move conservation in a more effective direction.

Discussion

Certain aspects of the interdisciplinary and interagency team were quite helpful in moving the SDM process from start to finish. First, a well-rounded team with representatives from responsible management agencies working together with species experts was important in providing both policy and science-based input. Second, the team participated fully in a series of multiple-day workshops; SDM is a deliberative process, and a team needs time to work through the problem thoroughly. Third, the participants exhibited the flexibility and open-mindedness needed to engage in a creative and analytical process. Although knowledgeable professionals and experts may have thought through a problem and have various solutions in mind, SDM is a value-focused, not an alternative-focused, process. Thus, the team members had to be willing to have their assumptions and ideas tested.

In retrospect, an SDM case study can appear to make direct and steady progress, but the actual path forward through the PrOACT steps (problem definition, objectives, alternatives, consequences, and trade-off analysis; see Runge and Bean, chapter 1, this volume) can be iterative, and the pace can be uneven. Systematically analyzing a decision can require confronting new ideas and challenging assumed constraints, which can be uncomfortable and even confusing to those involved. Flexibility on the part of the facilitators and patience among the team participants is important as the process unfolds. The facilitators need to have a sense of when to keep working on a decision component, when to move to the next step, and when to circle back to revisit and revise. The team must be prepared to trust the process during times of divergent thinking, when progress can be slow. Prototyping or decision sketching (i.e., iteratively working through multiple streamlined decision analyses, see Blomquist et al. 2010; Garrard et al. 2017; Gregory et al. 2012; and Smith, chapter 2, this volume) can be an essential tool for remaining flexible and experimenting with decision structures during the decision analysis.

Prototyping of the decision structure helped the team explore various approaches to developing a strategy without investing so much time that they would be reluctant to scrap an approach and start again. For example, the problem was initially framed as an effort to identify site-specific conservation targets, either habitat or demographic, that could be used to direct where management was needed. This conservation-targeting approach is consistent with and could be scaled up for range-wide recovery planning. However, the team realized the conservation targets, particularly for habitat, were alternatives rather than objectives, so they broadened the framing of the problem to focus on where and how to apply management to maximize persistence of Dwarf Wedgemussel in North Carolina. Prototyping provided a way to gain insights, learn what was important to the team, and develop alternative strategies creatively without getting too invested in an initial and tentative approach.

Ultimately, the SDM process was successful at helping the team think about conservation strategically and prospectively. The regulatory process tends to drive actions at the project level, which can result in reactive conservation. For example, a federally funded project with the potential to impact an imperiled species could result in funds being allocated to conservation measures. In the absence of a thoroughly analyzed conservation strategy, the allocation of funding could be inefficient, fragmentary, and suboptimal with respect to long-term or large-scale conservation objectives. The presence of an accepted conservation strategy helps counterbalance the reactive, project-by-project nature of conservation implementation.

ACKNOWLEDGMENTS

The authors are grateful to the other workshop participants who provided their time, data, and expertise to develop this decision analysis: Tom Augspurger, Judy Ratcliffe, Robert Nichols, Chris Eads, Tim Savidge, and Art Bogan. This effort benefitted greatly from advice from Mike Runge and Jonathan Daily on problem framing and population modeling. The authors thank Colin Shea, Mary Parkin, and Sarah Converse for helpful comments on early versions of this chapter.

LITERATURE CITED

Blomquist SM, Johnson TD, Smith DR, Call GP, Miller BN, Thurman WM, McFadden JE, Parkin MJ, Boomer GS. 2010. Structured decision-making and rapid prototyping to plan a management response to an invasive species. *Journal of Fish and Wildlife Management* 1:19–32.

Bottrill MC, Joseph LN, Carwardine J, Bode M, Cook C, Game ET, Grantham H, Kark S, Linke S, McDonald-Madden E, Pressey RL, Walker S, Wilson KA, Possingham HP. 2008. Is conservation triage just smart decision making? *Trends in Ecology and Evolution* 23:649–654.

Conroy MJ, Peterson JT. 2013. *Decision Making in Natural Resource Management: A Structured Adaptive Approach.* West Sussex, UK: Wiley-Blackwell.

Converse SJ, Shelley KJ, Morey S, Chan J, LaTier A, Scafidi C, Crouse DT, Runge MC. 2011. A decision-analytic approach to the optimal allocation of resources for endangered species consultation. *Biological Conservation* 144:319–329.

Courchamp F, Clutton-Brock T, Grenfell B. 1999. Inverse density dependence and the Allee effect. *Trends in Ecology & Evolution* 14:405–410.

Doremus H, Tarlock DA. 2005. Science, judgment, and controversy in natural resource regulation. *Public Land & Resources Law Review* 26:1–38.

Dunn H. 2004. Defining the ecological values of rivers: the views of Australian river scientists and managers. *Aquatic Conservation: Marine and Freshwater Ecosystems* 14:413–433.

Garrard GE, Rumpff L, Runge MC, Converse SJ. 2017. Rapid prototyping for decision structuring: an efficient approach to conservation decision analysis. Pages 46–64 in Bunnefeld N, Nicholson E, Milner-Gulland EJ, eds. *Decision-Making in Conservation and Natural Resource Management: Models for Interdisciplinary Approaches.* Cambridge, UK: Cambridge University Press.

Goodwin P, Wright G. 2007. *Decision Analysis for Management Judgment.* West Sussex, UK: John Wiley & Sons.

Gregory R, Arvai J, Gerber LR. 2013. Structuring decisions for managing threatened and endangered species in a changing climate. *Conservation Biology* 27:1212–1221.

Gregory R, Failing L, Harstone M, Long G, McDaniels T, Ohlson D. 2012. *Structured Decision Making: A Practical Guide to Environmental Management Choices.* West Sussex, UK: Wiley-Blackwell.

Gregory RS, Long G. 2009. Using structured decision making to help implement a precautionary approach to endangered species management. *Risk Analysis* 29:518–532.

Joseph LN, Maloney RF, Possingham HP. 2009. Optimal allocation of resources among threatened species: a project prioritization protocol. *Conservation Biology* 23:328–338.

Lankau R, Jorgensen PS, Harris DJ, Sih A. 2011. Incorporating evolutionary principles into environmental management and policy. *Evolutionary Applications* 4(2):315–325.

Linke S, Turak E, Nel J. 2011. Freshwater conservation planning: the case for systematic approaches. *Freshwater Biology* 56:6–20.

McLain D, Ross MR. 2005. Reproduction based on local patch size of *Alasmidonta heterodon* and dispersal by its darter host in Mill River, Massachusetts, USA. *Journal of the North American Benthological Society* 24:139–147.

Naidoo R, Balmford A, Ferraro PJ, Polasky S, Ricketts TH, Rouget M. 2006. Integrating economic costs into conservation planning. *Trends in Ecology and Evolution* 21:681–687.

Ricciardi A, Rasmussen JB. 1999. Extinction rates of North American freshwater fauna. *Conservation Biology* 13:1220–1222.

Runge MC, Converse SJ, Lyons JE. 2011. Which uncertainty? Using expert elicitation and expected value of information to design an adaptive program. *Biological Conservation* 144:1214–1223.

Smith DR, Allan NL, McGowan CP, Szymanski JA, Oetker SR, Bell HM. 2018. Development of a species status assessment process for decisions under the US Endangered Species Act. *Journal of Fish and Wildlife Management* 9(1):302–320.

Smith DR, Butler RS, Jones JW, Gatenby CM, Hylton R, Parkin M, Shultz C. 2017. Developing a landscape-scale, multi-species, and cost-efficient conservation strategy for imperiled aquatic species in the Upper Tennessee River Basin, USA. *Aquatic Conservation: Marine and Freshwater Ecosystems* 27:1224–1239.

Smith DR, McRae SE, Augspurger T, Ratcliffe JA, Nichols RB, Eads CB, Savidge T, Bogan A. 2015. Developing a conservation strategy to maximize persistence of an endangered freshwater mussel species while considering management effectiveness and cost. *Freshwater Science* 34:1324–1339.

Strayer DL, Sprague SJ, Claypool S. 1996. A range-wide assessment of populations of *Alasmidonta heterodon*, an endangered freshwater mussel (Bivalvia:Unionidae). *Journal of the North American Benthological Society* 15:308–317.

USFWS [US Fish and Wildlife Service]. 2013. *Dwarf Wedgemussel (Alasmidonta heterodon) 5-Year Review: Summary and Evaluation.* Concord, NH: New England Field Office, US Fish and Wildlife Service. http://ecos.fws.gov/speciesProfile/profile/speciesProfile.action?spcode=F029.

Vucetich JA, Nelson MP, Phillips MK. 2006. The normative dimension and legal meaning of *endangered* and *recovery* in the US Endangered Species Act. *Conservation Biology* 20:1383–1390.

7 Spawning Closures for Coral Reef Fin Fish

Terry Walshe and Stephanie Slade

The Australian state of Queensland has a legislative objective to provide for the use, conservation, and enhancement of the community's fisheries resources in a way that applies, balances, and promotes the principles of ecologically sustainable development. One of the tools used to achieve this purpose in the coral reef fin fish fishery is annual fishery closures during spawning periods. In addition to species protection, the decision maker must consider implications for other uses, values, and stakeholders when choosing a closure regime. The decision is difficult because of the multispecies nature of the fishery, uncertainty in scientific knowledge about fish responses to closures, and a history of value-based conflict among key stakeholders. This case study demonstrates a practical application of the simple multi-attribute rating technique (SMART) for multi-objective decision analysis that focuses on scientist and stakeholder participation in decision support. Within a pressing time constraint, we used naïve global value judgments to screen objectives, allowing us to focus on key trade-offs in the more formal SMART analysis. We illustrate how to use uncertain expert judgment and risk attitude in decision analysis to draw insight on robust compromise solutions.

Problem Background

Access to coral reef fin fish is important to a range of extractive users, including commercial, recreational, charter, and indigenous fishers, and nonextractive users, including tourists and conservationists on the Great Barrier Reef in Queensland, Australia. The spatial extent of the fishery coincides roughly with the boundaries of the Great Barrier Reef Marine Park. Together with other management tools including quotas and spatial zoning, seasonal closures are used to maintain harvest levels of key fishery species at sustainable levels. The periodic review of closures invites controversy, as fishers and conservationists contest their effectiveness and the costs of their implementation. For the period 2004–2008, 3 closures each of 9 days' duration were planned for the new moon periods in October, November, and December each year. But in 2008, based on advice suggesting December closures contribute little to the sustainability of fish stocks (Tobin et al. 2009) and in recognition of the high costs to fishers, the December closure was not implemented.

Against this background, a decision on the number, duration, and timing of closures for the period 2009–2013 was to be made quickly given the high sociopolitical interest in this management measure.

A 2-day workshop involving 13 participants encompassing a range of relevant expertise and stakeholder positions was convened to advance the decision. The brief of the workshop required that any closure regime implemented beyond 2008 needed to provide adequate protection for spawning coral reef fin fish species and that the imposition on commercial and recreational (including charter) fishing be no greater than for the period 2004–2008.

Decision Maker and Their Authority

Legislation subordinate to the Queensland Fisheries Act 1994 specifies spawning closures for coral reef fin fish. Ultimately, the Queensland government is the decision maker, but as is often the case in contested settings, the government is interested in the perspectives of affected stakeholders and whether a consensus view can be reached. This case study describes the structured decision-making approach we used in a 2-day workshop involving stakeholders and scientists who, in effect, acted as informal proxy decision makers. The outcomes were in no way binding on the government.

Ecological Background

The 2004–2008 closures were designed to provide protection to the key commercial target species, coral trout (*Plectropomus* spp. and *Variola* spp.), the peak spawning period of which is reasonably well known and strongly associated with the new moon in mid-spring to early summer. For other fin fish species, spawning period is relatively uncertain (Tobin et al. 2009). Closures targeting coral trout afford some protection to other coral reef fin fish species, although the magnitude of this effect is speculative. While critical to the commercial sector, coral trout is of less importance to recreational and charter-based fishing. A more comprehensive list of species of interest to all 3 fishing sectors would include

- Coral trout
- Red-throat emperor (*Lethrinus miniatus*)
- Red emperor (*Lutjanus sebae*)
- Large-mouth nannygai (*Lutjanus malabaricus*)
- Spangled emperor (*Lethrinus nebulosus*)
- Camouflage grouper (*Epinephelus polyphekadion*)
- Flowery cod (*Epinephelus fuscoguttatus*)
- Greasy rockcod (*Epinephelus tauvina*)
- Spanish flag (stripey; *Lutjanus carponotatus*)
- Tuskfish (*Choerodon* spp.)

Decision Structure

There are many potential configurations for spawning closures implying a potentially vast set of candidate alternatives, each of which may differ in

- Timing of closures (moon phase, month and duration);
- Species-specific closures; and
- Exemptions for specific sectors (e.g., the charter fleet).

It might be tempting for an analyst to approach the decision as a classic optimization problem, where the emphasis is on a computer-based search for the best alternative among many, with "best" defined by the user according to some specified objective function and a set of constraints. But in multi-objective, multi-stakeholder settings, optimization methods may be less useful because agreement on how to structure the objective function and accompanying constraints may be unattainable. Even if agreement on a preferred alternative can be reached, the underlying value judgments and trade-offs may be opaque.

Recognizing that a key obstacle to enduring policy on fishing closures is conflict stemming from the value-based positions of key stakeholders, we elected to place less emphasis on an exhaustive search of potential alternatives and greater emphasis on transparency in the value judgments underpinning trade-offs. This case study is a multi-objective decision problem involving a small set of discrete alternatives to which we applied the simple multi-attribute rating technique (SMART), an accessible shortcut to more rigorous and time-consuming methods available within the family of techniques known as multi-criteria decision analysis (Edwards and Barron 1994).

Decision Analysis

Objectives

Objectives were elicited using unstructured brainstorming. The central trade-off in this decision problem weighs the ecological benefits of closures against the costs of their imposition, although other considerations emerged in the workshop (see below). Ideally, we would account for all the ecological entities that may be impacted by alternative closure regimes, with concern for each entity captured through inclusion of a taxon- or community-specific objective. Although a *complete* set of objectives is a desirable property of sound problem formulation (Keeney 2007), it is usually impractical to include the many species and ecosystems that might plausibly be impacted in any natural resource management decision. In this case study, the constraint to complete the analysis within a 2-day workshop was especially limiting.

We were unable to include the full suite of species of importance to fishing and conservation interests. Each species to be included implied a corresponding objective, against which the consequences of each candidate alternative would need to be assessed. The task was simply too large for the time available. We therefore decided to shortlist species according to their perceived importance while recognizing that a management regime developed to protect shortlisted species would also provide unquantified benefits to other species.

We asked participants to assign *global* weights representing the importance of each species to the commercial, recreational, and charter fisheries, and to conservation. In formal multi-criteria decision analysis, the weights we assign to objectives should be *local*. Local weights are informed by 2 things—their importance and the range of the consequences associated with the best and worst alternatives under consideration (Edwards and Barron 1994; Fischer 1995). In making the range of consequences salient, the weights can be interpreted as compensatory judgments, which is the essential core of any trade-off. SMART makes the simplifying assumption that the value function for any single objective is linear. For example, when we assign the same weight to a 10% improvement in prospects for protection of a fish species and a 20-day loss of fishing access, we are saying that a 10% decline in protection is compensated by a 20-day gain in fishing access, or equivalently, that a 20-day loss in access is offset by a 10% gain in protection.

The task of assigning local weights is emotionally and cognitively demanding. In contrast, global weights are relatively easy to elicit, but their quantitative context is unspecified because the range of consequences is unspecified. Global weights may be colored by the experience, aspirations, or imagination of participants (Monat 2009). We elicited global weights in the early stages of our 2-day workshop, once participants were made aware of the specific decision setting of their judgments. Use of global weights in formal analysis is poor practice. We used them here as a screening tool, assuming that species having small global weights could be omitted because they were unlikely to have much bearing on the outcomes of a more detailed formal analysis that employed local weights.

Figure 7.1 records participants' pooled global weights describing the perceived importance of each species/species group to conservation and the commercial, recreational and charter fisheries. Based on qualitative interpretation of these graphical outcomes alone, participants agreed to the following shortlist of species for further consideration in the workshop:

- Coral trout
- Red-throat emperor
- Red emperor
- Large-mouth nannygai

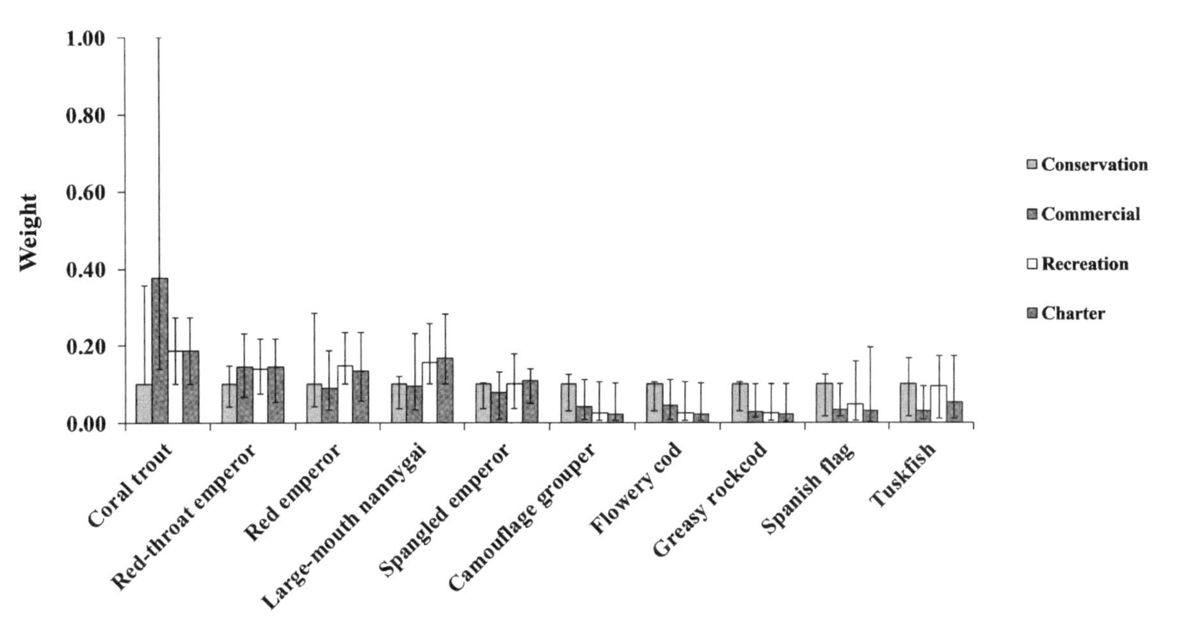

Figure 7.1. Median global weights assigned to coral reef fin fish species, describing the perceived importance of each species to conservation and to commercial, recreational, and charter fishing interests. Error bars capture the full range of responses among workshop participants.

- Spangled emperor
- Camouflage grouper / flowery cod

Beyond protection of the 6 fish species groupings listed above (where more protection is better), participants identified the following objectives:

- Costs to commercial fishers (less is better)
- Costs to recreational fishers (less is better)
- Costs to charter fishers (less is better)
- Broader ecosystem benefits (more is better)
- Ease of enforcement (more is better)
- Opportunities to learn the effect of spawning closures on fish protection and the broader ecosystem (more is better)

Alternatives

Participants were invited to identify candidate alternatives for closure regimes using unstructured brainstorming. After considering the time available in the workshop and what might be in the realm of political and operational feasibility, workshop participants agreed on 6 candidates for the period 2009–2013:

A1. No closures.
A2. No closures in 2009 and 2010, followed by 9-day closures in both October and November around the new moon in 2011, 2012, and 2013.
A3. 5-day closures in both October and November around the new moon each year.
A4. 9-day closures in both October and November around the new moon each year, with an exemption for the offshore charter fleet. No offshore charter fishing from December 20 to January 31 each year.
A5. 9-day closures in both October and November around the new moon each year.
A6. 5-day closures in both October and November around the new moon each year, plus a 5-day closure for coral trout only in September if the new moon occurred after September 15, and a 5-day December closure for coral trout if the new moon occurred before December 15.

Alternative 1 is a "do nothing" option, and alternative 5 described the regime in place at the time of the workshop. Where feasible, we suggest inclusion of a do-nothing option and the *status quo* in the set of alternatives in all decision problems for 2 reasons: they serve as important benchmarks against which the argument for change can be considered, and they may themselves be among the best alternatives available.

Consequences

Given sparse data and severe time constraints, we relied on expert judgment to estimate the consequences of each alternative against each objective. A common flaw in decision-support exercises is to inadvertently mix the experts' judgment about uncertain future events with personal or political preferences (Maguire 2004). Experts and stakeholders vary in their judgments of both elements. The explicit decoupling of causal judgments (consequences) and value judgments (trade-offs) as separate steps in structured decision-making provides partial insulation. Variation in biological judgments arises from uncertainty in scientific knowledge. Variation in value judgments reflects the priorities and preferences of individuals and organizations. The workshop integrated both elements in a structured decision-making framework using SMART (Edwards and Barron 1994).

Another common shortcoming in use of expert judgment is failure to treat language-based ambiguity (Regan et al. 2002). An important source of ambiguity in this case was variable interpretation of what might be meant by "protection" of fish species. To minimize this arbitrary source of disagreement, for each alternative closure regime and each species, participants were individually asked to estimate the probability (expressed as a percentage) that fish numbers encountered in Great Barrier Reef Marine Park general use and habitat protection zones ("blue zones") over the past 5 years would be maintained over the next 5 years. More specifically, given hypothetical implementation of alternative *a*, what is the chance the number of fish of species *y* would be maintained or increased over at least half of all blue-zone reefs? Participants were instructed to assume that all other management controls (e.g., quota, bag limits, and green zones) remained unchanged from the existing conditions.

The estimates of individuals are commonly compromised by overconfidence and motivational bias (Hammond et al. 2006; Morgan and Henrion 1990). To derive a plausible interval on probabilities, we pooled individual point estimate judgments of multiple experts and stakeholders (Armstrong 2001; Yaniv 2004). In an effort to curb motivational bias, prior to making their judgments participants were made aware that intervals would not include the 2 lowest and 2 highest estimates. Pooling *causal* judgments is strongly supported by psychological studies in expert judgment, which find that over many judgments the central tendency of the estimates of a group tend to outperform estimates by the best-credentialed individual within that group (Burgman 2015).

After providing initial estimates of the probability of maintaining the number of fish in blue zones over the next 5 years, participants were invited to justify and cross-examine each other's perspectives (Schultze et al. 2012). Estimates were then revised in the light of discussion, including exploration of the sparse data available. Results of revised estimates for each species are shown in table 7.1. This approach to elicitation was arguably best practice at the time of the workshop. Since then, similar but improved methods have been made available (Hanea et al. 2017).

The estimated performances of the alternatives against each objective are shown in table 7.1. Protection was described as "chance of successful maintenance of fish numbers," described above. The cost to each fishery sector was estimated by participants as aggregate fishing days lost over the 5-year regime. The 3 objectives dealing with broader ecosystem benefits, ease of enforcement, and prospects for learning were scored on a Likert scale from 1 (worst) to 4 (best). For judgments beyond those pertaining to the

Table 7.1. Estimated consequences of each alternative against shortlisted objectives

Objective	Preferred direction	A1	A2	A3	A4	A5	A6
Chance (%) of "protecting"							
Coral trout	More	29–80	35–85	50–85	66–90	69–90	62–90
Red-throat emperor	More	28–96	41–96	25–96	25–95	25–96	34–96
Red emperor	More	30–90	50–90	50–91	49–90	50–91	35–91
Large-mouth nannygai	More	30–90	50–90	50–90	48–90	50–90	35–90
Spangled emperor	More	44–95	50–95	50–95	50–95	50–95	44–91
Cam group/flow cod	More	29–90	39–90	49–95	49–95	50–95	40–90
Effective days lost to							
Commercial fishing	Less	0	60–66	50	100–110	100–110	75
Recreational fishing	Less	0	54	50	90	90	0–5
Charter fishing	Less	0	54–60	50	100	90–100	20–30
Ecosystem benefits	More	1	2	3	4	4	2–3
Ease of enforcement	More	4	3.4	3	2.8	3	1
Prospects for learning	More	2–4	4	2	2	2	3

Note: Intervals capture the judgments of 9 of the 13 participants (i.e., 4 outliers omitted). Cells reporting single point estimates represent agreed consensus judgments.

objective of fish protection, consensus point estimates were generally obtained. Where consensus could not be reached, an interval was provided to account for disagreement and uncertainty.

Among the candidate alternatives considered, coral trout was identified as the species most sensitive to spawning closures. It was also the most important to fishery users (fig. 7.1). Estimates ranged from 29–80% chance of maintaining fish numbers under no closures (alternative 1) to 69–90% under alternative 5. A more conservative (risk-averse) decision maker places greater emphasis on avoiding the possibility of poor outcomes (i.e., those alternatives with more extreme lower bounds). For coral trout considered in isolation, alternatives 4, 5, and 6 all appeal to a risk-averse decision maker because they have high lower bounds.

For the species of lesser (but nontrivial) importance—red emperor, large-mouth nannygai, spangled emperor, and camouflage grouper/flowery cod—alternatives 3 and 5 provide risk-averse options. Red-throat emperor is less sensitive to the 6 alternative spawning closures because it spawns mainly in winter (Tobin et al. 2009), outside the periods included in any of the alternatives.

Alternative 1, involving no closures, is plainly the best from the perspective of cost to fisheries over the 5-year period. It is also the easiest to enforce. For the commercial fishing sector, the next best is alternative 3, and for the recreational and charter sectors it is alternative 6. Ecosystem benefits beyond the protection of fin fish species are anticipated to be highest under alternatives 4 and 5. Prospects for learning are maximized under alternative 2.

There is clearly no dominant alternative that performs best across all 12 objectives. For individual workshop participants, personal value judgments were required to resolve trade-offs and identify preferred alternatives. Collective compromises seek to identify one or more alternatives that participants regard as broadly acceptable, rather than an alternative that is optimal for any subset of stakeholders or objectives. In the next section we describe how our use of SMART provided the structure for identifying individual and collective preferences.

Outcomes

Each of the 13 participants assigned local weights to the 12 objectives evaluated in the elicitation process. We used the swing weighting technique for elicitation (von Winterfeldt and Edwards 1986) where the weight apportioned to any objective reflected both the importance of the objective and the full range of consequences associated with alternatives (Steele et al. 2009). Decision scores summarizing the merit of alternatives across all objectives were obtained for each participant using SMART's simple weighted summation decision model (Edwards and Barron 1994). The decision score V for alternative i is

$$V_i = \sum_{j=1}^{n} w_j X_{ij}$$

where w_j is the weight for objective j, and X_{ij} is the linearly standardized consequence score for alternative i on attribute j, such that all consequences for any single objective range from 0 (worst) to 1 (best) across the full set of alternatives.

While there are good reasons to pool causal judgments, the same cannot be said for value judgments. It may be tempting to use the average weight for each objective across workshop participants in the SMART decision model to calculate a single decision score for each alternative, but outcomes from doing so can be misleading. Where weights across a group are distributed with multiple modes, decision scores obtained from pooled judgments may result in high scores being assigned to alternatives that few if any participants in fact support. Averaging also implies that workshop participants are in some way representative of the broader set of individuals with a stake in the decision problem, a claim that is rarely justified in multi-criteria decision analysis (Dobes and Bennett 2009). We used the 13 sets of decision scores corresponding to the 13 workshop participants as the basis for exploration of collective preferred alternatives.

The importance of uncertainty in consequences on decision scores can be evaluated with sensitivity analysis. A simple approach is to conduct and compare separate analyses on plausible lower and upper bounds based on the values reported in table 7.1 (Walshe and Burgman 2010). For each participant, we calculated decision scores for each alternative using lower bounds for all consequences in table 7.1 and then repeated the analysis using all upper bounds. Results of the range of decision scores obtained across all participants when using lower- and upper-bound estimates are shown in figure 7.2. In neither the lower-bound nor the upper-bound analysis is there a clear collective preference for any of the 6 candidate alternatives.

Any decision analysis is inevitably an abstraction and simplification of the many considerations that could be invoked in making a good decision. We therefore tend to use analysis as an exercise in *decision aiding* rather than a prescription for *decision making*. The role of workshop participants as prox-

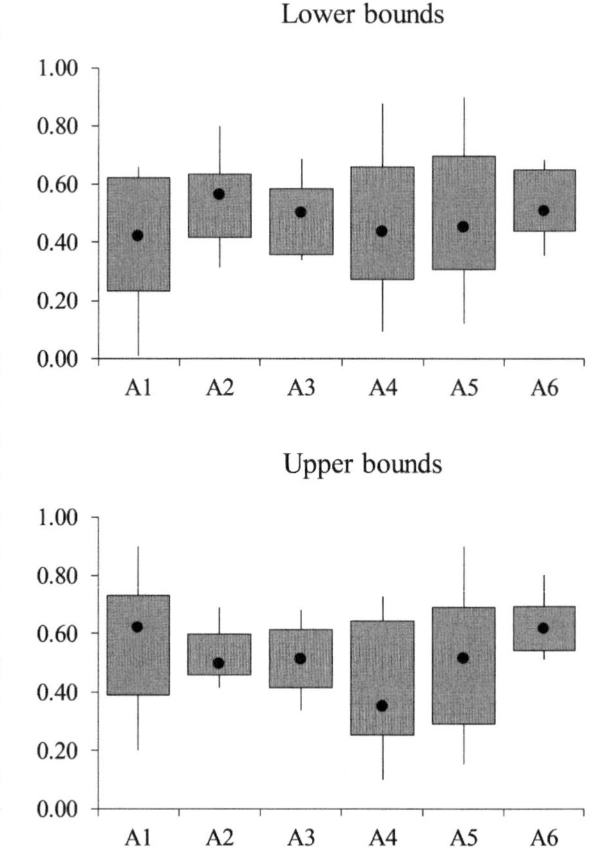

Figure 7.2. Participants' decision scores for the 6 alternatives, indicating the merit of each alternative aggregated over multiple objectives. Lower (and upper) bounds refer to decision scores calculated using the lower (and upper) bound of consequences. Median score is indicated by a dot, the box shows the range of 9 of the 13 participants, and whiskers indicate the full range of the 13 participants.

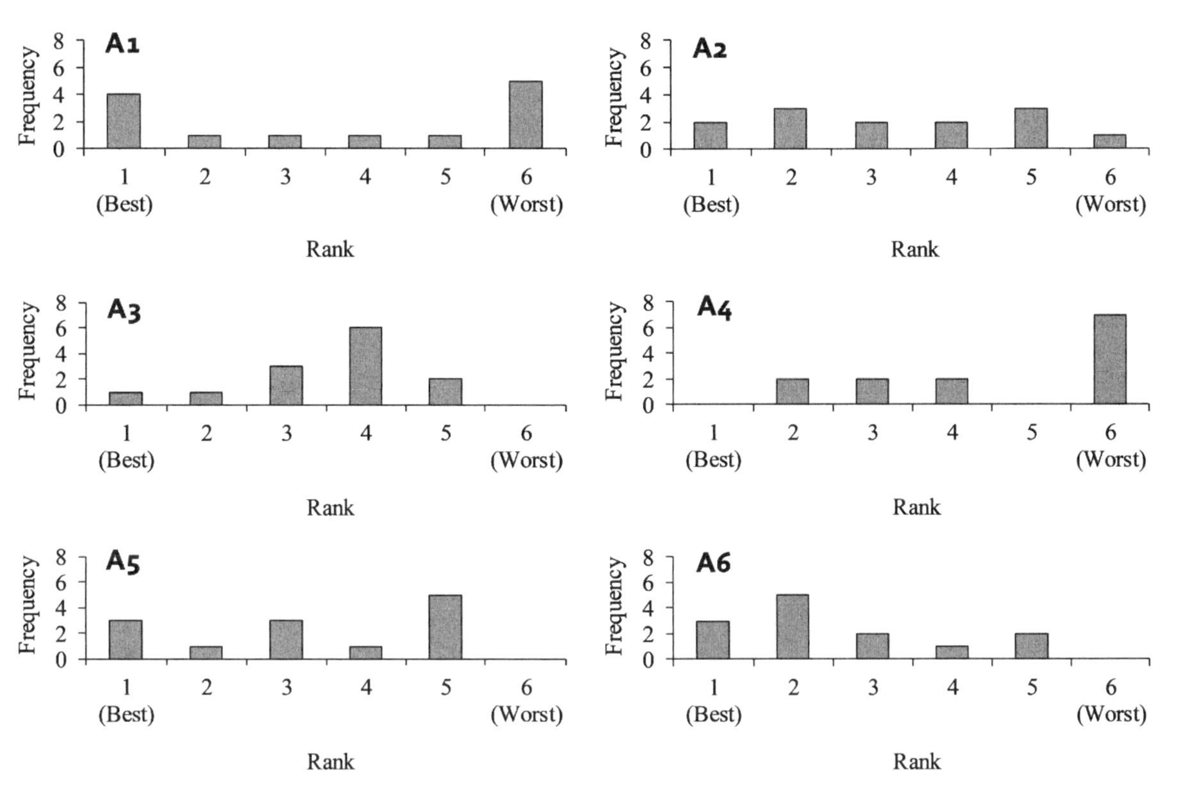

Figure 7.3. The frequency of rankings for each of the 6 alternatives among the 13 participants. Rankings were derived from the average of lower- and upper-bound calculations. A rank of 1 indicates the best alternative and 6 the worst.

ies for the decision maker, in this case the Queensland government, clearly places our analysis in the realm of decision aiding. On the basis of the dispersed outcomes summarized in figure 7.2, what advice can the analyst provide the decision maker?

The large range in decision scores associated with alternatives 1, 4, and 5 (for both lower and upper bound calculations) imply that although some participants strongly support them, others strongly oppose them. A less divisive approach is to consider implementing alternative 2, 3, or 6. The "maxi-min" strategy of decision making under uncertainty appeals to those who are risk averse (Luce and Raiffa 1957). The strategy involves focusing on the minimum outcome associated with each alternative and selecting the one with the largest minimum value (Morgan and Henrion 1990). The alternative with the largest minimum decision score is alternative 6, for both the lower- and upper-bound analyses.

The rank order of preference for each alternative among all participants is shown in figure 7.3. The distribution of rankings makes plain the divergence of views associated with alternative 1 (no closures). Four of the 13 participants considered it the best option. Five considered it the worst. Contrasts were driven essentially by different emphases on losses to fishers and protection of species or broader ecosystem benefits. The breadth of opinion regarding the merit of alternative 5 was likewise driven by these contrasting emphases. Alternative 4 involved special arrangements for the charter sector. Seven participants rated it the worst option. Alternative 6 had the broadest support: 10 of the 13 participants ranked it in their top 3. No one considered it the worst option. We note that our use and interpretation of figures 7.2 and 7.3 assume that the value judgments of all participants are equally relevant to the decision maker. This assumption is unlikely to be valid.

Decision Implementation

Outcomes of the workshop assisted the department in providing advice to the government, which announced 2 closures each of 5 days' duration for the new moon periods in October and November of 2009–2013, consistent with alternative 3. The government subsequently decided to provide certain exemptions for eligible long-range charter vessels that demonstrated disproportionate economic impact.

Exemptions for long-range charter vessels were implemented via a permit system involving application, assessment, and the development of conditions (including the species that could be taken) and additional reporting requirements. Despite this, the exemption was strongly opposed by other stakeholder groups on the grounds of equity, a perception of a "softening" of the government commitment to addressing sustainability concerns, issues of public perception, and compliance challenges.

Outcomes have led to enduring policy, with more recent announcements that the same closure regime would remain in place for the periods 2014–2018 and 2019–2023. The structured decision-making process employed at the 2009 workshop was well accepted by the range of stakeholders in attendance. There was not complete agreement among stakeholders in relation to either the consequence judgments or the value-based trade-off judgments; follow-up discussions with participants, however, confirmed that they identified with the logic of the structured decision-making process and understood the stepwise approach taken to reaching the recommendation. This understanding appears to have facilitated an enduring acceptance of the trade-offs associated with the recommended option, providing a sense of an outcome that effectively balances the complex policy drivers.

We note the following regarding alternative 3:

- It is among the subset of less divisive alternatives, alongside alternatives 2 and 6.
- It is among the best options for protection of red emperor, large-mouth nannygai, spangled emperor, and camouflage cod / flowery cod, noting that there was very little difference between the alternatives for red-throat emperor. There was also very little difference in alternatives 3 through 6 in protecting coral trout.
- Alternative 3 imposes a relatively low and equal impact on all fishery sectors, and the ecosystem benefits and ease of enforcement are moderately high.

We note the following additional qualifications regarding alternative 6:

- It provides a relatively high level of protection to coral trout.
- Its impact with respect to access to the fishing resource is highly inequitable, with commercial fishing bearing the largest cost and recreational fishing bearing negligible cost.
- Alternative 6 is the worst alternative with respect to ease of enforcement. The high costs of detecting noncompliance may need to be offset by improved deterrence through introduction of stronger penalties.

Outcomes of the workshop did not compel the government to any course of action. Nevertheless, we understand that the analysis and accompanying commentary were useful in arriving at a commitment to implement alternative 3. We speculate that this decision may have been shaped by additional consideration of the equity of costs borne by the 3 fishery sectors (an objective not included in our analysis) and a premium placed on ease of enforcement (a core consideration for the Queensland government).

Discussion

In many circumstances, informal and imprecise processes are used to adjudicate on decisions in natural resource management. The approach adopted here provides an extension to the SMART framework in which uncertainty may be characterized and carried through the logical steps that lead to a decision. A decision option that may yield a higher return might be declined in favor of an alternative with lower expected value but lesser uncertainty in the outcome (Morgan and Henrion 1990). Risk-averse decision makers may choose to minimize potential adverse outcomes. Such decision making under uncertainty can only be undertaken if the extent of uncertainty associated with alternatives is understood and communicated clearly.

Analyses were not without their limitations. Local weighting is a demanding task. It requires participants to clearly understand the merit of each alternative against each objective and then assign weights that reflect personal or organizational trade-offs. In a relative sense, the indicators used to characterize consequences associated with fish protection and costs to fishers were reasonably clear. The percentage chance of maintaining fish numbers and days lost to fishing are natural and accessible indicators. But the assignment of weights may have been problematic for the 3 criteria assessed on an arbitrary 4-point Likert scale (ecosystem benefits, ease of enforcement, and prospects for learning).

Inclusion of learning as a fundamental objective is not always appropriate. In many decision settings, learning is a means to improved understanding and precision in estimates of consequences associated with ends or fundamental objectives. Mixing means and ends objectives in a decision analysis invites confusion and double counting (Keeney 2007). In this instance, we elected to include learning because the decision will be revisited in future reviews of spawning closures, and workshop participants were keen to secure a firmer evidence base for future deliberations. McDaniels and Gregory (2004) provide further advice on when learning as an objective might usefully be included in structured decision making.

Much of the art of decision analysis involves the analyst's choice of approach or technique within the problem formulation stage. At the outset, the decision problem described here presented an array of challenges. They included the need for a collectively acceptable decision among a group of stakeholders holding divergent value-based positions, pervasive uncertainty around the extent to which spawning closures protect fin fish species other than coral trout, the implications of that uncertainty for risk attitude and preferred closure regimes, a potentially overwhelming number of alternatives, and, critically, a limited window of time in which the analysis was to be completed. Each of these challenges lends itself to different methods. In the "Decision Structure" section above, we noted the appeal of optimization techniques to search for better solutions among an otherwise bewilderingly large set of candidate alternatives. In the treatment of uncertainty and risk attitude, rigorous and detailed approaches are available from multi-attribute utility theory (von Winterfeldt and Edwards 1986), but their application is demanding, time-consuming, and difficult for participants to grasp. In particular, the elicitation of risk preferences and trade-offs under uncertainty can be turgid, with participants commonly providing inconsistent responses (Ruggeri and Coretti 2015).

In this case study, we judged the most pressing challenges to be mutual understanding of different stakeholder positions and the compressed time available for analysis. In response to these challenges we elected to use SMART. This simple approach to multi-objective problems is easy to explain, easy to implement, and readily engenders a sense of trust and transparency—qualities that are especially important in contested multi-stakeholder settings. We extended the standard approach to include uncertainty, whereby analyses were conducted on plausible lower and upper bounds for causal judgments. We discussed how risk attitude can be addressed under uncertainty without recourse to the turgid and difficult techniques in multi-attribute utility theory.

Here we believe we made the right choice for approach and problem formulation. Consistent with the idea of rapid prototyping within an iterative

PrOACT cycle (Garrard et al. 2017) and an emphasis on decision *aiding* rather than definitive decision making, our experience suggests that a simple technique provides the best prospects for ownership of the problem among decision makers or their proxies. Where obstacles are encountered that are inadequately addressed by a simple approach, more sophisticated methods can be deployed where time and resources allow.

ACKNOWLEDGMENTS

This work was supported by the Queensland Department of Agriculture and Fisheries. The authors thank Brigid Kerrigan for her energetic involvement in scoping the decision problem, Tracy Rout for able assistance in running the workshop, and all workshop participants for their insights and thoughtful contributions. This chapter benefited from comments on earlier drafts received from Claire Andersen, Sarah Converse, and 2 anonymous reviewers.

LITERATURE CITED

Armstrong JS. 2001. Combining forecasts. Pages 417–439 in Armstrong JS, ed. *Principles of Forecasting: A Handbook for Researchers and Practitioners*. Norwell, MA: Kluwer.

Burgman MA. 2015. *Trusting Judgments: How to Get the Best out of Experts*. Cambridge, UK: Cambridge University Press.

Dobes L, Bennett J. 2009. Multi-criteria analysis: "good enough" for government work? *Agenda* 16:7–29.

Edwards W, Barron FH. 1994. SMARTS and SMARTER: Improved simple methods for multiattribute utility measurement. *Organizational Behavior and Human Decision Processes* 60:306–325.

Fischer GW. 1995. Range sensitivity of attribute weights in multiattribute value models. *Organizational Behavior and Human Decision Processes* 62:252–266.

Garrard GE, Rumpff L, Runge MC, Converse SJ. 2017. Rapid prototyping for decision structuring: an efficient approach to conservation decision analysis. Pages 46–64 in Bunnefeld N, Nicholson E, Milner-Gulland E, eds. *Decision-Making in Conservation and Natural Resource Management: Models for Interdisciplinary Approaches*. Cambridge, UK: Cambridge University Press.

Hammond JS, Keeney RL, Raiffa H. 2006. The hidden traps in decision-making. *Harvard Business Review* 118:120–126.

Hanea A, McBride M, Burgman M, Wintle B, Fidler F, Flander L, Twardy C, Manning B, Mascaro S. 2017. Investigate D iscuss E stimate A ggregate for structured expert judgement. *International Journal of Forecasting* 33:267–279.

Keeney RL. 2007. Developing objectives and attributes. Pages 104–128 in Edwards W, Miles RFJ, von Winterfeldt D, eds. *Advances in Decision Analysis: From Foundations to Applications*. Cambridge, UK: Cambridge University Press.

Luce RD, Raiffa H. 1957. *Games and Decisions: Introduction and Critical Survey*. New York: Wiley.

Maguire LA. 2004. What can decision analysis do for invasive species management? *Risk Analysis* 24:859–868.

McDaniels TL, Gregory R. 2004. Learning as an objective within a structured risk management decision process. *Environmental Science & Technology* 38:1921–1926.

Monat JP. 2009. The benefits of global scaling in multi-criteria decision analysis. *Judgment and Decision Making* 4:492–508.

Morgan MG, Henrion M. 1990. *Uncertainty: A Guide to Dealing with Uncertainty in Quantitative Risk and Policy Analysis*. Cambridge, UK: Cambridge University Press.

Regan HM, Colyvan M, Burgman MA. 2002. A taxonomy and treatment of uncertainty for ecology and conservation biology. *Ecological Applications* 12:618–628.

Ruggeri M, Coretti S. 2015. Do probability and certainty equivalent techniques lead to inconsistent results? Evidence from gambles involving life-years and quality of life. *Value in Health* 18:413–424.

Schultze T, Mojzisch A, Schulz-Hardt S. 2012. Why groups perform better than individuals at quantitative judgment tasks: group-to-individual transfer as an alternative to differential weighting. *Organizational Behavior & Human Decision Processes* 118:24–36.

Steele K, Carmel Y, Cross J, Wilcox C. 2009. Uses and misuses of multi-criteria decision analysis (MCDA) in environmental decision-making. *Risk Analysis* 29:26–33.

Tobin RC, Simpfendorfer CA, Sutton SG, Goldman B, Muldoon G, Williams AJ, Ledee E. 2009. A review of the spawning closures in the Coral Reef Fin Fish Fishery Management Plan 2003. Report to the Queensland Department of Primary Industries and Fisheries. Fishing and Fisheries Research Centre, James Cook University, Townsville, Australia.

von Winterfeldt D, Edwards W. 1986. *Decision Analysis and Behavioral Research*. Cambridge, UK: Cambridge University Press.

Walshe T, Burgman M. 2010. A framework for assessing and managing risks posed by emerging diseases. *Risk Analysis* 30:236–249.

Yaniv I. 2004. The benefit of additional opinions. *Current Directions in Psychological Science* 13:76–79.

8 Managing Water for Oil Sands Mining

Dan W. Ohlson,
Andrew J. Paul, and
Graham E. Long

Oil (tar) sands mining in northeastern Alberta uses a water-based extraction process that relies primarily on the Lower Athabasca River. In this case study, we describe the use of structured decision making (SDM) to guide a multi-stakeholder process for developing the Lower Athabasca River Water Management Framework to meet oil sands mining, environmental flow, and traditional use requirements. We highlight how a consistent approach to establishing objectives and evaluation criteria across all values, even those that are conventionally difficult to quantify, served to directly involve a committee of stakeholders in the task of developing and evaluating alternatives. The result helped build trust and foster learning about what was feasible, about each other's interests, and about key trade-offs. Finally, we describe how achieving full consensus, while always the goal in a multi-stakeholder planning process, is not a prerequisite to providing the necessary information on which to develop public policy. The SDM process provided the information needed by government officials responsible for the final decision and gave them the confidence that they were making a defensible choice despite the highly controversial setting.

Problem Background

Regulatory and Historical Context

The Athabasca River emerges from the glaciers of Jasper National Park in Alberta and runs northeast to join the Peace-Athabasca Delta in the northernmost corner of the province. It is one of Canada's longest free-flowing rivers, famed for its scenery and cultural significance. In recent years, it has also become better known for its role as the primary source of process water required by oil sands mining companies as they extract bitumen from the tar-like, sandy deposits found along its banks and in the region surrounding the city of Fort McMurray. As the forecasted expansion of the oil sands industry rapidly grew in the 2000s, so too did controversy over this water extraction. Water removed from the river does not return; it is stored in massive tailings ponds, reducing instream flow in ways that have the potential to be harmful to the aquatic ecosystem and Aboriginal values.

Oil sands (also known as tar sands) mining in northeastern Alberta has always been contentious; government regulators struggle to balance significant economic development opportunities with environmental protection and social considerations. The Alberta provincial government created a regional sustainable development strategy in the late 1990s to provide a framework to achieve this balance. Since oil sands mining uses a water-based extraction process that relies primarily on withdrawals from the Lower Athabasca River, a key issue identified by stakeholders was the need to define the year-round instream flow requirements to protect the aquatic ecosystem of the Lower Athabasca River.

After undertaking a wide range of technical studies during the 2000s, federal and provincial governments jointly developed and released the interim Phase 1 Water Management Framework for the Lower Athabasca River in 2007 (Alberta Environment and Fisheries and Oceans Canada 2007). The Phase 1 Framework contained an interim management system that prescribed allowable water withdrawal rates associated with the amount of natural inflow in the river at different times of year. Although the interim system was based on a significant investment in the science of instream flow requirements (through aquatic field data collection, development of sophisticated hydraulic and habitat analysis methods, etc.), many stakeholders criticized shortcomings in both the framework and the process by which it was developed. No level of investment in the science of instream flow requirements can preclude the need to make difficult, value-based trade-off choices between water use for economic development and aquatic ecosystem requirements (Hatfield and Paul 2015). Recognizing this, the Phase 1 Framework specifically identified 2 key shortcomings to be addressed in a second phase of planning: (1) much greater attention to socioeconomic considerations was necessary; and (2) there was a need to further research an ecosystem base flow (EBF), described as a low-flow level where water withdrawals for industry would effectively stop to protect the aquatic ecosystem.

At the request of the provincial and federal regulatory agencies, the multi-stakeholder Cumulative Environmental Management Association (CEMA) established a Phase 2 Framework Committee (P2FC), a multi-stakeholder committee established to develop recommendations for a Phase 2 Water Management Framework that would prescribe long-term rules for cumulative oil sands mining water withdrawals from the Lower Athabasca River. Membership of the 20-person P2FC included the Alberta provincial government (Ministries of Environment, Sustainable Resource Development, and the Energy Resources Conservation Board), the Canadian federal government (Fisheries and Oceans Canada and Parks Canada), Aboriginal governments (Fort McKay First Nation, Fort Chipewyan Métis, and Fort McMurray Métis), oil sands mining companies (Canadian Natural Resources Limited, Imperial Oil Resources, Petro-Canada, Shell Canada, Suncor Energy, Syncrude Canada, and Total Canada), and nongovernmental organizations (Alberta Wilderness Association, South Peace Environmental Association, and World Wildlife Fund Canada). In 2008, 2 of the authors of this chapter (Ohlson and Long) were hired to design and implement a planning process using SDM to simultaneously enhance analytical rigor and improve the process of consultation/engagement. The third author (Paul) was the lead aquatic technical analyst for the Alberta provincial government. Terms of reference and process guidelines were developed to clearly prescribe the SDM-based Phase 2 process and the scope and mandate to deliver recommendations to federal and provincial regulators with an assigned deadline of December 2009. Under Alberta's Water Act and land use planning regime, the water management framework would become legally binding through government approval as part of the regional land use plan.

Ecological Background

The Athabasca River flow regime is driven by the northern climate of cold winters and warm summers. Naturally occurring low winter flows that average just 150 cubic meters per second (cms) are a limiting factor to overall aquatic ecological productivity. These are followed by very large spring and summer freshet flows that average 1,500 cms when snow and glacial melt waters combine with rainfall events. The Athabasca is one of the largest remaining free-flowing rivers in North America and is part of a rich aquatic and terrestrial ecosystem in northeastern Alberta. Relative to other rivers in the province, the Lower Athabasca has high fish diversity, with 31 species reported.

The fish community of the Lower Athabasca River shows complex and varied seasonal movement patterns to and from the mainstem river, Peace-Athabasca Delta, Lake Athabasca, and tributaries. For example, tens of thousands of lake whitefish (*Coregonus clupeaformis*) migrate from Lake Athabasca to the mainstem river to spawn in the fall and return to the lake during winter; walleye (*Sander vitreus*) and northern pike (*Esox lucius*) can migrate downstream to the delta in spring to spawn and return to the mainstem river during winter; burbot (*Lota lota*) migrate throughout the river and spawn in the winter; and Arctic grayling (*Thymallus arcticus*) use the mainstem river to migrate among tributaries where they spawn, feed, and overwinter.

Information on movement patterns of the fish community was gathered through traditional knowledge, local knowledge, scientific research related to oil sands development (e.g., the Alberta Oil Sands Environmental Research Program from 1975 to 1985 and the Regional Aquatics Monitoring Program from 1997 and ongoing), and research funded by CEMA (2003 to 2009) to specifically address information needs for development of a water quantity management framework.

High fish diversity results from a combination of factors including the region's postglacial history; diverse aquatic landscape of upland tributaries, mainstem river, delta, and lakes; and substantial movement of water through the area. Combined with the Peace and Slave Rivers, 82% of the entire province's water supply drains through the northeast. In an abstract sense, Alberta can be viewed as a dish pan tilted to the northeast with the province's lowest elevation (at about 200 meters above sea level) at the exit of the Slave River from Alberta.

The Peace-Athabasca Delta is one of the world's largest inland freshwater deltas. It is a Ramsar wetland of international significance (Ramsar Convention 2018) and is part of the Wood Buffalo National Park, recognized by UNESCO as a World Heritage Site (UNESCO 2018). It includes some of the largest undisturbed grass and sedge meadows in North America and is one of the most important waterfowl nesting and staging areas on the continent.

Aboriginal people of the Peace-Athabasca Delta are highly dependent on the aquatic environment, with fish historically forming the base of their food supply, muskrat trapping representing a significant part of the economy for centuries, and water providing the transport network for the region.

Decision Structure and Planning Process

This case study involved multiple competing objectives and multiple stakeholders. The competing objectives gave rise to difficult trade-offs in a contentious setting with highly polarized worldviews and legal constraints (e.g., existing water licenses).

From a process perspective, the P2FC, comprised of the senior decision makers or their proxies for each of the organizations involved, was supported by several task groups to support detailed development of information and assessment methods for various interest areas including instream flow and aquatic ecosystem function, socioeconomics, water storage engineering and mitigation, and climate change. The P2FC acted as the central point for discussions and delegated technical questions and analyses to the various task groups, which returned information for the P2FC's review and use. This committee met a total of 15 times over the course of 2008 and 2009 at a variety of locations in Fort McMurray, Calgary, and Edmonton.

An Instream Flow Needs Technical Task Group (IFNTTG) provided the biological expertise that underpinned the consideration of potential impacts of water withdrawals on the aquatic ecosystem. This group considered numerous potential impact hypotheses, developed the aquatic evaluation criteria, and conceptually designed and prioritized many of the adaptive management monitoring proposals. The

IFNTTG generally attempted to understand how the various flow alternatives might affect aquatic values and to communicate their assessments to non-biologists on the P2FC to help them evaluate competing management alternatives.

A Water Requirements Engineering Mitigation Task Group (WREMTG) was a pre-existing industry-led group with the task of exploring and characterizing potential engineering responses to future possible regulatory frameworks. WREMTG oversaw the development of cost, water storage footprint, and other estimates associated with engineering mitigation options that might be required to meet the draft Phase 2 Framework water management alternatives.

An Economics Task Group developed and commissioned studies on the potential social and economic impacts of water withdrawal alternatives. Its work focused primarily on assessing the potential impacts on local communities, especially First Nations and Métis.

An Ecosystem Base Flow Task Group was convened for a short period during the planning process to tackle specific questions around the development of an EBF for the Lower Athabasca River. Finally, a Climate Task Group was created during the Phase 2 process to develop and assess the potential flow implications for different climate change scenarios.

Decision Analysis

Objectives

The planning process guidelines, developed by the authors and a steering committee, clarified to all participants how the SDM approach would address social, environmental, and economic interests regarding water withdrawals from the Lower Athabasca River in a formal and consistent manner. After some initial screening, 3 broad objective categories—aquatic ecosystem health, traditional use, and economic development—and the approach to their analysis were developed and agreed upon, each using influence diagrams, data gathering, and evaluation (modeling) tools. Table 8.1 presents a summary of the objectives and evaluation

Table 8.1. List of final objectives and evaluation criteria used in the structured decision-making process for managing water withdrawals for oil sands mining on the Lower Athabasca River

Category	Objective	Evaluation Criteria
Aquatic ecosystem health	Fish habitat	Percent decrease in habitat availability from natural flow conditions. Assessed for 6 fish species, 4 life stages, open-water and ice-covered conditions, in 4 river segments using River2D hydraulic models and habitat suitability curves. Summarized into 2 fish-community level criteria: % of life stages affected and life stage with largest habitat loss.
	Mesohabitat	Percent decrease in mesohabitat availability from natural flow conditions. Assessed for 27 mesohabitat types based on rules for depth, velocity, and substrate, in 4 river segments using River2D hydraulic models. Summarized into 2 ecosystem level criteria: % of mesohabitat types affected and habitat loss for the most sensitive mesohabitat type.
	Whitefish spawning	Percent decrease in whitefish spawning habitat availability from natural flow conditions. Assessed in detail for 1 river segment during fall.
	Walleye recruitment	Percent decrease in walleye population from natural flow conditions. Assessed in detail for the delta and Lake Athabasca area for winter flow conditions.
Traditional use	Navigation suitability	Percent decrease in navigation suitability from natural flow conditions. Assessed for 3 periods (spring, summer, fall), in 4 river segments using River2D hydraulic models and navigation-depth suitability curves derived from community interviews with river boat users. Summarized by navigation suitability loss in the most sensitive season (fall).
Economic development	Cost for water storage	CAD millions for freshwater storage ponds and expanded tailings ponds.
	Water storage footprint	Area in square kilometers for freshwater storage ponds and expanded tailings ponds.

criteria, and the process of developing them is summarized below.

AQUATIC ECOSYSTEM HEALTH

Development of the aquatic ecosystem health objectives and evaluation criteria and analysis methods relied heavily on the many prior years of investment into instream flow studies. We began with an initial list of 29 potential impact hypotheses of how water withdrawals might influence overall aquatic ecosystem health spanning the full range from geomorphic processes (e.g., channel-forming flows, riparian impacts), through to mesohabitat conditions, side channel habitat connectivity, and microhabitat conditions (e.g., depth, velocity) for different fish species. These were initially screened by the IFNTTG through a series of workshops as being "accepted," "rejected," or "data deficient." Data gaps were immediately earmarked for future consideration in the adaptive management program; accepted hypotheses proceeded through a process of formal evaluation criteria development and analysis. This led to an identification of key uncertainties, which were also noted for later consideration in the adaptive management program.

In the end a concise set of objectives remained, detailed evaluation criteria definitions and modeling methods were developed, and influence diagrams were finalized to help communicate how water withdrawals influenced the overall aquatic ecosystem health objectives (figure 8.1).

TRADITIONAL USE

An innovative aspect of the work involved the development of a suitable approach to address the potential effects of water withdrawals on the traditional use objective. An independent study commissioned by the P2FC included interviews with Aboriginal community members regarding their use of the river. Using this information, the P2FC narrowed in on 4 specific components of the traditional use objective that could be influenced by water withdrawals: intergenerational knowledge transfer (i.e., grandparents and grandchildren out on the land and water), community diet and health (i.e., food gathering), use of the river for spiritual practices, and access to nearby river system (figure 8.1B). These components were all acknowledged as interconnected as well as influenced by other factors at the community level unrelated to river water withdrawals. Committee members agreed that it would be difficult, if not impossible, to assess the influence of water withdrawals on aspects like spiritual use of the river, although analogous attempts have been made in other contexts (e.g., Gregory et al. 2012; Runge et al. 2011). The influence diagram made it clear that the effect on the physical characteristics of the river (e.g., sandbars, riparian) and natural resources (e.g., fish and fish habitat) were being modeled in detail as part of the aquatic ecosystem health objective. Hence, the remaining requirement was to develop a criterion and methodology to measure the influence of water withdrawals on access and navigability. A model was developed to calculate the area of river suitable to navigation in different seasons using navigation-depth suitability curves and a River2D hydraulic model.

ECONOMIC DEVELOPMENT

From the oil sands industry's economic development perspective, any constraints on the availability of river water to meet processing needs, whether due to naturally occurring low flow periods or to additional requirements to protect instream flow, necessitates constructing off-stream storage to which water can be routed during periods of high river flow (i.e., spring/summer freshet) to be stored and later used during periods of reduced river flow (i.e., winter).

A base assumption for the P2FC process was that at some point in its future, the oil sands mining industry cumulatively may require up to an average of approximately 16 cms of fresh "make-up" water on a continuous basis. This make-up water is assumed to come entirely from the Lower Athabasca River. Assuming that instream flow rules would limit withdrawal during the winter months to a number less than 16 cms, then the balance would need to be made up from water storage. The evaluation criteria developed to

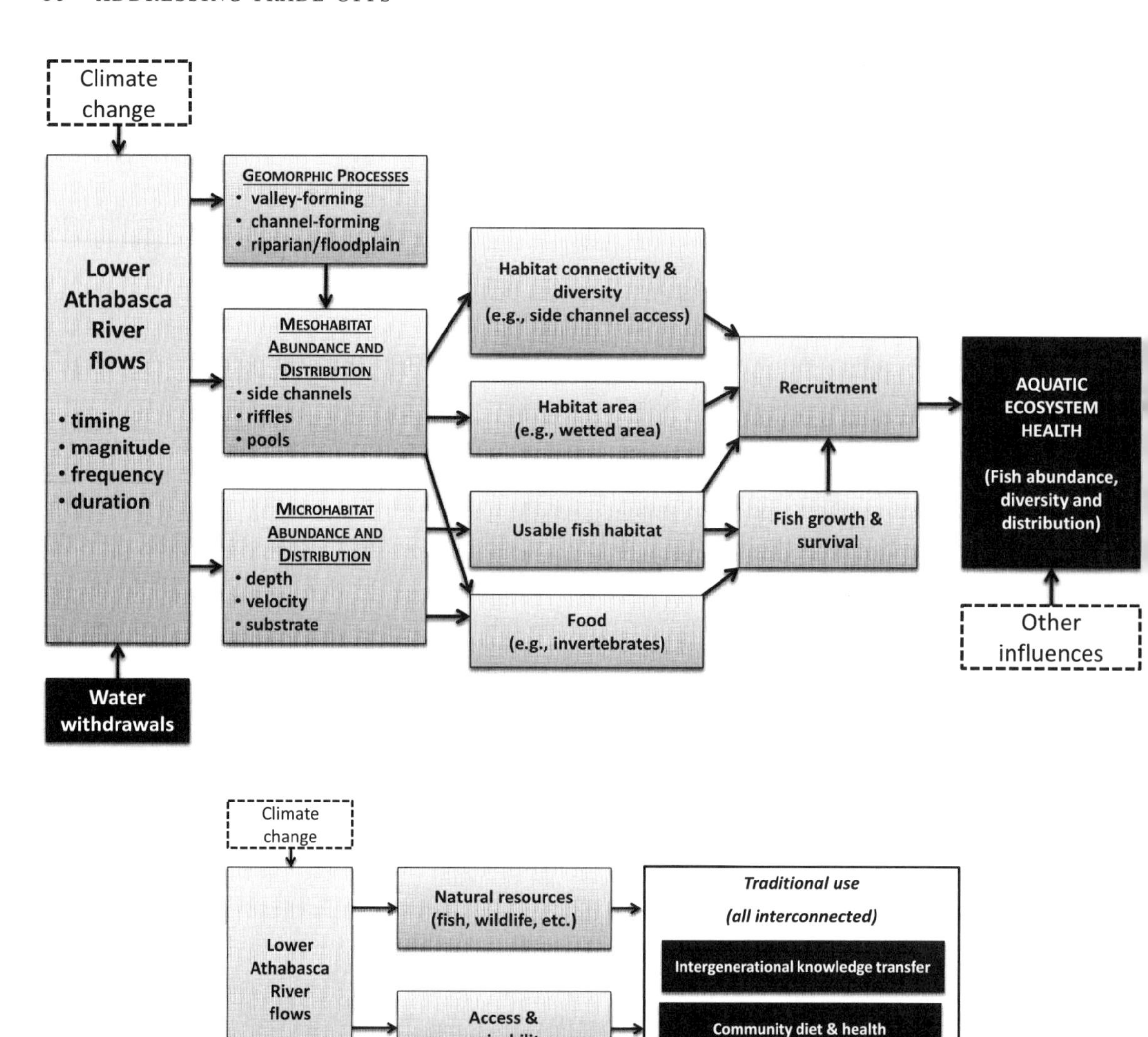

Figure 8.1. Influence diagrams linking the effects of water withdrawals from the Lower Athabasca River on stakeholder objectives including metrics of aquatic ecosystem health (*top*) and traditional use (*bottom*).

assess the impact on overall economic development of developing increased water storage under different water withdrawal alternatives included the capital cost and physical footprint of constructing new freshwater ponds or expanding existing tailings ponds.

Alternatives and Consequence Analysis

Conceptually, the alternatives to be considered were relatively simple. They involved consideration of what to do when naturally occurring flows were at average or high levels and what to do when naturally occurring flows were at low or even extreme drought levels. The task, therefore, was to develop alternative rule sets that prescribed how much water could be withdrawn from the river for oil sands mining at different times of the year given a range of potential natural inflows with significant variability from year to year and season to season. In practice, this was a very detailed exercise, following a process illustrated in figure 8.2.

In close collaboration with technical members of the P2FC, we developed a Microsoft Excel / Visual Basic tool that enabled the transparent creation and evaluation of alternatives by all participants. Base data inputs to this "flow calculator," as it became colloquially known, included a 50-year historical natural inflow record, synthetic 1:100 and 1:200 drought flow years, future climate change scenarios, and forecasted future industrial water withdrawal requirements for full build-out of oil sands reserves. The user-defined inputs were weekly water withdrawal rules. The flow calculator performed water balance calculations, and its outputs included a 50-year weekly time step series and accompanying statistics over the 50 inflow years for instream flow and wetted area, suitable navigation area, and industrial off-stream water storage requirements and cost. Instream flow outputs were also exported for use in the calculation of multiple fish and fish habitat criteria.

The tool was distributed to all participants and used extensively in real time during committee meetings to build understanding of the consequences of different flow management rules on each objective. After hearing of each other's needs through discussions, groups of participants representing a range of different interests would often huddle around computers attempting to find alternatives with mutually agreeable outcomes. This hands-on and transparent

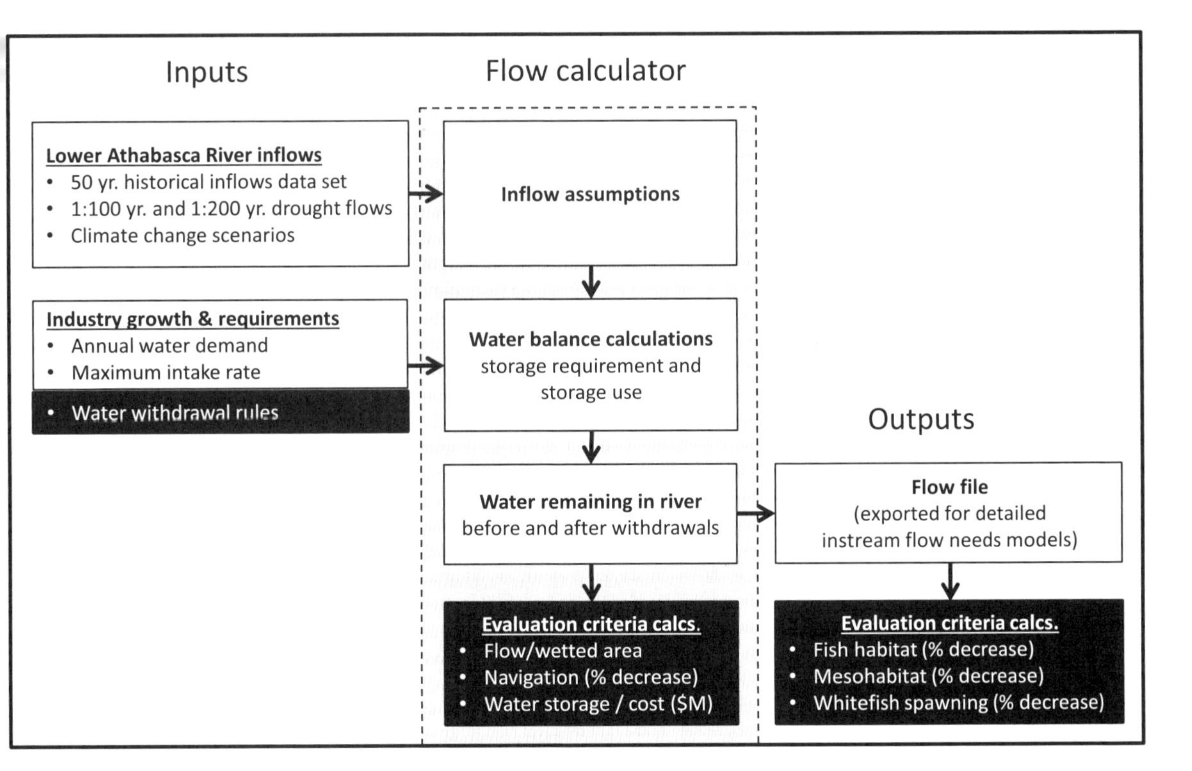

Figure 8.2. Structure of the flow calculator—the Excel / Visual Basic spreadsheet application used to input alternatives as a set of weekly water withdrawal rules and output results for evaluation criteria (data and charts) used in trade-off analysis discussions for the Lower Athabasca River structured decision-making process.

approach to alternatives development and evaluation established trust both in the analysis methods and among participants and promoted significant learning about the feasibility of proposed options.

At 4 key points in the process (generally referred to as "rounds"), results were summarized formally for different sets of alternatives in consequence tables to enable value-based trade-off discussions by the committee. As further described below, the alternatives developed during each round were

- Round 1: 7 broad bookend alternatives that represented the full (extreme) range of participant views on how to manage water withdrawals.
- Round 2: A set of 11 alternatives representing some participant proposals to seek a general balance across competing objectives.
- Round 3: 4 consistent alternatives based on a formal set of instream flow protection principles, while still presenting fundamental trade-offs between economic cost and environmental outcomes.
- Round 4: A final set of alternatives that further explored different alternatives for an ecosystem base flow and integrated climate change considerations.

Trade-Off Analysis

SDM facilitates the use of formal multi-attribute trade-off methods, including both quantitative methods (involving explicit weighting of objectives) and deliberative methods (involving structured discussions about values trade-offs). In a multi-stakeholder setting, we typically use a deliberative approach as a starting point, augmented by quantitative methods where we believe such methods will add insight to the choices at hand. A deliberative approach, though not quantitative, is still a structured and rigorous approach rooted in decision science methods and has the advantage of ensuring the continuous engagement of participants. Our approach involves

- Ensuring the objectives and criteria are well designed (essentially that they represent fundamental objectives and are complete and nonredundant) (Keeney 2007);
- Presenting the estimated consequences of alternatives in a consequences table, color-coded to focus attention on the trade-offs;
- Facilitating a discussion about key trade-offs;
- Eliminating dominated alternatives (alternatives that are outperformed or equaled on all criteria by at least one other alternative);
- Eliminating practically dominated alternatives (alternatives that all parties agree are inferior to another alternative, usually because of unacceptable trade-offs);
- Eliminating insensitive criteria (criteria that do not vary across the alternatives, a situation that often arises late in a process after a few rounds of refinement);
- Documenting the rationale for removing alternatives and/or criteria;
- Refining (or defining new) alternatives that provide a better balance across the objectives/criteria;
- Iterating this process until a preferred alternative is selected or the group reaches an impasse; and
- Using quantitative trade-off tools when needed, usually to help with reducing complexity, diagnosing sources of disagreement, or focusing limited meeting time on critical technical or value-based conflicts (Gregory et al. 2012, chapter 9; Locke et al. 2008, chapter 2).

In this process, we worked through 4 rounds of iteratively evaluating trade-offs and refining alternatives over a 9-month period.

ROUND 1 (ALTERNATIVES 1 TO 7)

The initial 7 alternatives included single-interest bookend alternatives (i.e., alternatives that served to maximize a single objective, the outcome of values-focused thinking; Keeney 1996). This round of alternatives facilitated the testing and refinement of modeling tools and enabled participants to learn about the key trade-offs and how to better develop new alternatives based on the insights gained.

ROUND 2 (ALTERNATIVES 8 TO 18)

This refined set of alternatives represented a spectrum of approaches put forward by participants to explore interests and seek a general balance between competing environmental, navigation, and industry objectives. Most alternatives employed rules that gradually increased protection as flows in the river decreased and applied less restrictive withdrawal rules during the summer when flows are higher to allow filling of off-stream storage in advance of the subsequent winter period. Several alternatives were eliminated through dominance assessment, and several criteria became insensitive across the range of remaining alternatives (including the navigation suitability criterion). As the number of agreeable alternatives was reduced, an efficiency frontier was identified when the performance of these alternatives was plotted across the remaining aquatic ecosystem health and storage/cost objectives. Alternatives close to the line signaled where fundamental trade-offs exist. Alternatives along the frontier could not be improved in terms of aquatic ecosystem health performance without increasing storage/cost, and vice versa. Participants agreed to explore a subset of alternatives lying on or near the efficiency frontier that had the potential to provide a mutually acceptable balance among objectives.

ROUND 3 (ALTERNATIVES 19 TO 22)

These 4 alternatives were developed using a formal set of instream flow protection principles and a targeted range for off-stream storage. Along the efficiency frontier, on one end alternative 19 required less storage and cost, and on the other end alternative 22 provided the highest amount of aquatic ecosystem protection at the highest storage and cost. As part of this round of trade-off analysis deliberations, given the keen interest by some committee members in the establishment of an EBF, sensitivity analyses were undertaken on low flow events and a range of ecosystem base flow water withdrawal exemptions ranging from "none" to "existing water license levels." Although all committee members acknowledged that these alternatives were "in the ballpark" of achieving an acceptable balance, a final decision regarding the EBF exemption level in combination with any of the remaining alternatives proved elusive.

At the end of this third round of trade-off analysis, and given the need for progress with a process deadline looming, the facilitators and analysts tabled a new compromise alternative reflecting our best judgment of the values expressed by committee members during the trade-off deliberations. Option A, as it became known and distinguished from the previous numbered alternatives, was based on the fundamental rules of alternative 21 (i.e., toward the aquatic protection end of the spectrum) coupled with a minimum water withdrawal rate of 8 cms that equaled the current existing water license holders, which reduced the total level of storage/cost impact on industry. This proposed solution acknowledged the P2FC principle in the committee's process guidelines to respect existing legal rights, including water rights.

ROUND 4 (OPTION A TO OPTION H)

Option A was rejected by some participants who felt that an 8 cms allowance would not adequately protect aquatic values at very low river flows during extreme droughts. A final round of alternative evaluations and trade-off analysis deliberations focused on exploring variations of the EBF withdrawal exemption levels at an agreed upon 1:100-year low flow level. In general, these alternatives represented a trade-off of decreased environmental protection in average inflow years (relative to that expected under option A) in exchange for increased protection in rare low inflow years (options B through H, in which the oil companies would voluntarily reduce some of their 8 cms water rights under these extreme conditions). Option H, with a 4.4 cms allowance, was the

Table 8.2. Consequence table (simplified) used in the fourth round of trade-off analysis discussions for the Lower Athabasca water management structured decision-making process

Category	Objectives	Evaluation criteria	Alt. 19	Alt. 20	Alt. 21	Alt. 22	Option A	Option H
Aquatic ecosystem health	Fish habitat	% loss (midwinter)	6	5	3	3	4	4
	Mesohabitat	% loss (midwinter)	24	18	13	10	15	17
	Whitefish spawning	% habitat loss	18	14	10	7	10	12
Traditional use	Navigation	% loss of suitability	2	2	2	3	2	2
Economic development	Cost for water storage	CAD millions	$700	$1,500	$2,100	$2,800	$1,900	$1,400

Note: Criteria in the form of percentages show the percent reductions from natural flow in years when the weekly flow in the river is below the Q80 exceedance level (i.e., the 20% lowest flow years).

final alternative proposed at this point, and an extensive set of sensitivity analyses was undertaken in which key model parameters were varied alone and in tandem (e.g., concerning climate change scenarios, operational flexibility, etc.) to provide participants with as much information as possible to take back to their constituents for review.

The consequence table for the fourth round of trade-off analysis discussions is shown in table 8.2. The final rule set for option H is illustrated in figure 8.3.

By the end of the SDM process, the P2FC had reached agreement on a suite of recommendations, including

- Water management rules for weekly cumulative withdrawals across the majority of inflow conditions;
- Implementation, including requirements for long-term status and trend monitoring and compliance reporting; and
- An adaptive management plan that provided the basis for effectiveness monitoring, addressed knowledge gaps and uncertainties through a research program, and specified management triggers.

The P2FC did not reach complete agreement on the final component on the water management framework, namely the water withdrawal rules under the EBF 1:100 drought year condition. Some committee members held the perspective that there should be a complete cutoff of all water withdrawals during extreme low flow events. Others thought that a complete cutoff would impose extraordinary cost and violate existing water license conditions, leading to legal uncertainty. These differing perspectives were documented as part of the final recommendations to inform further government-led consultations and decision making.

Despite the absence of complete agreement on the preferred alternative itself, at the conclusion of the planning process there was a clear sentiment that the progress made represented a remarkable success given the nature of the challenge. Government, environmental, Aboriginal, and industrial participants went on record as supporting the SDM process through which this recommendation was made.

Decision Implementation

The committee report and recommendations were delivered to Alberta Environment and Fisheries and Oceans Canada at the December 2009 deadline (Ohlson et al. 2010). Subsequently, government agencies took over to lead several review, consultation, and development actions.

First, a formal scientific peer review process was undertaken by the Canadian Science Advisory Secretariat (DFO 2011). Their conclusion was that the information and models used by the P2FC were the best currently available and that they could be used to provide guidance on potential ecosystem effects

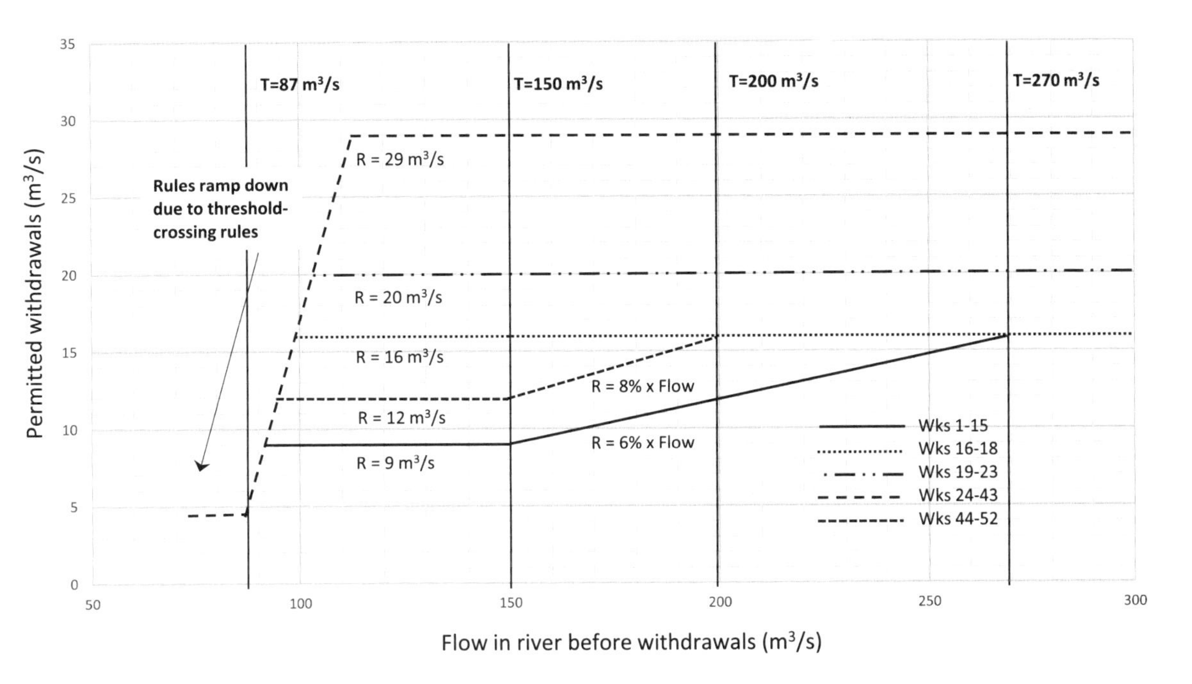

T = weekly flow trigger; R = cumulative water withdrawal limit
Source: Ohlson et al. 2010

Figure 8.3. Flow rules for option H, the preferred alternative for water management on the Lower Athabasca River, emerging from a structured decision-making process. The combined oil sands industry permitted maximum withdrawal of water from the Athabasca River as developed by the Phase 2 Framework Committee process. The permitted rate of water withdrawal (the y-axis) is a function of (1) the flow rate in the river before the withdrawal (the x-axis) and (2) the week of the year (each line in the chart represents a range of weeks in the year during which the rule applies). Some periods of the year have thresholds (or "triggers") in the value of the x-axis at which the nature of the permitted withdrawal changes. For example, for weeks 44 to 52, if the flow rate in the river before withdrawal is below the trigger line of 150 cms, the withdrawal limit is 12 cms. Above the trigger line of 150 cms but below 200 cms, the permitted withdrawal is 8% of the flow in the river. Threshold-crossing rules, not discussed here, ensure a smooth ramp-down to a maximum of 4.4 cms when the river flow rate before withdrawals is 88 cms or less.

from water withdrawals, while noting that uncertainty existed around model predictions and there was consequently a need for field monitoring.

During the review period, land use planning efforts for the overall Lower Athabasca Region advanced under Alberta's Land Stewardship Act, which aims to set long-term environmental management outcomes for air, land, biodiversity, and water. Unsurprisingly, oil sands mining was a central focus of planning and engagement in the region, and continued strong interest was expressed by many stakeholders in water management for the Lower Athabasca River. By August 2012, a land use plan for the Lower Athabasca Region was approved by the Alberta government along with management frameworks for air quality, water quality, and groundwater management. The Surface Water Quantity Management Framework for the Lower Athabasca River (Government of Alberta 2015) was approved in March 2015 after further consultation and development. The water withdrawal rules (now called weekly flow triggers and cumulative water use limits) were adopted directly from the P2FC's option H. The management framework also includes monitoring indices and adaptive management triggers for flow, water use, navigation, and aquatic ecosystem status.

The various management frameworks approved under the regional land use plan will guide the Alberta government in all future resource management decisions in the Lower Athabasca Region. Similarly, the federal government has acknowledged that it will use the water management framework for all future decisions within its regulatory authority under the Federal Fisheries Act.

Discussion

The SDM approach was undertaken against a background of extremely contentious political circumstances. Prior to this process, government faced significant criticism from some stakeholders that the then-current Phase 1 Water Management Framework did not make clear the balance being struck between protection of the aquatic environment and support for oil sands economic development, and that the framework did not address social and Aboriginal considerations. By bringing in neutral facilitators to facilitate an SDM approach for the Phase 2 planning cycle, several key advances were made.

First, the SDM approach mandated a consistent and principled approach to the treatment of all objectives, including Aboriginal interests, in a rigorous manner. All relevant interests were treated in a parallel and respectful manner, although inevitably the nature of the analysis differed in each case. This procedural rigor fostered a sense of mutual respect and collaboration in working toward the common goal of building a consequence table that could later be the focus of trade-off discussions.

Second, the SDM approach enabled the transparent development and evaluation of alternatives. Developing the flow calculator put the task of developing alternatives directly in the hands of participants. This inevitably sparked creativity as ideas emerged from radically different viewpoints. Since the tool provided immediate feedback on the performance of an alternative, new alternatives could be immediately tested. For participants, this triggered learning about what was feasible, about each other's interests, and about irreducible trade-offs. Toward the latter rounds of trade-off analysis, oil company managers were explicitly using the flow calculator to minimize aquatic habitat loss, and fish biologists were using it to minimize water storage requirements and cost, each searching for outcomes they hoped would be tolerable to themselves and to other parties. Clearly, the approach had led all parties to recognize that meeting their own priority objectives was inextricably tied to meeting others' as well.

Third, the SDM approach included explicit consideration of trade-offs as part of a contentious multi-stakeholder planning process. While exposing trade-offs could be considered risky in a public planning exercise, the process leading up to the discussion of trade-offs must be emphasized. The steps leading up to this point in the process over a period of 6 months included allowing all participants to express their interests, translating interests into objectives related to water withdrawals, developing understandable criteria and consequence assessment methods, and, of course, iteratively and collaboratively developing alternatives for consideration. In our experience, people are willing to make difficult trade-offs when they understand and accept the decision scope, recognize their objectives are being considered, and have had input into the alternatives under consideration.

Fourth, the SDM approach enabled the identification and prioritization of key uncertainties and data gaps to serve as the basis for compliance monitoring, further research, and adaptive management. Rather than stall the process, as important uncertainties and data gaps were identified at each step of the process they were either immediately actioned (if possible) or tracked for later discussion of implementation requirements. By the time the latter stages of the process came around, participants understood which uncertainties and data gaps mattered most for future implementation, making priority-setting a relatively easy task.

Based on this case study, several general conclusions can be drawn about multiple-objective problems. Decision framing at the outset is essential for success on problems with competing objectives in a highly contentious context. In this case, the scope for the P2FC planning process based on the principles of SDM was clearly described in terms of reference and planning process guidelines that all participants agreed to. Given the overall controversy over oil sands development in general, in order for some participants to commit to the process, everyone agreed first that discussing the merits of oil sands development per se was out of scope and second that participating on the committee did not imply support for oil sands development. Rather, the goal of the process was specifically stated as finding an acceptable balance between social, environmental, and economic interests regarding water withdrawals from the Lower Athabasca River within the context of a long-term, full build-out growth assumption.

From a process perspective, in classic SDM and values-focused form, it is imperative to get agreement on objectives (and criteria) early. The objectives must include all values relevant to the decision, and they must be put on a level playing field to keep all stakeholders engaged in the process. Means should be diligently separated from ends. Using influence diagrams explains the pathways of effect from alternatives to objectives, focuses discussion on appropriate evaluation criteria, develops creative alternatives, and facilitates the estimation of consequences and the identification of decision-relevant uncertainties. In this case study, focused attention on the traditional use objective early on and development of the navigation suitability criterion for comparing alternatives built early trust, such that it was not a concern in the latter stages of alternative evaluations when the decision became insensitive to the navigation objective.

Problems with competing objectives necessarily require the discussion of trade-offs, which introduces the need to choose an appropriate preference assessment technique. In this case, the problem reduced to a trade-off between 2 objectives—industry capital cost for water storage and winter fish habitat availability—with no opportunity to achieve further gains on one objective without creating losses on the other. Although the process was originally designed to conduct a multi-attribute trade-off weighting exercise, in the end we did not elicit weights. Trade-offs were addressed directly and explicitly but deliberatively. Weighting exercises provide a vehicle through which participants can express their views on the relative importance of the range of objectives and values in a consequence table. Often, when an impasse is reached in evaluating complex alternatives with a diverse group, weighting exercises can be powerful tools in breaking deadlocks: They help reveal areas of common agreement, and they help flag alternatives that might form the basis of compromise solutions. However, where a consequence table is simple enough for values trade-offs to be clear, such techniques are not always necessary and can lead to unnecessary distractions about the mechanics of the exercise itself. In this case, it is our belief that a weight elicitation exercise risked further and unnecessarily polarizing participants and was unlikely to lead to any new insight about the values of participants or the potential for a mutually agreeable solution.

Consensus is not essential for informed, broadly supported public decisions to emerge from a multistakeholder process. In this case study, all parties broadly agreed on most aspects of the water management solution but were unable to reach full consensus for reasons that were far beyond the scope of the process and control of those involved. For example, senior government officials and industry executives were not willing or able to address the legal ramifications of changing existing water licenses, while senior NGO representatives saw an EBF with a full cutoff (i.e., an absolute prohibition on any water withdrawal, including those already licensed, during extreme low flow circumstances) as a non-negotiable precedent for large river protection throughout North America. Nonetheless, the SDM process eliminated clearly inferior alternatives, substantially narrowed in on the majority of the required content and priorities for the water management framework, and clearly documented the diverse perspectives on the irreconcilable issue. This information was passed to provincial and federal government regulators who made final decisions. Good documentation of each step of the SDM process—from objectives, criteria, and assessment methods to the trade-off analyses and differing perspectives of alternatives—gave government officials the information they needed to make fully informed and defensible decisions.

ACKNOWLEDGMENTS

The authors are grateful to all committee members who provided generous contributions of time, knowledge, insight, and encouragement to the P2FC process. Significant technical and policy contributions were made by committee members Allan Locke, Patrick Marriott, Brian Makowecki, Chris Fordham, Ron Bothe, Stuart Lunn, Rick Courtney, and Mathieu Lebel. Todd Hatfield served as a key member of the consulting team throughout the process. Funding and secretariat services for the multi-year process were generously provided by the Cumulative Environmental Management Association.

LITERATURE CITED

Alberta Environment and Fisheries and Oceans Canada. 2007. *Water Management Framework: Instream Flow Needs and Water Management System for the Lower Athabasca River* (The Phase 1 Framework). http://aep.alberta.ca/water/programs-and-services/river-management-frameworks/athabasca-river-water-management-framework.aspx.

DFO. 2011. *Proceedings of a National Science Advisory Workshop on Instream Flow Needs for the Lower Athabasca River; May 31–June 4, 2010.* DFO Canadian Science Advisory Secretariat Proceedings Series 2010/056.

Government of Alberta. 2015. *Surface Water Quantity Management Framework for the Lower Athabasca River.* http://aep.alberta.ca/water/programs-and-services/river-management-frameworks/athabasca-river-water-management-framework.aspx.

Gregory R, Failing L, Harstone M, Long G, McDaniels T, Ohlson D. 2012. *Structured Decision Making: A Practical Guide to Environmental Management Choices.* Oxford, UK: Wiley-Blackwell.

Hatfield T, Paul AJ. 2015. A comparison of desktop hydrologic methods for determining environmental flows. *Canadian Water Resources Journal* 40:303–318.

Keeney RL. 1996. *Value-Focused Thinking: A Path to Creative Decisionmaking.* Boston, MA: Harvard University Press.

———. 2007. Developing objectives and attributes. Pages 104–128 in Edwards W, Miles RFJ, von Winterfeldt D, eds. *Advances in Decision Analysis: From Foundations to Applications.* Cambridge, UK: Cambridge University Press.

Locke A, Stalnaker C, Zellmer S, Williams K, Beecher H, Richards T, Robertson C, Wald A, Paul A, Annear T, eds. 2008. Campbell River, British Columbia. Pages 9–37 in *Integrated Approaches to Riverine Resource Management: Case Studies, Science, Law, People, and Policy.* Cheyenne, WY: Instream Flow Council.

Ohlson D, Long G, Hatfield T. 2010. *Phase 2 Framework Committee Report.* Cumulative Effects Management Association. http://library.cemaonline.ca/ckan/dataset/2008-0009.

Ramsar Convention. 2018. Peace-Athabasca Delta. https://rsis.ramsar.org/ris/241.

Runge MC, Bean E, Smith DR, Kokos S. 2011. *Non-native Fish Control below Glen Canyon Dam—Report from a Structured Decision-Making Project.* US Geological Survey Open-File Report 2011–2012. http://pubs.usgs.gov/of/2011/1012.

UNESCO. 2018. Wood Buffalo National Park. http://whc.unesco.org/en/list/256.

PART III ADDRESSING RESOURCE ALLOCATION

9 Introduction to Resource Allocation

James E. Lyons

With ongoing habitat loss and degradation, ever-increasing threats to biodiversity, and limited funding for conservation and management, nearly every natural resource manager routinely faces difficult resource allocation problems. Funding and capacity for natural resource management rarely meet the need, and informed resource allocations are increasingly important. The decision problems outlined in this chapter include not only habitat and species management but also a wide variety of administrative decisions. Ranking projects or plans by benefit-cost ratio is an intuitive, heuristic approach to resource allocation but may be inefficient. We present a general resource allocation framework in which these decision problems can be stated mathematically, making it relatively easy to find solutions using mathematical programming such as linear programming. Linear programming and other constrained optimization routines can be implemented in common software applications and used with a wide variety of decision problems, including project prioritization and portfolio decisions. Constrained optimization has advantages over intuitive benefit-cost ratios and can accommodate single- and multiple-objective problems. We also introduce the 3 case studies in this section, illustrating a variety of resource allocation problems: the first case study shows how to select cost-effective management actions for discrete management units such as wetlands or grassland patches; the second, how to use a patch dynamics model to allocate resources for a reserve network that protects habitat for multiple species of conservation concern; and the third, how to use stochastic simulation to determine allocation of resources in space and time for invasive species management.

Introduction

The challenge of resource allocation may be a universal dilemma faced by all organizations (Kleinmuntz 2007), including natural resource management agencies. Managers in all sectors—private, public, and nongovernmental—have more good ideas for species and habitat management projects than available resources will allow. With limited funding and capacity, conservation decision makers must choose actions that maximize the achievement of their management objectives: Which patches of remaining habitat should be selected to create a reserve network of protected lands? How should we allocate funding to multiple endangered species? Which recovery actions will maximize the probability of persistence? These resource allocation problems occur not only in the biological realm of habitat and species management but also in the administrative decisions that are part of natural resource management. The administrative decisions are also complex and varied: How should we allocate staff time to endangered species consultations? How to distribute a regional budget to field stations? Can we optimize workforce planning?

Resource allocation decisions that involve discrete choices are sometimes called portfolio problems: the decision maker must select a finite number of discrete "elements," say, projects or plans, from many potential elements that could be chosen if there were no constraints (Ando and Mallory 2012; Convertino and Valverde 2013; Lahtinen et al. 2017). One of the challenges in these problems is that the number of alternative ways that the discrete elements (e.g., projects or actions) could be combined increases exponentially with the number of elements and quickly becomes unwieldy, even with a small number of projects to choose from (Converse, chapter 11, this volume). Portfolio problems are common in natural resource management, and the capacity to recognize and solve this type of problem pays great dividends to structured decision making practitioners and analysts. The "knapsack problem" is a classic example of a portfolio decision: given a "knapsack" and multiple items, each with a certain weight and value, the decision problem is to fill the knapsack to maximize the total value of the items carried without exceeding the weight capacity of the knapsack. For example, when planning for conservation and recovery of threatened and endangered species, decision makers could use tools developed for knapsack problems to create a portfolio of recovery actions that will maximize the number of species recovered with available funds (Gerber et al. 2018).

A common refrain in the early stages of conservation planning is a call for a prioritized list of species, habitats, or research topics. Prioritization schemes are, in fact, another type of portfolio problem (Runge 2016). Implicit in most efforts to produce a list of priority species and habitats is a portfolio of management actions needed to address a conservation need. Therefore, managers could achieve more conservation benefit if the prioritization focused explicitly on management actions in the context of portfolio decision analysis (Game et al. 2013).

Resource allocation is increasingly important given ongoing habitat loss and degradation, increasing threats to biodiversity, and inadequate funding for conservation and management. Allocation decisions are difficult for a variety of reasons. Often there are myriad alternatives to choose from. A great deal of uncertainty about benefits, as well as costs, is not uncommon. And decision makers usually have multiple objectives to address with every decision. To help managers overcome these challenges, we present a general resource allocation framework that can be used to frame and solve a wide variety of decision problems. We also show how the framework can be extended to address multiple objectives. Finally, we discuss some helpful tools in resource allocation decisions, introduce the case studies in this section, and present challenging issues in resource allocation.

A Framework for Resource Allocation

Resource allocation includes all parts of the PrOACT cycle (Runge and Bean, chapter 1, this volume). Often at the heart of most resource allocation decisions, and an aspect which makes them difficult, are trade-offs among multiple objectives. Most, if not all, decision problems in natural resource management are multiple-objective problems; any decision involving financial resources certainly has multiple objectives (i.e., maximize natural resource benefits and minimize cost). As in all multiple-objective problems, the importance weights assigned to natural resource benefits and costs vary among decision problems as a function of stakeholder preferences (Converse, chapter 5, this volume). Here we review a general resource allocation framework that is not only helpful for framing resource allocation problems but also lends itself to a variety of optimization tools for tractable solutions and trade-offs in multiple-objective problems.

Some intuitive tools for resource allocation rely on a benefit-cost ratio, which in some cases may suffice. Joseph et al. (2009) provide a straightforward yet powerful tool for project selection. Their metric, "project efficiency," evaluates individual projects with a benefit-cost ratio tailored to the challenges commonly faced by managers. The project efficiency measure includes the probability of success for each project, a measure that incorporates not only uncertainty resulting from the stochastic nature of most management outcomes ("partial management control"; see Williams et al. 2007) but also uncertainty about the magnitude of benefits of the project given successful implementation (epistemic uncertainty, Runge and Converse, chapter 13, this volume). Joseph et al.'s (2009) project efficiency measure, which was designed in the context of threatened species management, also has the flexibility to assign importance weights to different species, an element of values-focused thinking (Runge and Bean, chapter 1, this volume) appreciated by many resource managers. Joseph et al. (2009) implement their project selection framework using a "greedy algorithm," which ranks projects using "project efficiency" and then selects projects until funds are exhausted.

Project efficiency, as defined by Joseph et al. (2009), has multiple benefits: it is intuitive, uses values-focused thinking, and provides a way to incorporate uncertainty when selecting cost-efficient actions. But in many cases, an intuitive approach based on a benefit-cost ratio will have shortcomings. Portfolio problems, for example, include a list of available projects or management actions, associated costs, and an available budget (Lahtinen et al. 2017). Again, the intuitive approach in this situation is to rank projects or actions by some measure of a benefit-cost ratio and select projects or actions until available funds are exhausted (Kleinmuntz 2007). However, this approach could leave some funds unspent and forgo additional value. There may be less costly projects that individually have lower benefit-cost ratios than the last funded project but, when taken together, could achieve greater value and still meet cost constraints. An alternative to the intuitive approach is a mathematical programming formulation (Williams 2013). A mathematical description of the problem is helpful because it facilitates combinatorial optimization to find the optimal portfolio.

A general resource allocation framework for a wide variety of conservation decisions can be summarized as follows (Kleinmuntz 2007). Suppose there are m discrete projects or management actions to choose for funding; as in the knapsack problem, each project or management action is either selected or not. Each project or management action being considered would provide certain benefits, b_i, and incur certain costs, c_i. Our optimization problem becomes:

$$\begin{aligned} &\textit{maximize} \sum\nolimits_{i=1}^{m} b_i x_i \\ &\textit{subject to} \sum\nolimits_{i=1}^{m} c_i x_i \le C \\ &x_i = \{1 \textit{ if item } i \textit{ is funded}, 0 \textit{ if item } i \textit{ is} \\ &\quad \textit{not funded}\}, i = 1, \ldots, m. \end{aligned} \tag{1}$$

where x_i is a binary decision variable for project i that indicates whether the project is selected for funding and C is the available budget. This linear model assumes that both costs and benefits are additive and that neither benefits nor costs depend on which other projects are selected (Kleinmuntz 2007). That is, we assume that there is no interaction among costs, such that selection of one project results in a cost savings for another project; similarly, the benefits of a project are the same regardless of what other projects are selected. This general formulation can accommodate a large variety of resource allocation problems, including portfolio, prioritization, and project selection. In portfolio problems, the number of ways that the elements (e.g., projects or actions) can be combined quickly becomes cumbersome (Converse, chapter 11, this volume). When framed in this way, however, the optimal solution can be found with optimization routines that readily search all the possible combinations of projects and actions to identify the combination with greatest benefits. When the decision variable is a binary "yes" or "no" decision for each project, the solution can be found using integer linear programming. Integer linear programming, as well as other optimization routines, are readily implemented in common software applications (e.g., Solver add-in for Excel; Kirkwood 1997) and in packages of the R computing environment (e.g., lpSolve package, Berkelaar et al. 2015). Conroy and Peterson (2013) provide a review of linear programming and other methods for optimization.

The mathematical programming formulation (eq. 1) has many advantages over a benefit-cost ranking (Kleinmuntz 2007). For example, suppose a decision maker at a regional office allocates funding to projects that are proposed by offices or stations in the region. The projects are of different types—habitat restoration, invasive species control, management of threatened and endangered species, and so on—and the decision maker would like to maintain a diversity of funded projects. It is relatively straightforward to add additional budget constraints similar to the one above that will result in a minimum number of each project type or a minimum proportion of the budget to each project type. Similarly, it is straightforward to add constraints that would result in a minimum amount to all stations if parity among offices is an administrative objective. The general formulation above can be extended in myriad ways with additional constraints to accomplish a wide variety of objectives related to management of biological resources and administration of financial resources.

Another extension of the general framework outlined above involves decisions about habitat management and restoration, which are often made in the context of specific management units (e.g., forest stands, managed wetlands, and grassland patches). When there are multiple, mutually exclusive management actions possible in each management unit, there is a form of project dependency among the available choices: the decision maker's choice is limited to 1 of the available actions, at most. In this situation, suppose that A represents the set of mutually exclusive management actions for a given management unit. Adding the constraint $\sum_{i \in A} x_i = 1$ allows 1 and only 1 action per unit (e.g., Lyons et al., chapter 10, this volume); similarly, the constraint $\sum_{i \in A} x_i \leq 1$ permits no more than 1 action from the set of available actions, with the possibility that none would be selected (Kleinmuntz 2007). In this way, the general resource allocation framework can be used to select a portfolio of management actions, treating decision making for habitat management as a portfolio analysis.

Many, if not all, resource allocation decisions in natural resource management include multiple objectives. For decisions with multiple objectives, multi-attribute value and multi-attribute utility models can be applied and the framework extended so that benefits are evaluated with respect to multiple, possible conflicting, objectives (Converse, chapter 5, this volume; Keeney 1992; Keeney and Raiffa 1993). If the full suite of objectives is represented by y_{ij} (outcome of project i with respect to objective j) and there are n performance measures reflecting performance toward those objectives, the benefit measure in the general framework above becomes a weighted average of the benefit contributed with respect to each preference measure (Kleinmuntz 2007):

$$b_i^* = \sum_{j=1}^{n} w_j v_j (y_{ij}) \qquad (2)$$

where w_j are the decision maker's preference weights (Converse, chapter 5, this volume) for each performance measure and v_j is a single-dimension value function (Kirkwood 1997). Value functions represent a decision maker's preference for various levels of performance on a single measure, normalized to a standard range for all measures, usually 0 to 1.

Value functions are a flexible and powerful aspect of multi-attribute value theory (Keeney 1992; Kirkwood 1997). These functions can take a variety of shapes to accurately reflect stakeholder preferences for each attribute. The value functions in figure 9.1 are from the case study in chapter 10 (Lyons et al., this volume) about decision making for managed wetlands with 3 objectives: non-breeding waterfowl, migrating red knots, and juvenile fish populations. The value function for non-breeding waterfowl is a straight line, indicating that every incremental improvement in numbers of non-breeding waterfowl (x-axis) between the worst and best outcomes is equally valuable to the decision maker; i.e., an increase from 5,000 to 10,000 birds is just as valuable as an increase from 20,000 to 25,000 birds. The value function for red knots is concave, indicating that the incremental improvements between the worst and best possible outcomes are not of equal value to the decision maker (fig. 9.1). In this case, an increase from 200 to 400 red knots is more valuable to the decision maker than an increase from 800 to 1,000 red knots. Finally, the value function for juvenile fish populations, which is based on the ratio of juvenile fish captured with a cast net inside and adjacent to the managed wetland, indicates that any ratio of fish inside to outside greater than 1 results in complete satisfaction of the decision maker and full value. The concave function for red knots and the step function for juvenile fish illustrate the flexibility of value functions to capture stakeholder preferences. There is some elicitation required with decision makers to understand their preferences, but the methods to elicit these value functions are straightforward (Goodwin and Wright 2004, chapter 3; Kirkwood 1997).

Value functions are used when uncertainty and risk are not a major concern. For decisions under uncertainty, when understanding the decision maker's attitude toward risk is important, similar functions are used but the elicitation procedures are slightly different, and the resulting functions are called "utility functions" (Runge and Converse, chapter 13, this volume).

In some resource allocation decisions, budget constraints may be flexible or not known exactly. In these cases, graphical (rather than mathematical)

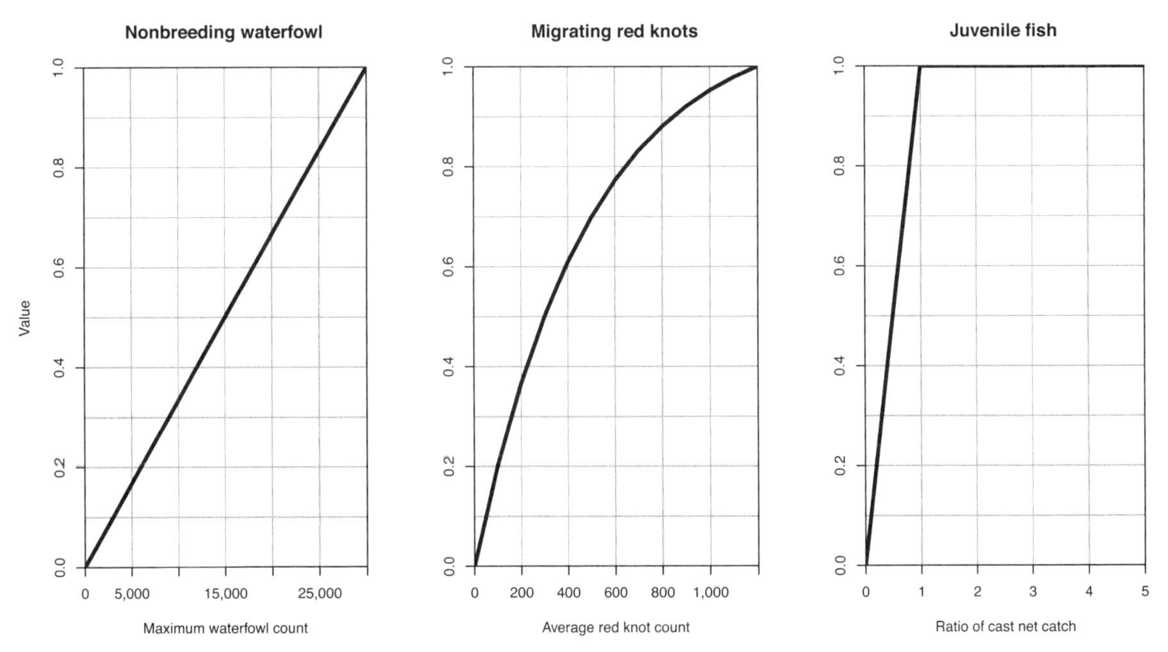

Figure 9.1. Single-dimension value functions reflect a decision maker's preferences for outcomes ranging from the worst option (minimum on x-axis) to best option (maximum on x-axis). Outcomes stated in terms of different objectives (i.e., scores on x-axis) are converted to a common scale ranging from 0 to 1 so that performance on multiple objectives can be aggregated. These value functions were used in decision analysis about wetland management units with 3 competing objectives: nonbreeding waterfowl, red knots, and juvenile fish abundance (Lyons et al., chapter 10, this volume). The straight-line value functions indicate that every increment on the x-axis is of equal value to the decision maker; the concave and step functions indicate that increments near the low end of the outcome scale are of more value than increments near the high end of the scale.

solutions are available, which allow decision makers to explore multiple options. A Pareto efficiency analysis is an intuitive graphical solution that can be used in a variety of situations. In a portfolio decision analysis, for example, a decision maker identifies candidate portfolios and calculates total benefits and costs for each. It is then possible to plot benefits and costs in 2 dimensions and identify the most efficient options over a range of budget constraints. Figure 9.2 shows such a Pareto plot from a portfolio analysis to identify a set of restoration practices for 34 different wetlands at Patuxent Research Refuge (Maryland, USA); each point represents a different portfolio of 34 restoration practices, one for each wetland (US Fish and Wildlife Service 2013). The line connecting the portfolios that offer the greatest benefit for a given level of cost (or the lowest cost for a given level of benefit) is the Pareto efficiency frontier. Portfolios on the efficiency front are "Pareto optimal," meaning it is not possible to find another allocation of resources (a different portfolio) that increases performance in one objective without sacrificing performance on another objective. The points that fall away from the efficiency front, such as portfolio 14 (fig. 9.2), are "dominated solutions": there are other portfolios, such as portfolio 23, that offer greater benefits for less cost. When costs and benefits are plotted in this way, the Pareto efficiency frontier will often have a concave shape showing the point of diminishing returns in benefit as costs increase. Decision makers could explore options near the efficiency frontier for a wide range of budgets if the exact budget constraint is not known, but the area near the "shoulder" of the efficiency front may be of interest if feasible. Decision makers could examine the elements of each of the portfolios near the shoulder on the efficiency frontier in more detail and select any of these options, knowing that each has the benefit of being Pareto optimal. Visual assessment of Pareto efficiency in more than 2 dimensions (2 objectives) is difficult, but various combinations of objectives can be explored using multiple plots (e.g., Joseph et al. 2009, fig. 1).

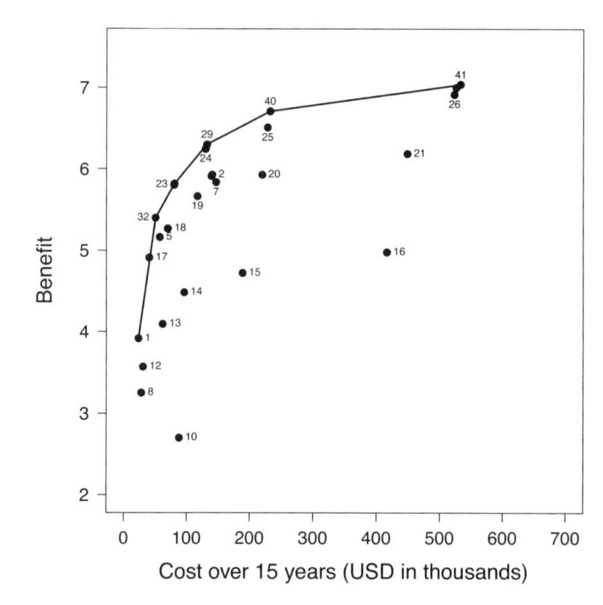

Figure 9.2. Pareto efficiency analysis to identify a set of restoration practices for 34 different wetlands at Patuxent Research Refuge (US Fish and Wildlife Service 2013). Each point represents a different portfolio of 34 restoration practices, 1 for each wetland. The analysis included 6 objectives related to forest land birds, waterbird abundance, waterbird species richness, fish abundance, Odonata species richness, and ecological integrity. "Benefit" (y-axis) is a standardized measure of performance for each portfolio with respect to these 6 objectives. The solid line ("Pareto efficiency frontier") shows the portfolios with greatest benefit for a given level of cost (or the lowest cost for a given level of benefit).

Case Studies

The next 3 chapters present case studies in which resource allocation was the focus of decision making. These case studies cover common challenges in conservation and management: habitat management under uncertainty about sea level rise, reserve design for multiple species in a world of diminishing habitat, and investments in control of invasive species.

In chapter 10, Jim Lyons and colleagues show how to select management actions for a set of discrete management units using a portfolio analysis with the general resource framework outlined above. The case study addresses a key uncertainty—the rate of sea level rise—while using a portfolio analysis to select management actions for multiple management units. The general resource allocation framework was implemented using integer linear programming with the Solver add-in in Excel, demonstrating that the approach is accessible to a wide variety of biologists and decision makers.

In chapter 11, Sarah Converse and colleagues show how selecting parcels of land to create a network of protected reserves can be approached as a resource allocation problem. The US Fish and Wildlife Service was considering a reserve network for 6 taxonomic entities—a bird, a butterfly, and 4 subspecies of pocket gopher—native to the prairie grasslands of the South Puget Sound region of Washington, USA. With limited resources it would not be possible to protect every remaining parcel of prairie habitat, so the decision was which of the available prairie parcels should be protected to maintain viable populations of all 6 taxonomic groups. Converse and colleagues used a patch dynamics model to simulate population change and determine which combination of parcels would result in an acceptably low threshold for the probability of extinction for all 6 taxa. Converse and colleagues integrate population processes for multiple species as part of the reserve design problem (Nicholson et al. 2006), rather than treating reserve design as a minimum set problem that relies on presence-absence data at the time of the decision (a minimum set problem is one in which the decision maker seeks the smallest set of sites that will retain the target number of species or taxonomic groups). The authors also show how judicious problem framing can make a complex and difficult decision with multiple objectives more tractable. Finally, Converse and colleagues found that the prototyping approach (Smith, chapter 2, this volume) resulted in useful insights and lead to new, creative management alternatives not previously considered.

In chapter 12, Joslin Moore and Charlie Pascoe present an analysis that focused on management of an invasive species, grey sallow willow, in the alpine and subalpine peatlands of Victoria, Australia. Peatland communities are normally resistant to willow invasion because their dense vegetation cover limits invasion, but wildfire and other disturbances remove the native vegetation cover and exacerbate the threat of invasion by willows. Managers in this case recognized a dual threat: current invasion, represented by the seedlings and small willows in the peatlands, and future invasion, represented by the mature willows (seed sources) outside the peatlands. The resource allocation challenge was to balance control of the current invasion, by removing seedlings from within peatlands, with control of the seed sources, by removing mature willows from areas adjacent to the peatlands. This decision was complicated by uncertainty about the effectiveness of management actions and differential costs of management to areas within and adjacent to the peatlands. The authors used a dynamic, stochastic simulation model to evaluate different strategies of resource allocation in space and time. Furthermore, Moore and Pascoe used a sensitivity analysis and a value of information analysis (Smith, chapter 17, this volume) to evaluate the impact of uncertainty in the model on the optimal resource allocation. This case study will be especially helpful to practitioners faced with uncertainty not only about the resource allocation but also about how the system responds to management actions and the benefits of reducing that uncertainty.

Open Questions and Challenging Issues in Resource Allocation

One aspect of portfolio analysis that may be underappreciated in natural resource management is the concept of "baseline conditions" (Clemen and Smith 2009). Habitat managers might use a portfolio analysis to select a set of habitat enhancements for multiple management units under their control (Convertino and Valverde 2013; Neckles et al. 2015; Stralberg et al. 2009). Similarly, decision makers in a regional office might use a portfolio analysis to allocate financial resources to projects at the different field offices or stations in their region. In these and similar portfolio analyses, we often use value functions and multiattribute utility theory to measure total benefit of a portfolio (e.g., Kirkwood 1997). This approach relies on a strong assumption about the cost of not doing a project (Clemen and Smith 2009). That is, when we define value functions using 0, the baseline for measurement, as the value of the worst outcome, we implicitly assume that the value of not selecting a project is 0, when in fact selecting a particular management practice often results in an incremental increase in resource benefits from a current level that is not zero.

Clemen and Smith (2009) suggest that we should measure benefits—for example, the results of selecting a particular set of habitat management projects—using the sum of incremental improvements. Kirkwood recommends this solution (see updates to Kirkwood 1997 at www.public.asu.edu/~kirkwood/SDMBook/sdmadd.htm). Morton (2015), however, suggests that this solution may not always be possible or desirable. The appropriate baseline depends on the way the problem is framed and how the objectives will be measured; structured decision-making practitioners and analysts should be aware of this issue and determine the appropriate baseline on a case-by-case basis.

Many resource allocation decisions are dynamic decision problems (Runge, chapter 21, this volume): Decision makers often choose how to allocate money, time, and effort on a regular and repeating basis, e.g., annually. The repeated nature of these decisions can present additional challenges to decision makers and analysts; for example, the actions taken in one decision cycle may affect the options available or outcomes in future decision cycles. Converse et al. (chapter 11, this volume) described the challenges of a dynamic reserve design problem and why they chose to frame the problem as a one-time decision; the benefits from technical tractability of the one-time decision outweighed the benefits of including the details of the dynamic approach. Nevertheless, given the pace of change in natural resource systems, the dynamic aspects of land protection will be increasingly important for natural resource managers (Bonneau et al. 2018; McDonald-Madden et al. 2008; Oetting et al. 2006).

LITERATURE CITED

Ando AW, Mallory ML. 2012. Optimal portfolio design to reduce climate-related conservation uncertainty in the Prairie Pothole Region. *Proceedings of the National Academy of Sciences* 109:6484–6489.

Berkelaar M, others. 2019. *LpSolve: Interface to "Lp_solve" v. 5.5 to Solve Linear/Integer Programs. R Package Version 5.6.13.* https://CRAN.R-project.org/package=lpSolve.

Bonneau M, Sabbadin R, Johnson FA, Stith B. 2018. Dynamic minimum set problem for reserve design: heuristic solutions for large problems. *PLOS ONE* 13:e0193093.

Clemen RT, Smith JE. 2009. On the choice of baselines in multiattribute portfolio analysis: a cautionary note. *Decision Analysis* 6:256–262.

Conroy MJ, Peterson JT. 2013. *Decision Making in Natural Resource Management: A Structured, Adaptive Approach.* Hoboken, NJ: Wiley.

Convertino M, Valverde LJ, Jr. 2013. Portfolio decision analysis framework for value-focused ecosystem management. *PLOS ONE* 8:e65056–e65056.

Game ET, Kareiva P, Possingham HP. 2013. Six common mistakes in conservation priority setting. *Conservation Biology* 27:480–485.

Gerber LR, Runge MC, Maloney RF, Iacona GD, Drew CA, Avery-Gomm S, Brazill-Boast J, Crouse D, Epanchin-Niell RS, Hall SB, Maguire LA, Male T, Morgan D, Newman J, Possingham HP, Rumpff L, Weiss KCB, Wilson RS, Zablan MA. 2018. Endangered species recovery: a resource allocation problem. *Science* 362:284–286.

Goodwin P, Wright G. 2004. *Decision Analysis for Management Judgment.* 3rd ed. Hoboken, NJ: Wiley.

Joseph LN, Maloney RF, Possingham HP. 2009. Optimal allocation of resources among threatened species: a project prioritization protocol. *Conservation Biology* 23:328–338.

Keeney RL. 1992. *Value-Focused Thinking: A Path to Creative Decisionmaking.* Cambridge, MA: Harvard University Press.

Keeney RL, Raiffa H. 1993. *Decisions with Multiple Objectives: Preferences and Value Tradeoffs.* New York: Cambridge University Press.

Kirkwood CW. 1997. *Strategic Decision Making: Multiobjective Decision Analysis with Spreadsheets.* Belmont, MA: Duxbury Press.

Kleinmuntz DN. 2007. Resource allocation decisions. Pages 400–418 in Edwards W, Miles RF, Jr., von

Winterfeldt D, eds. *Advances in Decision Analysis: From Foundations to Applications*. New York: Cambridge University Press.

Lahtinen TJ, Hamalainen RP, Liesio J. 2017. Portfolio decision analysis methods in environmental decision making. *Environmental Modelling & Software* 94:73–86.

McDonald-Madden E, Bode M, Game ET, Grantham H, Possingham HP. 2008. The need for speed: informed land acquisitions for conservation in a dynamic property market. *Ecology Letters* 11:1169–1177.

Morton A. 2015. Measurement issues in the evaluation of projects in a project portfolio. *European Journal of Operational Research* 245:789–796.

Neckles HA, Lyons JE, Guntenspergen GR, Shriver WG, Adamowicz SC. 2015. Use of structured decision making to identify monitoring variables and management priorities for salt marsh ecosystems. *Estuaries and Coasts* 38:1215–1232.

Nicholson E, Westphal MI, Frank K, Rochester WA, Pressey RL, Lindenmayer DB, Possingham HP. 2006. A new method for conservation planning for the persistence of multiple species. *Ecology Letters* 9:1049–1060.

Oetting JB, Knight AL, Knight GR. 2006. Systematic reserve design as a dynamic process: F-TRAC and the Florida Forever program. *Biological Conservation* 128:37–46.

Runge MC. 2016. Portfolio problems. Pages 10.1–10.3 in Runge MC, Romito AM, Breese G, Cochrane JF, Converse SJ, Eaton MJ, Larson MA, Lyons JE, Smith DR, Isham AF, eds. *Introduction to Structured Decision Making, 2016 edition*. Shepherdstown, WV: US Fish and Wildlife Service, National Conservation Training Center. https://nctc.fws.gov/courses/descriptions/ALC3171-Introduction-to-Structured-Decision-Making.pdf.

Stralberg D, Applegate DL, Phillips SJ, Herzog MP, Nur N, Warnock N. 2009. Optimizing wetland restoration and management for avian communities using a mixed integer programming approach. *Biological Conservation* 142:94–109.

US Fish and Wildlife Service. 2013. Patuxent Research Refuge impoundment structured decision-making summary report (appendix G). Pages G1–G25 in *Comprehensive Conservation Plan, Patuxent Research Refuge*. Hadley, MA: US Department of the Interior, US Fish and Wildlife Service.

Williams BK, Szaro RC, Shapiro CD, US Department of the Interior Adaptive Management Working Group. 2007. *Adaptive Management: The US Department of the Interior Technical Guide*. Washington, DC: US Department of the Interior, Adaptive Management Working Group.

Williams HP. 2013. *Model Building in Mathematical Programming*. 5th ed. Hoboken, NJ: Wiley.

10 Resource Allocation for Coastal Wetland Management

Confronting Uncertainty about Sea Level Rise

James E. Lyons,
Kevin S. Kalasz,
Gregory Breese, and
Clint W. Boal

Coastal wetlands are rich and diverse ecosystems with a wide variety of birdlife and other natural resources. Decision making for coastal wetland management is difficult given the complex nature of these ecological systems and the frequent need to meet multiple objectives for varied resources. Management challenges in the coastal zone are exacerbated by uncertainty about sea level rise and impacts on infrastructure, particularly the levees and structures which managers use to manipulate water levels in managed wetlands and create high-quality habitat for birds and other wildlife. The most challenging decisions in coastal wetland management involve resource allocation for habitat manipulations and longer-term investments to maintain management control in wetlands that are increasingly compromised by sea level rise and increasing storm frequency and intensity associated with a changing climate.

We used multi-criteria decision analysis to create a resource allocation framework for managed wetlands that identifies the most effective and efficient management strategies that are robust to uncertainty about sea level rise. The prototype framework includes a small number of managed wetlands, for which subject matter experts articulated potential management and restoration actions. The consequences of these actions were predicted using expert elicitation with the subject matter experts; furthermore, expert judgment was used to articulate expected outcomes with 2 hypotheses about the rate of sea level rise. We used a constrained optimization (integer linear programming) to find optimal resource allocation strategies given a range of budget constraints; we also used a Pareto efficiency analysis for a graphical solution to the problem if the exact budget constraint was not known. Finally, given the importance of preference weights in a multi-criteria decision analysis, we evaluated sensitivity to objective weights. With this resource allocation framework, we showed how to identify optimal combinations of management and restoration actions to maximize benefits in terms of stated objectives. We show how multiple working hypotheses about sea level rise can be incorporated into decisions for coastal wetland management. Our resource allocation approach can be modified for a wide variety of natural resource management settings.

Problem Background
Collaborative Wetland Management in the Coastal Zone

In coastal Delaware, there are 22 managed wetlands (approximately 4,050 hectares) under the stewardship of the Delaware Department of Natural Resources and Environmental Control (DNREC) and the United States Fish and Wildlife Service (USFWS). The types and sources of water in these wetlands, and the ways that water levels are manipulated, vary somewhat among these managed lands. Each wetland is managed independently of the others, and while all of them have the capacity to be managed for multiple objectives, wintering waterfowl and hunting tend to be the driving force behind annual decisions about habitat management. High-quality habitats for waterfowl and other waterbirds can be created by manipulating water levels at key points in the annual cycle to promote growth of beneficial food plants and create shallow water and mudflats. Managers recognized that there would be ecological benefits to managing impoundments collectively rather than individually: collective management could increase available habitat for a variety of taxonomic groups as well as recreational opportunities for multiple stakeholders. Some of the wetlands are more vulnerable to coastal environmental stressors and sea level rise than others: Some are directly on the bay shore, whereas others are more inland with a salt marsh to buffer them from the effects of severe storms. In 2010, the USFWS and DNREC decided to explore ways to collaboratively address resource allocation to achieve multiple objectives of both agencies and other stakeholders during the next 40 years and collectively identify best management practices for their impoundments across the landscape.

One of the management challenges in the coastal zone is deciding when and where to make infrastructure investments to maintain management control in wetlands compromised by rising seas and increasingly frequent and intense storms. To maintain control of water levels in managed wetlands damaged by sea level rise, managers may consider raising dikes, replacing or enlarging water control structures, creating openings in dikes to increase water exchange and tidal flow, and building new impoundments; alternatively, managers may abandon impoundment maintenance and opt to restore natural salt marsh habitat. Our decision analysis therefore included not only short-term resource allocations made in the context of annual decisions about water level manipulations but also long-term investments that will improve habitat conditions for waterbirds in the coastal zone and prevent degradation of resources caused by system change.

The purpose of our study was to develop decision guidance for wetland management in the coastal zone of Delaware that could be used by DNREC and USFWS. We conducted a week-long decision analysis workshop attended by subject matter experts and decision makers. Two of the workshop participants were decision makers; one was a DNREC wetland management program administrator responsible for management decisions for the state's wetlands, and the other was the USFWS refuge project leader responsible for land management decisions. These 2 individuals, along with the 7 additional workshop participants (see Acknowledgments), also served as subject matter experts. Given the relatively short time frame available with the expert panel, our goal was to produce a relatively simple prototype decision process that could be easily extended to represent all dimensions of the actual problem in the future (e.g., additional wetlands and additional objectives). Therefore, we selected a subset of the impoundments in the coastal zone under management of our stakeholders to include in the prototype. The wetlands selected were Raymond Pool at Bombay Hook National Wildlife Refuge (NWR; freshwater, 40 ha); Little Creek North (saline, 187 ha); Logan Lane South (saline, 173 ha); and Unit 3 at Prime Hook National Wildlife Refuge (hereafter "Prime Hook", freshwater, 1742 ha). We selected this subset because these impoundments have a great deal of ecological importance to waterbirds in the coastal zone of Delaware and so that our prototype would include areas managed by both DNREC and USFWS. We also selected a subset of all the objectives of interest to our stakeholders for the prototype. These simplifications made the decision problem more tractable for the expert panel while maintaining all critical elements of the decision, especially uncertainty about sea level rise. Although we limited the number of impoundments used in our prototype so that it could be completed in the time available with the expert panel, there is no practical limit on the number of management units (or objectives) that could be included when the framework is implemented for all managed wetlands in the coastal zone of Delaware.

The resource allocation framework we present below allows for flexible management scenarios to determine where in the landscape impoundments will be constructed, restored, maintained, or allowed to revert to tidal systems as sea level rise makes it too costly to maintain and manage them. The decision problem has multiple objectives and uses a portfolio analysis to identify resource allocation strategies. We demonstrate 2 methods—constrained optimization and graphical solutions—to find the best resource allocation for combinations of management actions that maximize multiple objectives under uncertainty caused by sea level rise.

Ecological Background

Coastal wetlands are among the most productive and ecologically valuable natural ecosystems (Bildstein et al. 1991). Largely because of their managed coastal wetlands, Bombay Hook, Prime Hook, and Chincoteague NWRs have more waterfowl, wading birds, and shorebirds than any other refuge in the USFWS's Northeast or Southeast region (Aagaard et al. 2016). Waterfowl numbers typically peak during migrations and winter. Impoundment management for waterfowl usually involves lowering water levels during the growing season to promote germination and growth of waterfowl food plants and then maintaining water levels at appropriate depths throughout the nonbreeding season to make plant and invertebrate food resources available to waterfowl. Delaware Bay is a globally important stopover area for red knots (*Calidris canutus*) and other migratory shorebirds: hundreds of thousands of shorebirds converge on the shores of the bay each spring to feed on the eggs of horseshoe crabs (*Limulus polyphemus*), an energy-rich and easily digestible food source which fuels their migration to Arctic and sub-Arctic breeding areas in Canada. Red knots in eastern North America (*C. c. rufa*) were listed as threatened under the Endangered Species Act in the United States in 2014 (USFWS 2014). Although red knots and other shorebirds forage primarily at the sandy beaches of the bay shore, they also use the managed wetlands and other marsh habitats for roosting during diurnal high tides and at night. Shorebirds prefer large, open, unvegetated areas for roosting, and wetland managers can promote appropriate habitat conditions using water level manipulations and vegetation management practices to maintain open mud and sand flats with sparse and low vegetation. Finally, managed wetlands can provide important nursery habitat for fish populations by offering shelter and foraging areas, especially for juvenile life stages under water regimes that maintain favorable aquatic environments and accommodate fish passage. The wetlands at the coastal refuges in Delaware, as in other states, are also economically important: Waterfowl hunting is estimated to bring in several million dollars to Delaware each year. However, maintaining impoundments so that they provide these ecological and economic benefits incurs costs, and managing for multiple objectives (e.g., waterfowl, shorebirds, and fish populations) involves ecological trade-offs that are difficult to quantify and are plagued with uncertainty.

Decision Analysis

Objectives

The decision makers on our expert panel selected 3 objectives for our prototype: (1) maximize numbers of migrating and wintering waterfowl, (2) maximize numbers of roosting red knots during migration, and (3) maximize juvenile fish populations (fig. 10.1). Quantitative performance measures were chosen for each objective: (1) maximum number of waterfowl between October and March, (2) mean number of red knots roosting in each wetland (average of

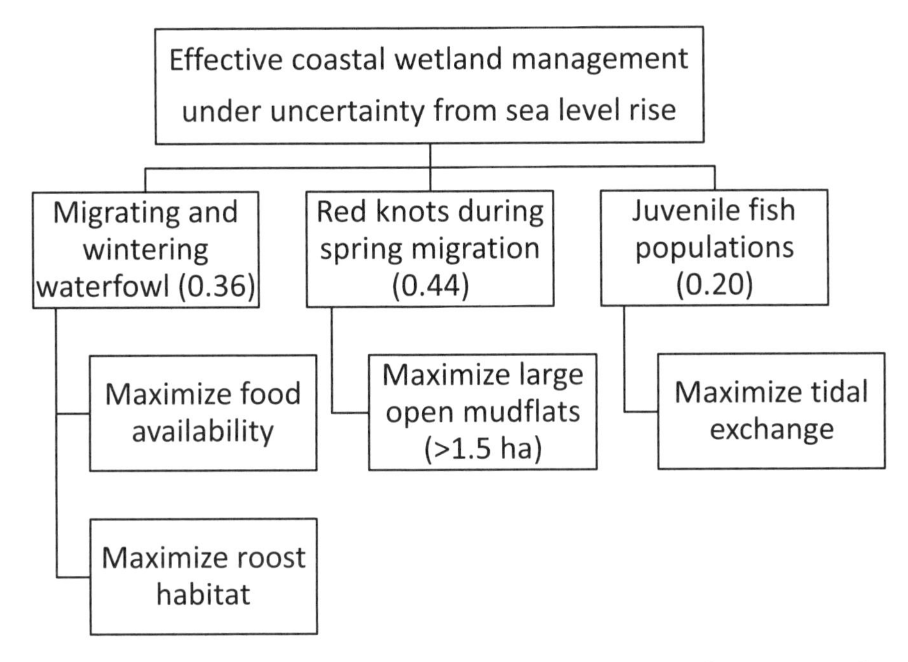

Figure 10.1. Objectives hierarchy for coastal wetland management in Delaware, including a strategic objective at the top. Fundamental objectives are shown just below the strategic objective and include weights determined by stakeholder preferences in parentheses; means objectives are included in lower levels of the hierarchy. See text for quantitative performance measures associated with each of the fundamental objectives.

4–5 weekly counts during the stopover period), and (3) ratio of juvenile fish caught with a cast net inside a water control structure to natural habitat outside the water control structure.

Management Actions

Managers choose between different water regimes, which are defined by timing and intensity of flushing, drawdown, and flooding. These actions are determined by the impoundment management plan, which prescribes various flood stages to meet annual and seasonal management objectives. For example, an impoundment may be maintained at full pool during winter, gradually drawn down to provide food for migrating waterfowl in the spring, kept low to maximize plant growth during the summer, and slowly flooded in late summer through fall to spread food availability for waterfowl over the fall season. This type of water regime also maximizes diversity of water levels. No one water regime can maximize benefits for all objectives; each species group tends to have its own optimal annual water regime.

We selected 3–6 feasible management actions for each impoundment (table 10.1). We first identified 3 management actions in the form of 3 different annual schedules for water level manipulations (i.e., annual water regimes), which were motivated by our 3 objectives. Management action 1 was an annual water regime considered beneficial to migrating and wintering waterfowl (fig. 10.2). Action 1 specified "full pool" during winter, gradual drawdown for migrants in the spring to increase waterfowl food plant growth during the summer, and slow flooding from late summer through fall to spread food availability over the fall season. Action 2 was tailored to provide roosting habitat for red knots during spring migration (fig. 10.2). This water regime maximized open mudflats during May and included relatively quick filling again after May to minimize plant growth on mudflats and maintain open conditions. Action 3, a compromise solution to achieve multiple benefits, including fish

Table 10.1. Potential management actions and costs for each managed wetland

Management action	Cost USD (thousands)
Bombay Hook NWR, Raymond Pool	
Action 1. Waterfowl water regime	$2
Action 2. Shorebird water regime	$2
Action 3. DE saline water regime	$1
Little Creek North	
Action 1. Waterfowl water regime	$5
Action 2. Shorebird water regime	$5
Action 3. DE saline water regime	$7
Action 4. Replace wcs, repair dike, sediment control, & action 1 (waterfowl water regime)	$800
Action 5. Replace wcs, repair dike, sediment control, & action 2 (shorebird water regime)	$800
Action 6. Replace wcs, repair dike, sediment control, & action 3 (DE saline water regime)	$800
Logan Lane South	
Action 1. Waterfowl water regime	$10
Action 2. Shorebird water regime	$10
Action 3. DE saline water regime	$13
Action 4. Construct new 61 ha (150 ac) impoundment, pumping, & action 1 (waterfowl water regime)	$700
Action 5. Construct new 61 ha (150 ac) impoundment, pumping, & action 2 (shorebird water regime)	$700
Action 6. Construct new 61 ha (150 ac) impoundment, pumping, & action 3 (DE saline water regime)	$700
Prime Hook NWR (Unit 3)	
Action 1. Waterfowl water regime	$23
Action 2. Shorebird water regime	$15
Action 3. DE saline water regime	$10
Action 4. Raise levee, replace wcs, & action 1 (waterfowl water regime)	$1,200
Action 5. Raise levee, replace wcs, & action 2 (shorebird water regime)	$1,200
Action 6. Raise levee, replace wcs, & action 3 (DE saline water regime)	$1,200

Note: Water regimes are described in "Management Actions" and depicted in figure 10.2. Wcs = water control structure. "Delaware (DE) saline" is a water regime developed by the Delaware Department of Fish and Wildlife to meet multiple objectives.

nursery habitat, was referred to by habitat managers as "Delaware saline" (fig. 10.2); managers believed that this action, by its intermediate nature, provided habitat for shorebirds during the partial drawdown and for waterfowl and juvenile fish during much of the year. Actions 1–3 could be applied in any of the impoundments. Note also that these water regimes often result in some level of benefits for each of the objectives, not just the motivating objective. For example, action 1 is designed to maximize benefits for nonbreeding waterfowl, but red knots and other shorebirds are still expected to use the impoundment, just not as often and in the numbers as expected under action 2 (shorebird water regime).

In addition to the 3 general management actions described above, we also identified restoration actions specific to each impoundment that were designed to improve management control of water levels and continue achieving multiple objectives despite rising sea level. For example, action 4 for Prime Hook was to raise the levee, replace the water control structure, and then manipulate water levels according to action 1, the "waterfowl water regime" (table 10.1). Action 4 is thus expected to restore and maintain water level control in the management unit as sea level continues to rise. We included additional restoration and management actions specific to Prime Hook, Little Creek North, and Logan Lane South designed to reduce impacts of sea level rise (table 10.1). Because it sits above the current and anticipated tidal range, Raymond Pool at Bombay Hook NWR is not expected to be susceptible to loss of management control over time due to sea level rise.

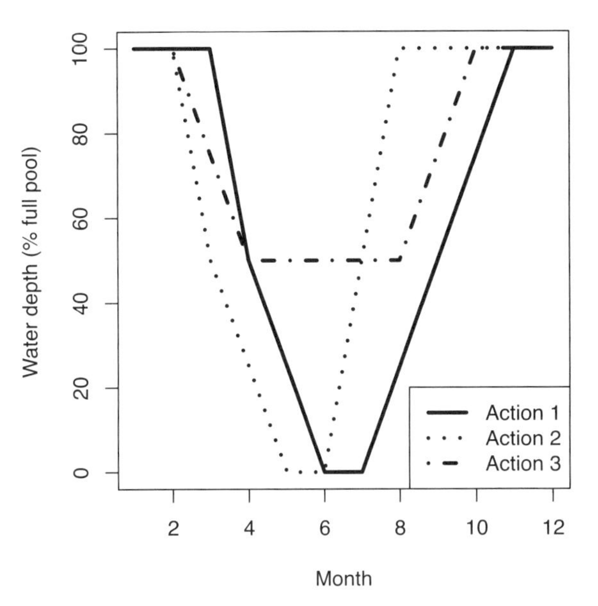

Figure 10.2. Schematic showing 3 alternative management actions (annual water regimes). Action 1 (solid line) would be most beneficial to migrating and wintering waterfowl; action 2 (dotted line) would be beneficial to migrating shorebirds; and action 3 (dash-dotted line) would benefit multiple taxa, including juvenile fish populations.

Portfolios of Management Actions

No single water regime will be optimal for all objectives, but it may be possible to develop a portfolio approach in which each impoundment is managed such that the combination of actions across the landscape (i.e., across all impoundments) maximizes total benefit for all objectives combined. Multiple considerations impact the optimal combinations of actions, including the number of existing impoundments, their locations on the landscape, effectiveness of current and future water level control, and availability of areas for new impoundments. Using a portfolio approach to resource allocation, decision makers choose a management action (table 10.1) for each of the impoundments in the allocation framework; the collection of management actions for all impoundments is the "management action portfolio." Using a constrained optimization routine (e.g., integer linear programming), it is possible to identify a management action portfolio that maximizes the sum of benefits from all impoundments while meeting cost and other constraints of interest. In addition to the optimal solution identified by constrained optimization, in the type of resource allocation problem described here, it is also possible to identify a subset of action portfolios that are Pareto-optimal (i.e., minimizing investments beyond a point of diminishing returns identified by graphical solutions) and to provide decision makers with a rich set of efficient portfolios for a variety of budget constraints (see "Pareto Efficiency Frontier" below).

Consequence Analysis

EXPERT ELICITATION AND SELECTING THE EXPERT PANEL

While the management agencies involved in this study had monitoring data on waterfowl, shorebird, and fish populations in their managed wetlands, there were insufficient data to make predictions about the responses of these populations to the specific management actions of this study. Therefore, we relied on expert judgment (Martin et al. 2005; Runge et al. 2011; Smith, chapter 17, this volume) to predict the responses of these systems to specific management and restoration actions, with the participants in our decision analysis workshop serving as our panel of experts. Our expert panel included 2 individuals who will themselves either select management actions or make recommendations that will be heavily weighed by additional decision makers. In some cases it was necessary for our expert panel to represent the preferences and values of additional decision makers who were not present and to consider multiple regulatory frameworks when selecting objectives and outlining potential management actions. The 2 National Wildlife Refuges in Delaware—Bombay Hook and Prime Hook—were established with explicit purposes under specific legislation. In addition, the USFWS National Wildlife Refuge System has a regulatory framework for deciding upon management alternatives. Similarly, DNREC has a regulatory framework for managing impoundments.

Some of the DNREC impoundments are held in partial ownership with Delaware Department of Transportation (DELDOT), in which case the regulatory framework of DELDOT also impacts decision making in collaborative efforts. Finally, adjacent landowners have interest in the management of impoundments given potential flooding issues, and there are several special interest groups concerned about impoundment management for reasons that include mosquito control, hunting, birding, fishing, and other recreational activities.

PREDICTIVE MODEL

The expert panel predicted the outcomes in response to management actions using a combination of graphical conceptual models and expert judgment. Our expert panel first constructed an influence diagram for each objective (fig. 10.3). Influence diagrams are conceptual models that link management actions with objectives or threats (Clemen 1996). Once influence diagrams were completed and verified, we used expert elicitation techniques with our expert panel to generate predicted outcomes for each of the management actions (table 10.2) with respect to each objective (fig. 10.1). The expert panel made predictions in terms of the specific performance measures identified for each objective. During this elicitation, the panel used the influence diagrams and prior management experience with these wetlands to guide predictions about management outcomes. For example, an impoundment with greater food availability would be predicted to attract more waterfowl (fig. 10.3a). Rather than making a single prediction for each management action, our experts provided multiple predicted outcomes for each of the objectives; multiple judgments were used to capture a range of possible outcomes. Predicting outcomes with a single value overestimates confidence in the predictions and does not accommodate uncertainty related to environmental variation or to the conceptual models (Burgman 2005; Morgan and Henrion 1992). Therefore, our panel provided multiple estimates—usually 3, representing low, moderate, and high outcomes—and associated probabilities that summed to 1. The point estimates for low, moderate, and high outcomes were multiplied by their associated probability and summed for an expected value of the outcome. In some cases, experts provide only 2 outcomes (low and high), and these were treated in a similar manner. For example, if the predicted numbers of waterfowl in response to a management action were 10,000, 20,000, and 30,000 with associated probabilities of 0.1, 0.6, and 0.3, the expected value (EV) of the management action (in number of waterfowl) would be 22,000.

Habitat conditions are expected to deteriorate over time as sea level rise and other stressors damage coastal impoundments and reduce management control of water levels. To capture these anticipated changes in benefits over time and incorporate this uncertainty into our resource allocation, our expert panel made predictions for 3 points in time: 10, 20, and 30 years into the future. This allows explicit incorporation of anticipated changes at the impoundments into management decisions to best allocate financial resources for long-term benefits. The 30-year time frame was selected as a period we could reasonably expect existing impoundments to be maintained and to function given the stresses of sea level rise.

We accounted for epistemic uncertainty (Runge and Converse, chapter 13, this volume) by creating 2 models representing 2 working hypotheses about the rate of sea level rise. Some wetlands will be impacted sooner than others because of their elevation and current infrastructure, and there will be differential costs and benefits to investments in infrastructure depending on the actual rate of sea level rise. Where to first replace or enhance infrastructure could be determined by expected longevity of management control and resource benefits as well as expected costs and benefits of restoration. The first model was formed under the hypothesis that sea level rise will be 0.5 m over 100 years, which was the estimated rate at the time of the workshop ("current SLR"; Williams et al. 2009). Our second model was created

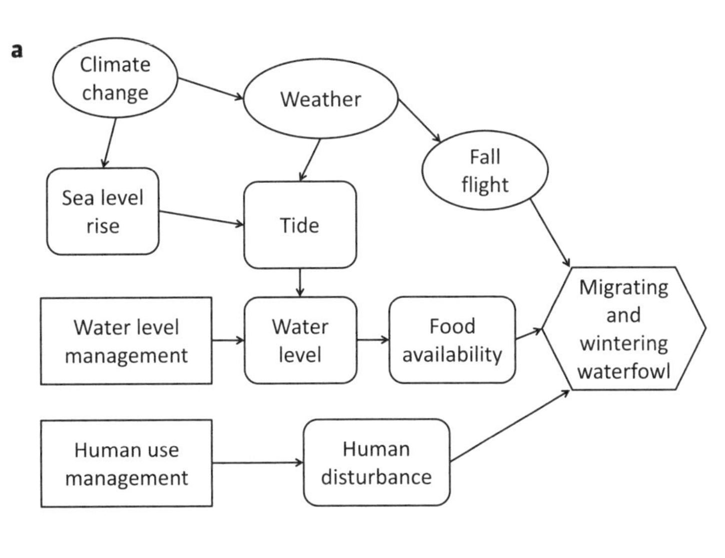

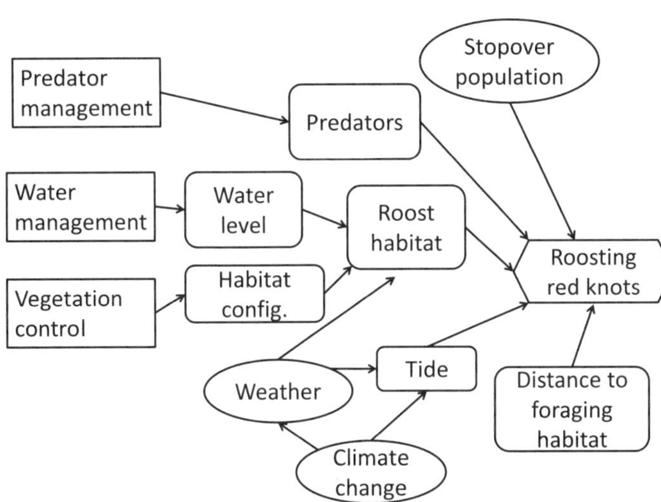

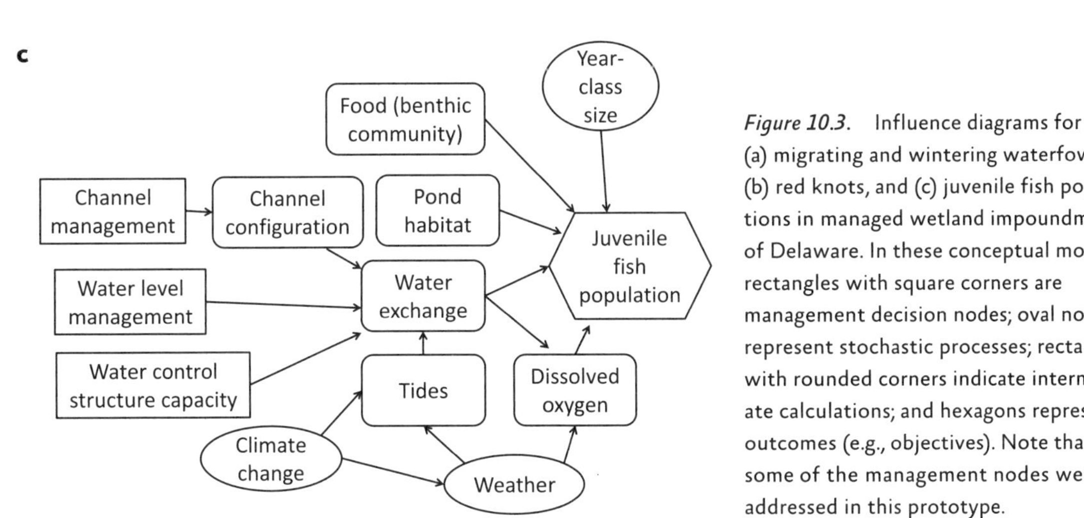

Figure 10.3. Influence diagrams for (a) migrating and wintering waterfowl, (b) red knots, and (c) juvenile fish populations in managed wetland impoundments of Delaware. In these conceptual models, rectangles with square corners are management decision nodes; oval nodes represent stochastic processes; rectangles with rounded corners indicate intermediate calculations; and hexagons represent outcomes (e.g., objectives). Note that some of the management nodes were not addressed in this prototype.

Table 10.2. Costs and benefits for all portfolios evaluated in the prototype

Portfolio	Description	Management benefit (ΣWEV_i)	Cost USD (thousands)
A	Action 1 (waterfowl water regime) in all impoundments	1.81	$39.5
B	Action 2 (shorebird water regime) in all impoundments	2.29	$32
C	Action 3 (DE saline water regime) in all impoundments	1.02	$30.5
D	Minimum-cost action for each impoundment	1.48	$26
E	Maximum-cost action for each impoundment	2.63	$2,702
F	Cost constraint $100K	2.29	$32
G	Cost constraint $1M	2.40	$827
H	Cost constraint $1.5M	2.50	$1,217
I	Cost constraint $2M	2.52	$1,907
J	Cost constraint $2.5M	2.61	$2,012
K	Custom portfolio	1.33	$821
L	Custom portfolio	2.00	$1,220
M	Custom portfolio	1.86	$27
N	Custom portfolio	1.95	$2,015
O	Custom portfolio	1.62	$1,517
P	Custom portfolio	1.81	$2,701
Q	Custom portfolio	1.36	$2,701
R	Custom portfolio	2.18	$2,015
S	Custom portfolio	1.60	$2,012
T	Custom portfolio	1.99	$1,512

Note: Portfolios A–E and K–T were custom portfolios, with A–E specifically designed for management objectives or cost extremes. Management benefit WEV_i is the weighted expected value of each management action in the portfolio. Water regimes are described in "Management Actions" and depicted in figure 10.2. "Delaware (DE) saline" is a water regime developed by the Delaware Department of Fish and Wildlife to meet multiple objectives.

under the hypothesis that sea level rise would be 1 m over 100 years ("accelerated SLR"). Model weights for our 2 hypotheses about sea level rise, which sum to 1, reflected credibility or degree of support for each hypothesis and were assigned by group consensus of our expert panel. The current SLR hypothesis received a model weight of 0.7; the accelerated SLR hypothesis received a weight of 0.3. These weights can be adjusted over time (Runge, chapter 21, this volume) with appropriate monitoring of sea level rise.

Using our conceptual models of the system (fig. 10.3) and our evaluation of differential impacts of sea level rise on management capabilities over time, our expert panel made predictions under both the current SLR (model 1) and the accelerated SLR (model 2), at 10, 20, and 30 years in the future. For example, under model 1, expected waterfowl abundance at Little Creek North with action 1 was 680, 485, and 330 waterfowl at 10, 20, and 30 years, respectively. We calculated expected value of action i over the entire planning period as the sum of expected values at period $j = 10$, 20, and 30 years:

$$EV_i = \sum_{j=1}^{3} EV_{i,j}, \quad (1)$$

where $EV_{i,j}$ is expected value of action i at time period j. Finally, we calculated expected value under model uncertainty as

$$EV_i^* = \sum_{k=1}^{2} \theta_k EV_{i,k}, \quad (2)$$

where θ_k is model weight for model k and $EV_{i,k}$ is expected value of action i under model k. Figure 10.4 shows predicted changes in waterfowl abundance

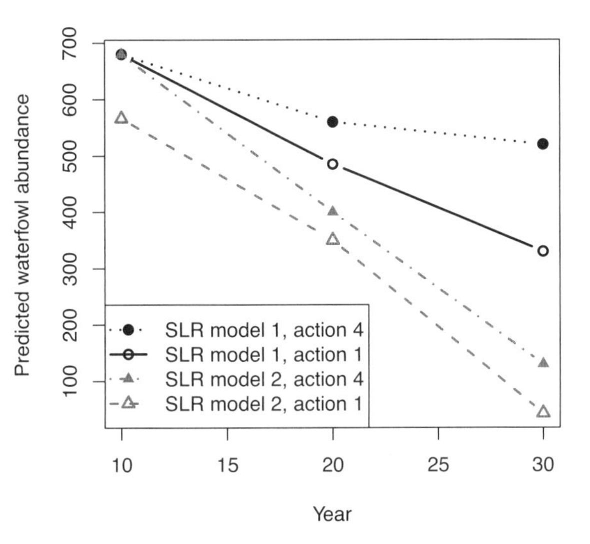

Figure 10.4. Predicted waterfowl abundance at Little Creek North for 2 different management actions and 2 models of sea level rise. The predicted abundances at 10, 20, and 30 years are the expected values reflecting parametric uncertainty. An expert panel identified a discrete probability distribution for waterfowl abundance in each scenario that accounted for incertitude in the conceptual model.

over time for 2 different management actions and under 2 different models of sea level rise.

Decision Solution

TRADE-OFFS USING SINGLE-DIMENSION VALUE FUNCTIONS

The expert panel predicted the outcome of each potential management action in terms of the performance measures for each objective: maximum waterfowl count during October–March surveys; average number of roosting red knots during spring migration; and a ratio of cast net catch inside and outside of the impoundment. Clearly, it is not possible to simply sum these measurements, either as a measure of overall benefit or a means to make trade-offs among objectives. To make trade-offs among objectives in a multi-objective problem, it is necessary to convert predicted responses to a common scale (i.e., "apples to apples"). We used a value model composed of single-dimension value functions (Kirkwood 1997) and converted predicted biological outcomes to a common 0–1 scale:

$$v(x) = \begin{cases} \dfrac{1-e^{[-(x-Low)/\rho]}}{1-e^{[-(High-Low)/\rho]}}, \rho \neq Infinity \\[2ex] \dfrac{x-Low}{High-Low}, otherwise \end{cases}$$

where x is a predicted outcome, $v(x)$ is the associated value between 0 and 1 of the outcome, *Low* is the lowest score for the objective (i.e., low end of the range of biological outcomes), *High* is the highest score for the objective, and ρ is an exponential constant. The value function above applies where higher scores are more desirable (i.e., an increasing value function; decreasing value functions are similar, see Kirkwood 1997). We created a single-dimension value function for each of our objectives and elicited the shape parameters (ρ) from our expert panel. The shape of the value function allows decision makers to express relative preference of values across the range of outcomes (Kirkwood 1997).

The final step in preparation to make trade-offs among objectives and identify optimal resource allocations is to set objective weights. Objective weights, which sum to 1, reflect the relative importance assigned to objectives by decision makers in multi-objective decision problems. Weights for objectives were elicited from the expert panel using a modified swing weighting technique (Conroy and Peterson 2013). After pairwise comparisons of worst and best outcomes for each objective, our expert panel decided that equal weight on all 3 objectives was not appropriate and that red knots during spring migration was the most important objective for this prototype. Objective weights for waterfowl and fish objectives were assigned relative to the red knot objective with final weights of 0.44, 0.36, and 0.20 for red knots, waterfowl, and fish populations, respectively (fig. 10.1).

CONSTRAINED OPTIMIZATION

The optimization routine is designed to maximize the sum of management benefits for all actions included in the portfolio, while meeting constraints (e.g., available budget). Management benefit for action i is defined as the weighted sum of expected value for all 3 objectives:

$$WEV_i = \sum_{l=1}^{3} \omega_l EV_{i,l}^{*}, \qquad (3)$$

where ω_l is objective weight for objective l and $EV_{i,l}^{*}$ is expected value under model uncertainty of action i with respect to objective l.

We used linear programming as implemented in the Solver add-in to Microsoft Excel to maximize our objective function

$$max\left(\sum_{i=1}^{n_{actions}} WEV_i I_i\right)$$

subject to constraints

$$\textit{Cost of Management Portfolio} = \sum_{i=1}^{n_{actions}} y_i I_i \leq \textit{Available budget},\ I_i \in \{0,1\},$$

where y_i is cost in dollars of management action i; I_i is a decision variable equal to 1 if action i is included in portfolio and 0 otherwise; and $n_{actions}$ is the total number of potential management actions for all impoundments. Additional constraints were used to ensure that only one management action was chosen for each impoundment.

PORTFOLIO EVALUATION

We identified optimal portfolios with the constrained optimization routine and a range of cost constraints: \$50,000 and from \$1,000,000 to \$2,500,000 in increments of \$500,000 (5 different constraints in all, resulting in 5 different optimal portfolios). We also created and evaluated 15 custom portfolios by selecting actions of interest for each impoundment and examining costs and benefits based on our objective function. Evaluating custom portfolios in this way relies on equations 1–3 but does not require constrained optimization. The framework is extremely flexible and can incorporate a wide variety of constraints, in addition to or in place of cost, to tailor portfolios to any combination of objectives and constraints desired by decision makers. All 20 portfolios (5 identified with cost constraints and 15 custom portfolios) are described in table 10.2.

PARETO EFFICIENCY FRONTIER

Optimization based on integer linear programming identifies a single optimal portfolio given a set of constraints (e.g., cost and other constraints of interest). If the budgets available to decision makers are not known precisely, in many cases a graphical solution such as a Pareto efficiency frontier is helpful for decision guidance. A Pareto efficiency frontier is identified by plotting the costs (x-axis) and benefits (y-axis) of portfolios of interest. Portfolios in the upper left-hand area of the graph are the most effective and efficient options available, whereas portfolios in the lower right-hand area tend to have low benefits and high costs. Options found along the Pareto frontier cannot be improved upon without giving up something in terms of benefits or cost.

OPTIMIZATION AND GRAPHICAL SOLUTIONS

The management and restoration actions (table 10.1) show a range of costs (\$1,000 to \$1.2 million) because actions may range from relatively inexpensive habitat manipulations occurring every year (annual actions) to expensive restoration actions occurring once in the next 30 years (one-time actions), as well as combinations of annual and one-time actions. Note that there are no restoration actions listed for Raymond Pool because this wetland is not subject to impacts of sea level rise. The purpose of the optimization routine and graphical solution is to identify optimal and near-optimal resource allocations, including where to invest first in the more expensive wetland restoration practices if funds are available. At the low end of budget constraints, we evaluated a scenario in which there are insufficient funds for infrastructure changes (restoration) in any of the wetlands. With a $50,000 cost constraint, the optimization results indicate that portfolio B (action 2, "shorebird water regime"; see fig. 10.2) is the optimal allocation of available resources. Portfolio B is optimal because action 2 returns not only the greatest benefits for roosting red knots but also waterfowl benefits comparable to action 1, which is designed specifically for waterfowl only. If the budget is increased to $1 million, portfolio G, which includes restoration (replace the water control structure, repair dike, control sediments) at Little Creek North, is optimal. Given that a $1 million budget would constrain restoration actions to only one wetland, this analysis indicates that restoration at Little Creek North would provide greatest benefits for all objectives. If the budget is increased to $1.5 million, it is still possible to restore only one wetland, but in this case portfolio H with restoration of Prime Hook is optimal and Little Creek North reverts to action 2 and is not restored. With a budget of $2.5 million, restoration is possible in more than one wetland; the optimal portfolio (I) included restoration of Prime Hook (followed by action 1) and Logan Lane South (followed by action 2).

In our example, portfolios B, D, and M lie on the Pareto efficiency frontier at the low end of the range in costs of portfolios considered (fig. 10.5). For a given budget (x-axis), these portfolios provide maximum benefits across the landscape (total value of approximately 1.5–2.3). Portfolios A and C provide inferior benefits for a given budget and are suboptimal choices (value of approximately 1.0–1.8). Portfolio C (action 3, "Delaware saline," in all wetlands) is most beneficial to fish populations (fig. 10.6), but this portfolio is suboptimal in part because it is less effective at meeting multiple objectives.

It is instructive to evaluate management benefits toward each of our objectives individually for portfolios of interest (fig. 10.6). Portfolios A (action 1 in

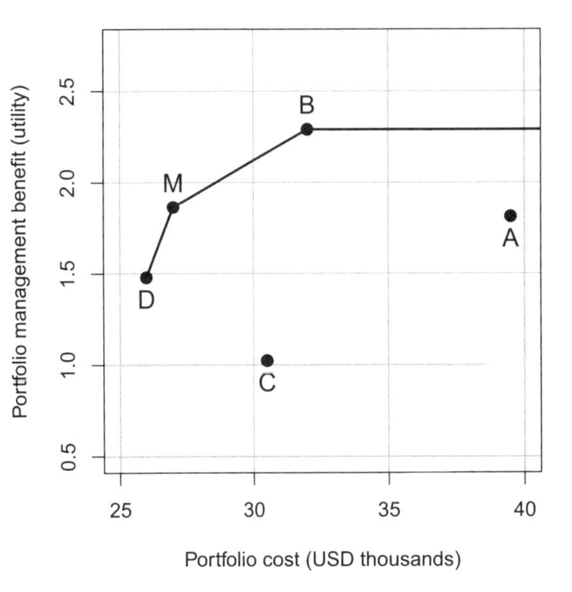

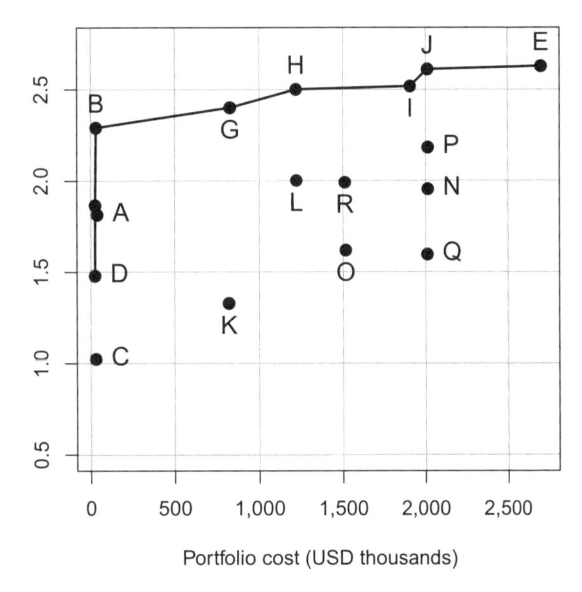

Figure 10.5. (*Left*) Pareto efficiency frontier for management portfolios with cost <$40K. Portfolio A, action 1 in all impoundments; portfolio B, action 2 in all impoundments; portfolio C, action 3 in all impoundments; portfolio D, minimum-cost actions in each impoundment; portfolio M, action 3 in Prime Hook and action 2 in all other impoundments. (*Right*) All 20 portfolios included in the decision analysis. Some portfolios are hidden by others with identical costs and benefits (i.e., when cost constraints were binding, a portfolio with a higher cost constraint may result in the same optimal portfolio as when using a lower cost constraint). See figure 10.2 for schematic showing the water regime for actions 1–3.

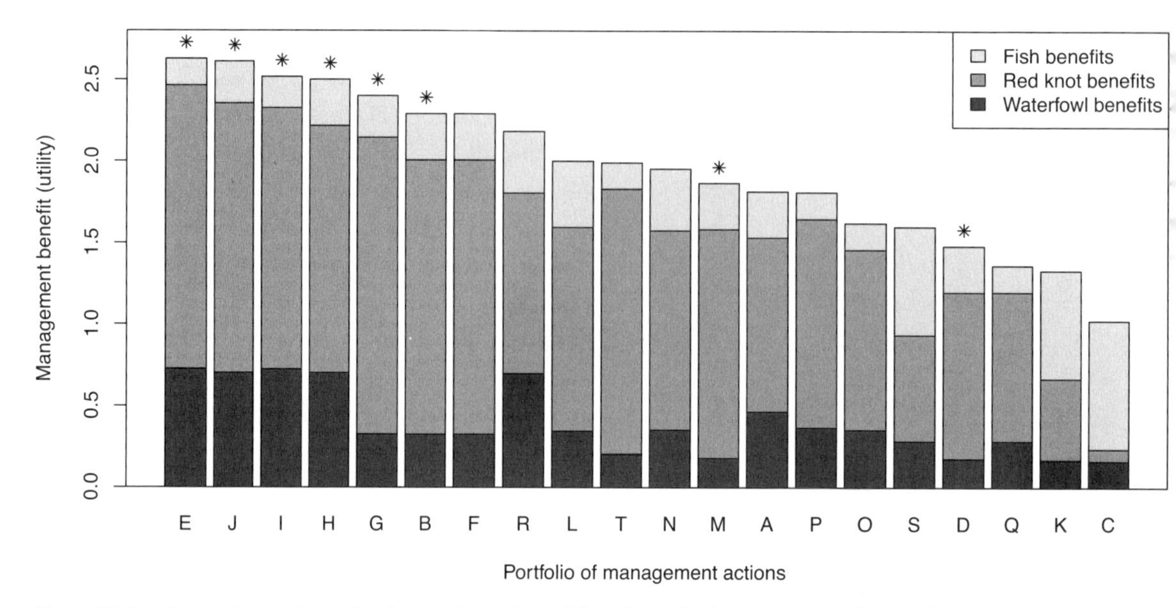

Figure 10.6. Comparison of benefits from selected portfolios for each objective (waterfowl, red knot, and fish populations). Each bar shows expected management benefit for a given portfolio. Portfolio B is optimal for budgets <$100K, and portfolio G is optimal for budgets <$1M. Pareto-optimal portfolios are indicated with an asterisk (*).

all wetlands) and B (action 2 in all wetlands) provide a similar distribution of benefits for all 3 objectives. Most of the management benefit of portfolio C, however, results from benefits to fish populations. Given that fish populations receive less weight (0.20) than other objectives (fig. 10.1), this portfolio is not as effective as portfolios B or D (fig. 10.6). This result was unexpected because action 3 was thought to achieve multiple objectives and had been implemented in the past with this reasoning. When our experts combined inference from a conceptual model of the system and clearly articulated management actions, predicted benefits for multiple objectives with this management strategy were smallest of all the management actions considered, which provided insights and demonstrated trade-offs that had been overlooked previously.

SENSITIVITY TO OBJECTIVE WEIGHTS

Given the importance of preference weights in any decision analysis, we conducted a sensitivity analysis to determine if the preferred alternative indicated with this analysis was robust to the weights chosen by our expert panel. We conducted the sensitivity analysis with portfolios A–C because these portfolios include the same management action in all impoundments. We evaluated changes in total management benefit (utility) as the weight assigned to the management objective for red knots varied from 0 to 1 because this objective received the highest weight of our 3 objectives. As the weight assigned to red knots was varied, the relative proportions for the other objectives were held constant. Overall, our portfolio analysis seems robust to the changes in the weight assigned to red knots (fig. 10.7). Portfolio B, which was identified as optimal for budgets less than $50,000 (fig. 10.5), had greatest management benefits for any amount of weight ≥0.17. The weight assigned to red knots by our expert panel was 0.44; this weight would have to be reduced to 0.16 before one of the other portfolios had greater total benefit (fig. 10.7).

Decision Implementation

Results of this study were presented at a regional conference about climate change and coastal wet-

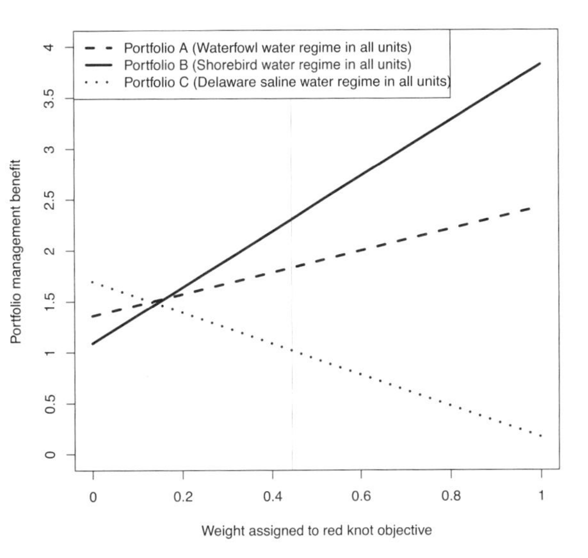

Figure 10.7. Sensitivity analysis for weight assigned to red knots for 3 portfolios (see table 10.2 for portfolio definitions). Portfolio B is optimal until weight assigned to red knots is reduced to 0.17. Portfolio A is optimal when weight for red knots is 0.14–0.16. Portfolio C is optimal when weight assigned to red knots ≤0.13. Dashed vertical line shows weight for red knot objective (0.44) assigned by expert panel representing decision makers and stakeholders.

land management as an example of National Wildlife Federation's Climate-Smart Conservation (Kane 2011). In addition, the authors and members of our expert panel that were staff with the DNREC shared the prototype with decision makers in the agency and other stakeholders. DNREC then refined and began implementing restoration options developed at the workshop at Little Creek North and Logan Lane South to maintain their function for the next 20 to 30 years in the face of sea level rise. The state has upgraded the Logan Lane South impoundment by improving the dikes and water control structures. The ability of Prime Hook NWR to manage its impoundments was further compromised by Hurricane Sandy in 2012. The impacts from Hurricane Sandy demonstrated the importance of explicitly incorporating a robust range of uncertainty into decision making, especially when predicting future conditions from SLR, storm impacts, and climate change. Following this storm, an unprecedented opportunity to restore habitat was provided in the form of recovery and resiliency funding. The refuge took this opportunity to make a major shift in management from coastal impoundments to tidal marsh. A dune system was restored, impoundments were linked and opened to tidal flow, and channels were cut into the marsh, restoring tidal flow. A major objective of the restoration, which was completed in 2017, was improving habitat resiliency for SLR and storm events likely to occur in the future, thus providing flood protection for adjacent landowners.

Discussion

Decision making for coastal wetland management is complicated by uncertainty about sea level rise and the relative benefits of infrastructure investments to maintain management control of water levels within the boundaries of managed wetlands. Resource allocation that is determined by the best choice of a management action or strategy in multiple wetlands uses a "project selection" framework. Even when the numbers of management objectives and options are small, the number of ways that those projects or actions can be combined quickly becomes overwhelming and impossible to address without numerical optimization. Portfolio analysis via integer linear programming or a similar constrained optimization provides a powerful and flexible analytical approach.

We provide a resource allocation framework to achieve the maximum benefits over a 30-year time horizon from coastal wetlands in Delaware that are managed under climate change uncertainty for multiple competing objectives. We incorporated uncertainty about the rate of sea level rise using 2 scenarios, 0.5 and 1.0 meters over the next 100 years. Our flexible approach to resource allocation thus provides decision guidance that is robust to uncertainty surrounding rates of sea level rise. The flexibility of this approach could be used to frame this decision problem in other ways (see part 1, this volume) or to include alternative objectives. For example, a related objective of coastal wetland managers may be to maximize the extent of tidal salt marsh habitat as a means objective that provides greater resiliency to the coastal zone, including adjacent terrestrial habitats and developed areas, from sea level rise and storm surge. Including the extent of salt marsh habitat as an objective could help identify which impoundments should be allowed to revert to natural tidal systems to maximize resiliency. A portfolio analysis for resource allocation can easily accommodate changes to the objectives or their framing (see part 1, this volume).

When accounting for uncertainty about sea level rise using 2 different rates of sea level rise, we used best available estimates of sea level rise at the time of this study. More recent predictions suggest that sea level rise will be faster than predicted and closer to our accelerated model. Our expected benefits and predictions in terms of absolute numbers of birds and fish populations may be optimistic; nevertheless, our conclusions about resource allocation would likely be unchanged because the relative degradation of conditions and management control, and the relative benefits of impoundment restoration, would be similar with a different expected rate of sea level rise.

Several aspects of environmental variation influence this allocation framework. For example, weather and vegetative response are not known completely. There is also uncertainty about how long a given impoundment can provide sustained benefits under a given management regime. Experience of our expert panel suggests that there may be a need to occasionally allow full flooding and draining to reinvigorate the impoundment. Similarly, our expert panel suggests that there are clear disadvantages of using the same management strategy year after year; rather, there is a need to rotate management actions periodically. Our conceptual model and predictions do not account for any temporal dynamics and likely do not capture important aspects of these systems. A dynamic model and optimization (e.g., Martin et al. 2011), or a framework that encompasses, say, a 3-year rotation of management actions, may have advantages for decision making.

Our consequence analysis and predicted outcomes are based on expert judgment of a panel with decades of combined experience in all aspects of wetland management for wildlife. These experts relied on their experience with the wetlands in our prototype and the influence diagrams we created to capture important aspects of how these systems respond to management. With 3 objectives, 2 models of sea level rise, and 3 time periods of interest (10, 20, and 30 years), our study required a great deal of expert elicitation. Eliciting so many parameters and avoiding expert fatigue was one of the facilitation challenges in this study (Speirs-Bridge et al. 2010). To avoid expert fatigue as much as possible, we assigned certain experts to small groups according to their areas of expertise to work on certain parameters, rather than asking every expert to provide judgment on every parameter. This division of labor, and our participatory model building before the elicitation began, helped maintain consistency in our expert judgment.

Future management decisions would benefit from targeted monitoring data to estimate the parameters in our system models and improve our understanding of system dynamics; the influence diagrams could help identify the fundamental and means objectives that would be most valuable as monitoring targets (Lyons et al. 2008). Furthermore, it is relatively straightforward to create a more quantitative model like a Bayes decision net from the influence diagrams, thus creating a model of the system that is readily updated with monitoring data. Finally, a digital elevation model or hydrogeomorphic studies could help us make better predictions about the loss of management control over time and degradation of management benefits as a result of compromised management control.

ACKNOWLEDGMENTS

The authors thank the following individuals for participation in our expert panel: G. Breese, J. Clark, M. DiBona, R. Hossler, B. Jones, K. Kalasz, B. Meadows, M. Stroeh, and B. Wilson. They are grateful to Donna Brewer, USFWS National Conservation Training

Center, for organizing the workshop. R. Katz and H. Neckles provided helpful comments that improved the manuscript. The findings and conclusions in this article are those of the authors and do not necessarily represent the views of the US Fish and Wildlife Service. Any use of trade, firm, or product names is for descriptive purposes only and does not imply endorsement by the US government.

LITERATURE CITED

Aagaard KJ, Lyons JE, Loges BW, Thogmartin WE. 2016. Summary of abundance data collected during the pilot phase of the Integrated Waterbird Management and Monitoring program (IWMM), 2010 to 2014. Brussels, IL: US Fish and Wildlife Service. https://ecos.fws.gov/ServCat/Reference/Profile/68100.

Bildstein KL, Bancroft GT, Dugan PJ, Gordon DH, Erwin RM, Nol E, Laura X. Payne, Stanley E. Senner. 1991. Approaches to the conservation of coastal wetlands in the Western Hemisphere. *The Wilson Bulletin* 103:218–254.

Burgman MA. 2005. *Risks and Decisions for Conservation and Environmental Management*. Cambridge, UK: Cambridge University Press.

Clemen RT. 1996. *Making Hard Decisions: An Introduction to Decision Analysis*. Pacific Grove, CA: Duxbury Press.

Conroy MJ, Peterson JT. 2013. *Decision Making in Natural Resource Management: A Structured, Adaptive Approach*. Hoboken, NJ: Wiley.

Kane A. 2011. *Practical Guidance for Coastal Climate-Smart Conservation Projects in the Northeast: Case Examples for Coastal Impoundments and Living Shorelines*. Merrifield, VA: National Wildlife Federation.

Kirkwood CW. 1997. *Strategic Decision Making: Multiobjective Decision Analysis with Spreadsheets*. Belmont, MA: Duxbury Press.

Lyons JE, Runge MC, Laskowski HP, Kendall WL. 2008. Monitoring in the context of structured decision-making and adaptive management. *Journal of Wildlife Management* 72:1683–1692.

Martin J, Fackler PL, Nichols JD, Lubow BC, Eaton MJ, Runge MC, Stith BM, Langtimm CA. 2011. Structured decision making as a proactive approach to dealing with sea level rise in Florida. *Climatic Change* 107:185–202.

Martin TG, Kuhnert PM, Mengersen K, Possingham HP. 2005. The power of expert opinion in ecological models using Bayesian methods: impact of grazing on birds. *Ecological Applications* 15:266–280.

Morgan MG, Henrion M. 1992. *Uncertainty: A Guide to Dealing with Uncertainty in Quantitative Risk and Policy Analysis*. Cambridge, UK: Cambridge University Press.

Runge MC, Converse SJ, Lyons JE. 2011. Which uncertainty? Using expert elicitation and expected value of information to design an adaptive program. *Biological Conservation* 144:1214–1223.

Speirs-Bridge A, Fidler F, McBride M, Flander L, Cumming G, Burgman M. 2010. Reducing overconfidence in the interval judgments of experts. *Risk Analysis* 30:512–523.

United States Fish and Wildlife Service [USFWS]. 2014. Fish and Wildlife Service. *Federal Register* 79:73706–73748.

Williams SJ, Gutierrez J, Titus JG, Gill SK, Cahoon DR, Thielere ER, Anderson KE, FitzGerald D, Burkett V, Samenow J. 2009. Sea-level rise and its effects on the coast. Pages 11–24 in Titus JG, ed. *Coastal Sensitivity to Sea-Level Rise: A Focus on the Mid-Atlantic Region*. Washington, DC: US Environmental Protection Agency.

11 Reserve Network Design for Prairie-Dependent Taxa in South Puget Sound

Sarah J. Converse, Beth Gardner, and Steven Morey

Conserving species requires managing threats, including habitat loss. One approach to managing habitat loss is to identify and protect habitat in networks of reserves. Reserve network design is a type of resource allocation problem: How can we choose the most effective reserve network design given available resources? We developed and implemented a patch dynamics model to evaluate proposed reserve networks in terms of ability to sustain populations of several taxa that are dependent on native prairie in the South Puget Sound region of Washington, USA. With expert input, we built a patch dynamics model for each taxon and used the model to examine probability of persistence in 50 years under a variety of reserve network designs, including the existing reserve network. Results suggest that the existing reserve network offers varying levels of protection for the different taxa, from desirable (>90% certain that the probability of persistence is ≥75% in 50 years) to negligible. We identified a reserve network that was >90% certain to protect all 6 taxa of interest, which would require a combination of land protection and translocations of taxa to new or existing reserves. Post hoc, we also identified possible hybrid alternatives, involving addition of new reserves and growth of existing reserves, that protected all 6 taxa without translocations. The approach we demonstrate is technically tractable and allows for the evaluation of any proposed reserve network design, thereby allowing a decision maker to evaluate a set of reserve networks that meet resource constraints and determine which of those best meets conservation objectives.

Problem Background

In the face of severe habitat loss, we undertook an effort to identify a network of reserves to be protected and managed for 6 animal taxa associated with a relict prairie ecosystem in the South Puget Sound (SPS) region of the State of Washington, USA. The SPS prairie ecosystem has declined substantially since Euro-American settlement (Crawford and Hall 1997; Dunwiddie and Bakker 2011). These prairies were maintained historically by Native American use of fire (Dunwiddie and Bakker 2011). Since Euro-American settlement, expanding agriculture and urbanization, reduction in fire frequency, and invasive plants have contributed to a severe reduction in native prairie habitat. It has been estimated that only 2–3% of the historical grasslands of western Washington are still dominated by native grassland vegetation (Crawford and Hall 1997; Chappell et al. 2001). At the outset of our effort, the existing reserve network in SPS was small (<2,300 hectares), and its adequacy for protecting taxa of concern was in doubt.

Given the threat posed by continuing habitat loss, the project leader of the US Fish and Wildlife Service's Washington Fish and Wildlife Office (WFWO) was interested in identifying areas to be protected in SPS to conserve taxa of interest. We took a reserve network design approach to this problem. In reserve network design, contiguous patches of land constitute reserves, and multiple discrete reserves form a reserve network. When designing a reserve network, a decision maker selects some set of reserves to achieve a conservation objective, such as improving persistence probability of species. The decision maker must allocate resources to achieve protection of the reserve network. Therefore, reserve network design is a resource allocation problem, where the interest is in identifying the most effective reserve network design given available resources.

A challenge in reserve network design is that there are typically many alternative reserve networks. Reserves do not represent alternatives; instead, it is a set of reserves (i.e., a reserve network) that represents a discrete alternative. In other words, we are interested in selecting a portfolio, or collection, of reserves. For Y available reserves, we have 2^Y possible reserve networks. As Y becomes even moderately large, direct evaluation of all possible reserve networks becomes technically infeasible.

Golovin et al. (2011) presented a quasi-optimization approach to solving a dynamic version of the reserve network design problem in SPS. In the dynamic version, patches of land and the budget to allocate to protection of land become available and are selected sequentially (e.g., annually). Golovin et al. (2011) showed that, under certain conditions, a greedy algorithm can produce a near-optimal solution to this problem. With a greedy algorithm applied to reserve network design, the single reserve that by itself most improves performance on the objective function (e.g., by itself most improves probability of species persistence) is chosen, and then the next best reserve is chosen, and so on, until the budget is exhausted. In the dynamic approach, this process would be repeated with each sequential (e.g., annual) set of available reserves and budgets. A near-optimal solution is one that approximates the optimal solution but is not guaranteed to be globally optimal. While comprehensive, the approach demonstrated by Golovin et al. (2011) is mathematically and computationally intensive.

Here, we present a less computationally intensive approach that differs from the previous approach in that it (1) is designed to evaluate a restricted set of reserve networks and so does not produce an optimal or near-optimal solution and (2) treats the reserve network design problem as a static problem, whereby the entire reserve network is selected at a single time point. Our approach is feasible for quantitatively trained ecologists, whereas the approach presented by Golovin et al. (2011) required collaboration between quantitative ecologists and computer scientists.

We also added a new dimension to the decision problem by considering not just acquisition of reserves but translocations of taxa to reserves to increase their distribution, which should in turn increase probability of persistence. In order to utilize the greedy algorithm, it was necessary for Golovin et al. (2011) to assume that no colonization could take place between reserves in the reserve network. This is not unreasonable in the SPS context (see "Consequences"), which highlights the underlying challenge that building reserves is not adequate if species cannot access them. By evaluating not just reserve networks but translocation as well, we provide a broader set of alternatives that could be more effective at protecting the taxa of interest.

Under our approach, a decision maker would be able to evaluate any specific reserve network. If the budget were available to protect $\leq X$ hectares, the decision maker could evaluate a set of alternative reserve networks of $\leq X$ hectares and choose the one that best achieved the conservation objectives. Alternatively, a threshold could be set for performance on conservation objectives (e.g., a minimum probability of persistence of the taxa over a given time frame), and the smallest reserve that achieves that objective could be identified. In either case, the financial resources could be allocated most effectively conditional on the set of alternatives evaluated. If the cost of a reserve network design were not strongly correlated with area of the network, another metric for cost could be devised (e.g., based on estimated market value of the land). Other costs could also be integrated, such as those of managing the reserve network or translocating species into the network.

Decision Maker and Their Authority

The US Fish and Wildlife Service has regulatory authority for terrestrial species that are listed under the US Endangered Species Act and engages with partners to implement conservation actions for species that are candidates for listing under the Endangered Species Act. Our effort began in 2009, when the taxa of interest were candidates for listing. Ultimately each of the taxa was listed as threatened (5 taxa) or endangered (1 taxon).

The project leader of the WFWO was interested in developing a tool that could assist in decision making about habitat protection actions. The mechanism by which land would receive protection was not specified and did not constrain our approach. Multiple mechanisms could lead to protection of lands in a reserve network designed for conservation of species that are candidates for or are listed under the Endangered Species Act. Outright purchase of land by the US Fish and Wildlife Service is uncommon, but protection could be achieved through formal conservation agreements with landowners, by working with state and nonprofit partners to purchase land or place it in a conservation easement, or by designating critical habitat under the Endangered Species Act. In any of these cases, some amount of money, work, or political capital would need to be allocated to achieve protection of the reserve network. As noted above, we evaluated the reserve network design problem as a static, one-time decision and did not consider temporal aspects of how patches would become available or would be protected. In fact, the reserve network would likely be constructed over time as resources and mechanisms for protecting land became available.

Ecological Background

The taxa of interest in SPS included the Taylor's checkerspot (*Euphydryas editha taylori*; TCS), the streaked horned lark (*Eremophila alpestris strigata*; SHL), and 4 geographically distinct subspecies of the Mazama pocket gopher (Olympia pocket gopher, MPG-OLY, *Thomomys mazama pugetensis*; Roy Prairie pocket gopher, MPG-ROY, *T. m. glacialis*; Tenino pocket gopher, MPG-TEN, *T. m. tumuli*; and Yelm pocket gopher, MPG-YLM, *T. m. yelmensis*). Each is strongly associated with the SPS prairie ecosystem.

The TCS is a subspecies of Edith's checkerspot with a distribution in western Washington, Oregon, and British Columbia (Stinson 2005) although the subspecies is now represented in Oregon and British Columbia by only a small number of populations. TCS is reliant on grasslands with short grasses, appropriate larval host plants, and spring nectar sources (Stinson 2005). TCS have poor dispersal ability, especially across forested habitat (Duggan et al. 2015 and references therein). Translocations of TCS have met with some success (Schultz et al. 2011).

The SHL is a subspecies of the horned lark found only in western Washington and western Oregon (Pearson and Altman 2005 and references therein). In Washington, SHL select nesting areas with short grasses and bare ground while avoiding shrub cover (Pearson and Hopey 2005). Because of these habitat preferences, airports have become a relative stronghold for SHL. SHL exhibit high site fidelity, and efforts to encourage colonization of vacant habitat may include physical translocation or conspecific attraction (e.g., call playbacks; Duggan et al. 2015).

The distribution of the MPG extends through western Washington, western Oregon, and into northern California (Stinson 2005). We evaluated 4 subspecies, all of which are endemic to Washington with highly restricted ranges. The presence of MPG appears to be most strongly determined by soils that facilitate their fossorial habits and forbs that provide critical forage (Stinson 2005 and references therein). MPG dispersal is rare beyond 300 m from the natal area (Daly and Patton 1990; Stinson 2005). A small number of reintroductions have been successful in establishing MPG in new sites (e.g., Olson 2012).

Decision Analysis

Objectives

The primary objective, articulated by management of the WFWO, was to achieve a minimum of 75% probability of persistence (≥1 occupied reserve) after 50 years, for each of the target taxa; we considered meeting this level of protection with 90% confidence to be adequate. A secondary objective identified by management of the WFWO was to minimize restoration effort (the effort required to restore appropriate habitat for these taxa). This second objective was handled implicitly because the patch dynamics model included effects associated with vegetation composition. Vegetation communities that would require more restoration had lower habitat value (e.g., forested sites).

We further assumed that costs would scale to 2 aspects of the alternatives: the area of the reserve network and the number of translocations undertaken. As discussed above, it would be possible to develop other metrics for cost, but here we assumed that minimizing the area of the network and the number of translocations would be beneficial.

Alternatives

Alternative reserve networks were constructed of individually platted pieces of real estate (parcels). Parcels could form reserves by themselves or in conjunction with adjacent parcels. Reserves constituted the patches on which the patch dynamics model operated (see "Consequences"). A set of reserves formed a reserve network. Our first step in defining alternatives was to identify the set of admissible parcels for consideration.

We identified the set of land parcels in appropriate portions of the Washington counties of Grays Harbor, Lewis, Mason, Pierce, and Thurston, as well as Ft. Lewis Army Base (within Pierce County), that met the following criteria: (1) located at least partially on appropriate soil types (soil types that can support prairies); (2) land use classified by county surveyor's office as undeveloped, agriculture, open space, or forest; (3) at least 2.02 ha (5 acres) in size; and (4) could be combined with adjacent qualifying parcels to assemble a contiguous reserve of ≥40.47 ha (≥100 acres). This last requirement eliminated small and isolated parcels that could not be formed into a contiguous reserve. This set of criteria yielded 5,274 candidate parcels for consideration (fig. 11.1).

We considered a restricted subset of the full set of possible reserve networks that consisted of 16 alternatives (table 11.1). These alternatives were chosen to illustrate a variety of approaches to building a reserve network. These included a "cheap" alternative, whereby the decision maker would rely on the existing reserve network; an "expensive" alternative, whereby the decision maker would protect all occupied parcels and build a reserve network with a relatively large area; and several moderate alternatives in terms of area protected. We included several moderately sized alternatives with the same total area, which allowed us to specifically consider the question of resource allocation. For a fixed reserve area, we sought alternatives that performed better than others and so represented a better allocation of resources. Many more reserve networks could be selected for evaluation; we present here a small but illustrative subset.

The alternatives were paired such that Alternative *x* included the same protected parcels as Alternative *x*T, with T indicating the addition of translocations. In all cases, translocations were made to any reserve in the reserve network that did not have a full complement of the taxa. Only one MPG subspecies was assumed to be translocated per reserve, based on the subspecies range in which a given reserve was located. The alternatives, in terms of the reserve network designs, were

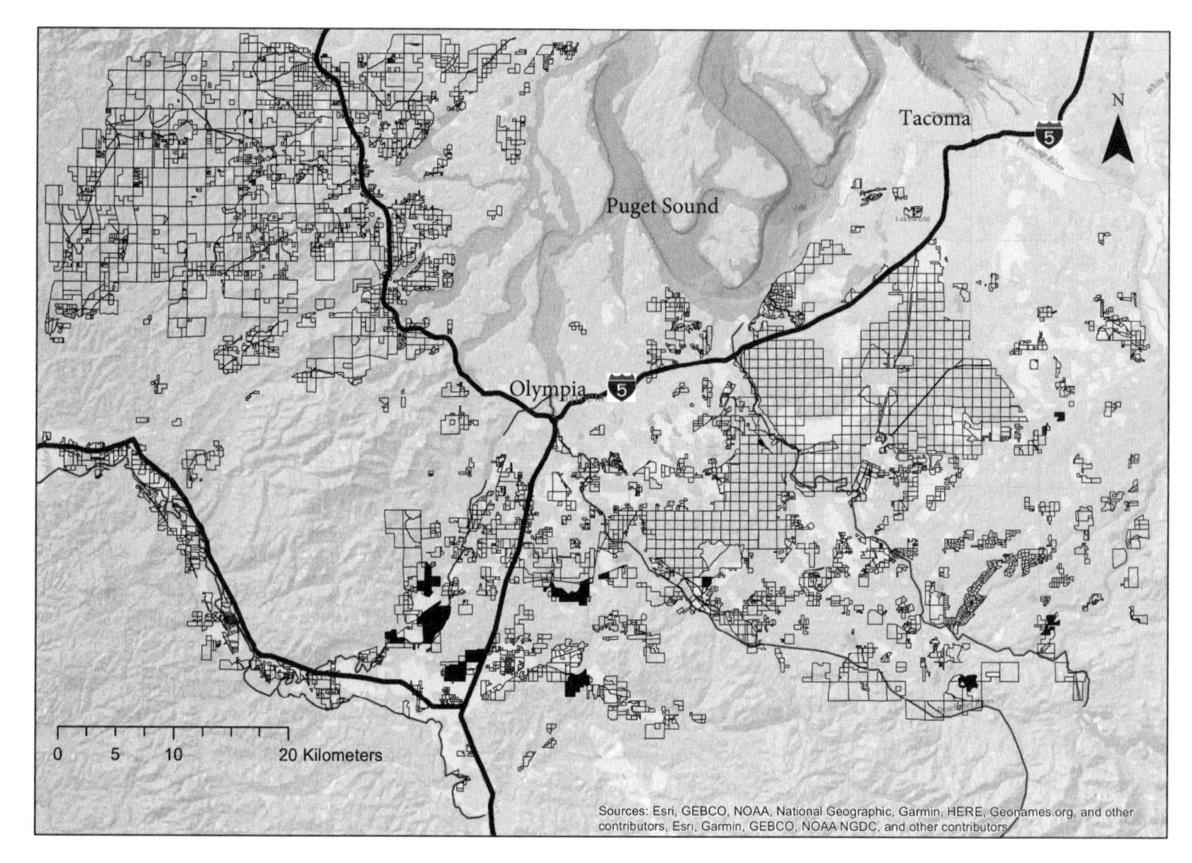

Figure 11.1. The South Puget Sound study area, with the outline of admissible parcels and currently protected reserves represented by filled parcels.

Table 11.1. Characteristics of alternative reserve network designs, including paired spatial designs with and without translocations, for protecting taxa of concern in the South Puget Sound prairies ecosystem

Alt.	Description[a]	Parcels/reserves[b]	Area (ha)	Area (ha) on prairie soils	Trans.[c]
1, 1T	Existing reserves	17/17	2,267	1,369	5, 16, 13
2, 2T	Existing + 1 smallest occupied	20/20	2,328	1,423	5, 18, 16
3, 3T	Existing + 2 smallest occupied	23/23	2,439	1,530	6, 20, 19
4-1, 4T-1	Existing + random occupied	38/29	3,762	2,759	6, 24, 25
4-2, 4T-2	Existing + random occupied	38/28	3,762	2,783	6, 23, 24
4-3, 4T-3	Existing + random occupied	38/27	3,762	2,777	6, 22, 23
5, 5T	Existing + largest neighbors	30/16[d]	3,762	2,547	5, 15, 12
6, 6T	Existing + all occupied parcels	82/33	6,881	5,624	7, 27, 26

Note: For each paired set of alternatives, we provide the number of parcels protected, the number of (contiguous) reserves, the area of the reserve network, the area of the reserve network on prairie soils, and for the translocation alternatives, the number of translocations for Mazama pocket gopher (MPG), streaked horned lark (SHL), and Taylor's checkerspot (TCS).

[a] See text for detailed description of alternatives.

[b] Number of parcels is the number of individually platted pieces of real estate contributing to the reserve network. Number of reserves is the number of discrete reserves, each composed of contiguous patches of land.

[c] Number of translocations to all unoccupied reserves for MPG, SHL, and TCS, respectively.

[d] Only 13 of the 17 existing reserves had neighbors that were in the candidate set of parcels. Adding these 13 resulted in spatial joining of 2 existing reserves (see alt. 1), thus the resulting reserve network had only 16 reserves rather than 17.

- Alternative 1, 1T: Existing reserve network only (composed of 17 parcels comprising 2,267 ha).
- Alternative 2, 2T: Existing reserve network plus the smallest available occupied parcel for any taxon that was not protected adequately under Alternative 1.
- Alternative 3, 3T: Existing reserve network plus the 2 smallest available occupied parcels for any taxon that was not protected adequately under Alternative 1.
- Alternative 4-1, 4T-1: Existing reserve network plus randomly selected parcels occupied by any taxon that was not protected adequately under Alternative 1, such that the total area was equal to the area protected under Alternative 5.
- Alternative 4-2, 4T-2: Same as Alternative 4-1 with different random selection.
- Alternative 4-3, 4T-3: Same as Alternative 4-1 with different random selection.
- Alternative 5, 5T: Existing reserve network plus the largest parcel adjacent to each existing reserve.
- Alternative 6, 6T: Existing reserve network plus all parcels known to be occupied by at least 1 of the taxa.

Consequences

The purpose of modeling in decision analysis is to provide a link between alternatives and objectives, such that the consequences of any given alternative can be evaluated in terms of the objectives. In this case, we needed to be able to make predictions about persistence probability of species under alternative reserve network designs. We developed a patch dynamics model to predict the probability of persistence in reserves. A patch dynamics model assembles a matrix of 0s and 1s ($Z_{i,t}$) that describes the occupancy status (0 = unoccupied, 1 = occupied) of each of i reserves at time period t. We assume each of the $Z_{i,t}$ arises as a function of a stochastic process such that:

$$Z_{i,t} \sim Bernoulli(\psi_{i,t}).$$

Typically, we would define the probability of occupancy for $t \geq 2$ as a function of the previous time step such that:

$$\psi_{i,t} = \varphi(Z_{i,t-1}) + \gamma(1 - Z_{i,t-1})$$

where φ (patch survival) is the probability that a patch is occupied in year t given that it was occupied in year $t-1$, and γ (colonization) is the probability that a patch is occupied in year t given that it was unoccupied in year $t-1$. However, in this case, we set the colonization parameter to 0 in all cases. Therefore, the patch dynamics model simplified to:

$$\psi_{i,t} = \varphi(Z_{i,t-1}).$$

We assumed no colonization between reserves for several reasons. First, for both MPG and TCS, the probability of colonization was extremely restricted, based on the ecology of the species and as predicted by expert judgment, dropping to near 0 at 1 km (MPG) or 2 km (TCS) from an occupied source patch. Second, for SHL, strong site fidelity suggests that colonization events are rare. This assumption is conservative and so will tend to result in underestimation of persistence probability in a given reserve network. Furthermore, this assumption made the problem much more technically tractable, as it did for Golovin et al. (2011; see "Problem Background") in the quasi-optimization approach.

Under the patch dynamics model, we assumed that any place that was not protected could not be occupied by the taxa. In other words, the patch dynamics model only considered dynamics of protected patches. We refer to this as the "hostile matrix" assumption. This conservative approach was warranted because the future land use changes in the SPS are difficult to project but likely to be substantial.

In the absence of empirical data to build a patch dynamics model for the taxa of interest, we worked with an expert panel to determine the input parameters and the structure of models for patch survival. We held workshops with a group of biologists with expertise on the target taxa and the SPS prairie ecosystem. The goal of these workshops was to parameterize patch dynamics models for each of the taxa. Substantial uncertainty existed about the ecological processes governing population dynamics. Our intent during the expert workshops was to formally capture this uncertainty so that it could be reflected in the predictive patch dynamics models.

We do not describe the model completely here; further detail on model structure, parameterization, and execution can be obtained by contacting the first author. Briefly, for all taxa, probability of patch survival was modeled as a function of habitat in the patch (land cover, slope, and soil type) and patch size. The exact shape of relationships and parameter values differed between TCS, SHL, and MPG but not among subspecies of MPG.

We also considered translocations. The probability of executing a successful translocation is uncertain; we assumed that all translocations had a success probability of 0.5. This is the probability that, if a translocation were attempted for a given taxon, the reserve would become occupied by that taxon and thereafter would be subject to the patch survival process.

To determine the set of sites occupied by the taxa, we used a database provided by the WFWO. Accuracy of the database was unknown, and at a minimum the database was likely missing some locations. However, in the absence of better information, we assumed these locations were complete and correct. Violations of this assumption would render predictions less accurate. Ascertaining occupancy status would be an important step before going ahead with protection of sites that were assumed to be occupied.

We implemented the patch dynamics model in the R programming environment (R Development Core Team 2004). We executed the patch dynamics model by running 500 stochastic simulations of the model and recording whether ≥375 of the 500 simulations (i.e., ≥75%) resulted in the taxon being extant after 50 years. Any taxon meeting this threshold we considered "protected." We repeated this for each of 5,000 sets of values sampled independently from the sampling distributions—reflecting parametric uncertainty—for each of the model parameters. We thereby built sampling distributions describing the probability that each taxon was protected under each alternative.

Decision Solution

The existing reserve network (Alternative 1) consisted of 2,267 ha in 17 reserves, with 1,369 ha on prairie soils (table 11.1). Assuming the initial occupancy status of the taxa was accurate, 4 of the existing reserves were occupied by TCS, none by MPG-OLY or MPG-ROY, 1 by MPG-TEN, 2 by MPG-YLM, and 1 by SHL. Under this alternative, the desired level of protection (>90% certainty that probability of persistence was ≥75% at 50 years in future) was not met for MPG-OLY, MPG-ROY, or MPG-TEN (table 11.2). It was clear that neither MPG-OLY nor MPG-ROY would be protected under Alternative 1, as these 2 subspecies did not occupy any of the existing reserves. For TCS, the desired level of protection was also not met; we were 72.4% certain that persistence probability was ≥75%. For both MPG-YLM and SHL, we were >90% certain that persistence probability was ≥75%.

If we examine all 16 alternatives considered, we find that only 8 of the alternatives are non-dominated based on the number of taxa protected, reserve area, and number of translocations (table 11.3). A dominated alternative is one that performs worse than another alternative on at least one objective and equally on all other objectives. A dominated alternative does not represent a rational choice, because the alternative that dominates it is at least as good or better on all objectives. By contrast, any non-dominated alternative could be chosen by a rational decision maker who preferred the trade-offs it represented. Of the 8 non-dominated alternatives, only 1 (4T-3) protected all 6 taxa at the desired level. This alternative involved protection of 3,762 ha and 51 total translocations. However, likely fewer translocations would be required, as some of the taxa were

Table 11.2. The predicted probability of meeting the management objective for each taxon under reserve network design and translocation alternatives in the South Puget Sound prairies

	Non-translocation alternatives							
Taxa	1	2	3	4-1	4-2	4-3	5	6
MPG-OLY	0.000	0.000	0.000	0.979	0.988	0.987	0.000	0.986
MPG-ROY	0.000	0.000	0.000	0.975	0.984	0.947	0.000	0.997
MPG-TEN	0.082	0.080	0.079	0.084	0.084	0.081	0.938	0.082
MPG-YLM	0.995	0.995	0.994	0.994	0.995	0.994	0.996	0.999
SHL	0.992	0.995	0.994	0.997	0.998	0.997	0.998	0.999
TCS	0.724	0.716	0.726	0.715	0.709	0.717	0.828	0.913
	Translocation alternatives							
	1T	2T	3T	4T-1	4T-2	4T-3	5T	6T
MPG-OLY	0.000	0.000	0.001	0.978	0.986	0.988	0.972	0.987
MPG-ROY	0.000	0.001	0.001	0.977	0.984	0.947	0.000	0.996
MPG-TEN	0.992	0.994	0.991	0.992	0.990	0.992	0.997	0.995
MPG-YLM	0.995	0.995	0.994	0.994	0.994	0.996	0.997	0.998
SHL	0.996	0.998	0.996	0.998	0.998	0.998	0.999	0.999
TCS	0.832	0.851	0.833	0.911	0.911	0.914	0.936	0.967

Note: The objective for each taxon was ≥75% probability of persistence 50 years in the future. Results are based on simulation runs of a patch dynamics model for prairie-dependent taxa of interest under the 16 management alternatives (described in text). Taxa include 4 subspecies of Mazama pocket gopher (MPG), streaked horned lark (SHL), and Taylor's checkerspot (TCS).

Table 11.3. Summary of performance of 16 reserve network design and translocation alternatives for South Puget Sound prairies

Alternatives	Taxa protected	Area (ha)	Total trans.	Dominated? By which alternative?
1	2	2,267	0	No
2	2	2,328	0	Alternative 1
3	2	2,439	0	Alternative 1
4-1	4	3,762	0	No
4-2	4	3,762	0	No
4-3	4	3,762	0	No
5	3	3,762	0	Alternatives 4-1, 4-2, 4-3
6	5	6,881	0	No
1T	3	2,267	34	No
2T	3	2,328	39	Alternative 1T
3T	3	2,439	45	Alternative 1T
4T-1	6	3,762	55	Alternative 4T-3
4T-2	6	3,762	53	Alternative 4T-3
4T-3	6	3,762	51	No
5T	5	3,762	32	No
6T	6	6,881	60	Alternatives 4T-1, 4T-2, 4T-3

Note: Evaluated by a patch dynamics model, including number of taxa protected (protection is defined as ≥ 90% confidence that probability of persistence at 50 years in the future is ≥75%), the area protected, and the total number of translocation actions required, across all taxa (4 subspecies of Mazama pocket gopher, streaked horned lark, and Taylor's checkerspot). Also indicated is whether the alternative is dominated, and if so, by which alternative. A dominated alternative is one that performs worse than another alternative on at least 1 objective and equal on all other objectives.

protected without translocations (for MPG-OLY, MPG-TEN, MPG-YLM, and SHL we had >98% confidence that probability of persistence was ≥75%). Eliminating translocations for all taxa which were adequately protected under Alternative 4-3 (SHL, MPG-OLY, MPG-ROY) would reduce the number of translocations from 51 to 25 (leaving 2 translocations for the MPG-TEN subspecies, plus 23 translocations for TCS; table 11.1).

To identify a preferred reserve network, a decision maker could select from any of the alternatives that are non-dominated by choosing the one that represents an acceptable trade-off of costs versus benefits. It would also be possible, building on the results from those alternatives that were formally considered, to devise additional hybrid alternatives. For example, an alternative composed of the largest parcel adjacent to the single reserve occupied by MPG-TEN (as in Alternative 5), all parcels added under Alternatives 4-1, 4-2, or 4-3, and additional parcels occupied by TCS (as in Alternative 6) would be adequate to protect all taxa with >90% certainty, and without any translocations. An iterative approach to constructing alternatives could be used to build progressively more targeted alternatives that combine attractive features of the tested alternatives (see also Ohlson et al., chapter 8, this volume).

Discussion

In our approach to the SPS reserve network design problem, we constructed a patch survival model for threatened taxa on SPS prairies and used the model to evaluate the existing reserve network and a suite of alternatives involving various additions to that network along with translocations. Our results indicated that it was possible to achieve a high probability of meeting the conservation objectives (≥75% probability of persistence in 50 years) for each taxon with a combination of land protection and translocations. We also identified, through examination of simulation results, that protection could be achieved without translocations through a combination of adding to existing reserves and protecting occupied parcels.

While we only examined a restricted set of alternatives, we used a technically tractable approach to evaluate alternatives, and also included translocations, which are likely to be an important tool in the fragmented SPS prairies (Duggan et al. 2015). While our approach does not provide an optimal or quasi-optimal solution, decision makers are likely to be interested in evaluating only a restricted set of options, e.g., when an opportunity arises to add to the existing reserve network. We did not evaluate a dynamic version of this problem per se, but a heuristic arising from the work of Golovin et al. (2011) indicates that a reasonably good solution can be obtained by simply protecting the best parcels available given the budget available at any point in time. Another benefit of our approach is that new information can be integrated into the modeling framework, and available alternatives can be reanalyzed with this new information.

Any decision analysis involves a series of decisions *within* the decision about how to conduct the analysis. These smaller decisions may be made in more or less formal ways. In this case, our decision to develop a simulation-based approach was based on several factors evaluated informally. The analysis of Golovin et al. (2011) taught us a great deal about the problem and about the pros and cons of framing it as a dynamic problem and using a quasi-optimization approach to solve it. One value of this simulation approach was that we were able to directly consider reintroductions. Second, the simulation approach did not require advanced skills in computer science, which often will not be available to decision makers. We also found that the simulation approach facilitated development of a deeper understanding of the process model because we could focus on running a wide variety of simulations. Finally, the simulation approach facilitated evaluation of "what-if" scenarios about alternative conservation strategies. For example, would the decision maker need to utilize reintroductions to meet management objectives, or would land protection be enough? Which taxa most required land protection, and which were relatively safe under the status quo? In this case, we expected that broad strategies, rather than particular reserve designs, would be of greater interest to the decision maker.

Most reserve network design approaches do not consider dynamics of populations but instead consider presence/absence of species as selection criteria, though exceptions exist (e.g., Moilanen and Cabeza 2002). The benefit of our approach is that the principals of island biogeography, where larger reserves with larger areas are more likely to sustain taxa, are explicitly accounted for. This approach is transferrable to any reserve network design problem in which patch dynamics models can be constructed for the taxa of interest. While we used an expert elicitation process to build our model, a promising empirical approach to parameterizing a model for reserve network design is use of a dynamic occupancy model (MacKenzie et al. 2003), which is a statistical model that is structurally similar to our patch dynamics model but integrates an observation component to account for imperfect detection. A hierarchical implementation of a dynamic occupancy model (e.g., Royle and Kéry 2007) would facilitate inclusion of process uncertainty and spatial autocorrelation, which we also included in our patch dynamics model.

We made several important assumptions, including about the probability of translocation success, the veracity of the initial occupancy data, and that no colonization could occur between reserves. Violations of these assumptions could introduce bias into the evaluation of alternatives. For example, if translocation success probability is <0.5, our model results for probability of persistence would likely be biased high, and the estimated number of translocations needed would be biased low.

Further testing of these assumptions through data collection and analysis could improve our approach. The reliance on expert judgment in building our patch dynamics model further argues for the importance of continued monitoring and updating. We accounted for uncertainty in expert judgments by developing sampling distributions for model parameters based on the elicited expert judgment and sampling from those distributions in the model simulations. This modeling approach provides a basic framework into which future data could be integrated and used to further improve decisions about land protection actions in SPS.

ACKNOWLEDGMENTS

Funding for this project was provided by the US Fish and Wildlife Service Washington Fish and Wildlife Office (WFWO). K. Berg and J. Bush provided valuable management direction. Other contributors from the WFWO included M. Jensen, C. Langston, and T. Thomas. The authors thank members of the expert panel, including J. Bakker, T. Kaye, J. Kenagy, S. Pearson, M. Singer, and D. Stinson. K. Flotlin of the USFWS WFWO contributed her expertise on Mazama pocket gophers, and D. Stokes gave input in the setup phase. Reviews by S. Morrison, G. Olson, and J. Lyons improved the manuscript.

LITERATURE CITED

Chappell CB, Mohn Gee MS, Stephens B, Crawford R, Farone S. 2001. Distribution and decline of native grasslands and oak woodlands in the Puget Lowland and Willamette Valley ecoregions, Washington. Pages 124–139 in Reichard S, Dunwiddie P, Gamon J, Kruckeberg A, Salstrom D, eds. *Conservation of Washington's Rare Plants and Ecosystems*. Seattle, WA: Washington Native Plant Society.

Crawford R, Hall H. 1997. Changes in the south Puget Sound prairie landscape. Pages 11–15 in Dunn P, Ewing K, eds. *Ecology and Conservation of the South Puget Sound Prairie Landscape*. Seattle, WA: The Nature Conservancy of Washington.

Daly JC, Patton JL. 1990. Dispersal, gene flow, and allelic diversity between local populations of *Thomomys bottae* pocket gophers in the coastal ranges of California. *Evolution* 44:1283–1294.

Duggan JM, Eichelberger BA, Ma S, Lawler JJ, Ziv G. 2015. Informing management of rare species with an approach combining scenario modeling and spatially explicit risk assessment. *Ecosystem Health and Sustainability* 1:22.

Dunwiddie P, Bakker JD. 2011. The future of restoration and management of prairie-oak ecosystems in the Pacific Northwest. *Northwest Science* 85:83–92.

Golovin D, Krause A, Gardner B, Converse SJ, Morey S. 2011. Dynamic resource allocation in conservation planning. *Proceedings of the AAAI Conference on Artificial Intelligence* 25:1331–1336.

MacKenzie DI, Nichols JD, Hines JE, Knutson MG, Franklin AB. 2003. Estimating site occupancy, colonization, and local extinction when a species is detected imperfectly. *Ecology* 84:2200–2207.

Moilanen A, Cabeza M. 2002. Single-species dynamic site selection. *Ecological Applications* 12:913–926.

Olson GS. 2012. Mazama pocket gopher translocation study: progress report. Cooperative Agreement #13410-B-J023. Olympia, WA: Washington Department of Fish and Wildlife.

Pearson SF, Altman B. 2005. *Range-Wide Streaked Horned Lark (*Eremophila alpestris strigata*) Assessment and Preliminary Conservation Strategy.* Olympia, WA: Washington Department of Fish and Wildlife.

Pearson SF, Hopey M. 2005. *Streaked Horned Lark Nest Success, Habitat Selection, and Habitat Enhancement Experiments for the Puget Lowlands, Coastal Washington and Columbia River Islands.* Natural Areas Program Report 2005-1. Olympia, WA: Washington Department of Natural Resources.

R Development Core Team. 2004. *R: A Language and Environment for Statistical Computing*. Vienna, Austria: R Foundation for Statistical Computing.

Royle JA, Kéry M. 2007. A Bayesian state-space formulation of dynamic occupancy models. *Ecology* 88:1813–1823.

Schultz CB, Henry E, Carleton A, Hicks T, Thomas R, Potter A, Collins M, Linders M, Fimbel C, Black S, Anderson HE, Diehl G, Hamman S, Gilbert R, Foster J, Hays D, Wilderman D, Davenport R, Steel E, Page N, Lilley PL, Heron J, Kroeker N, Webb C, Reader B. 2011. Conservation of prairie-oak butterflies in Oregon, Washington, and British Columbia. *Northwest Science* 85:361–388.

Stinson DW. 2005. *Status Report for the Mazama Pocket Gopher, Streaked Horned Lark, and Taylor's Checkerspot.* Olympia, WA: Washington Department of Fish and Wildlife.

12 Optimizing Resource Allocation for Managing a Shrub Invading Alpine Peatlands in Australia

Joslin L. Moore and Charlie Pascoe

The Bogong High Plains contain the largest area of alpine and subalpine vegetation in Victoria, Australia, including a significant proportion of Victoria's alpine peatlands. Alpine peatlands are listed as an endangered ecological community and protected by the Australian government. One of the major threats to the alpine peatlands on the Bogong High Plains is invasion by *Salix cinerea* (gray sallow willow). Mass germination of willow commenced in 2003 and is associated with disturbance of peatlands by wildfire and the presence of substantial seed-producing populations in nearby waterways. While the main mitigation strategy is to control willow individuals, it was unclear whether effort should be focused on removing the young willow plants establishing in peatlands, or if some effort should be allocated to removing mature willow individuals outside peatlands that may act as a seed source in the future. This chapter addresses the common problem of allocating a fixed budget among potential management alternatives. Simulations using a model of willow spread, establishment, and control were used to identify optimal allocations across a range of possible budgets. Given uncertainty regarding exactly how the system worked (reflected as uncertainty in the model parameters), we used value of information and a sensitivity analysis to show that the optimal resource allocation was consistent across a wide range of potential parameter values, which increased confidence that the optimal resource allocation would be effective in the real world.

Problem Background

The Bogong High Plains is the largest contiguous alpine and subalpine area in Victoria, Australia, comprising approximately 120 square km of alpine and subalpine grassland, heathland, peatland, and snow gum (*Eucalyptus pauciflora*) woodland (McDougall 1982). Annual rainfall varies from 1,200 to 2,400 mm, falling as snow above 1,400 m for 3 months of the year (McDougall 1982).

The Bogong High Plains contain 32% of Victoria's alpine peatlands, which are listed as an endangered community under national legislation (Department of the Environment, Water, Heritage and the Arts 2009). *Salix cinerea* L. (gray sallow willow) was introduced to the region in the 1950s to stabilize hydrological work associated with the Bogong hydroelectric scheme (Carr et al. 1994), both on the high plains and along lowland waterways. Susceptibility of peatlands to willow invasion is linked to disturbance, particularly fire (McDougall and Walsh 2007); in an undisturbed state, peatlands have dense vegetation cover, rendering them largely unsuitable for willow germination.

Salix cinerea is a multistemmed dioecious shrub willow and is one of the most invasive willows introduced to Australia (Cremer et al. 1995; Holland-Clift and Davies 2007). It is the only willow known to have invaded Australia's relatively weed-free alpine and subalpine regions (McDougall et al. 2005); although restricted to moist environments, it is not limited to strictly riparian areas. Reproducing predominantly by seed, a light pappus facilitates long-distance seed dispersal, thereby aiding the rapid colonization of disparate areas, although no persistent seed bank is formed in Australia (Cremer 2003). A ready colonizer of disturbed environments, the species establishes in seminatural and natural environments (Cremer 2003) and regenerates and resprouts after fire (Karrenberg and Suter 2003).

Wildfires in Australia's alpine and subalpine areas are relatively infrequent (Zylstra 2006). However, in early 2003, during a severe drought, severe wildfires ignited by lightning burnt over 20,000 square km of southeastern Australia's mountain and alpine region, including the majority of the Bogong High Plains, laying bare moist sediments and peats previously densely vegetated by peatland and wet heath communities. On these newly exposed substrates, mass germination of willow seedlings occurred, commencing in the first summer after the fire. Extensive areas of substrate were expected to remain bare for many years.

The management of this invader on the Bogong High Plains was a high priority for Parks Victoria, the state government agency responsible for managing most of the area. Parks Victoria considers willow invasion to be one of the major threats to protected alpine peatland communities on the Bogong High Plains. Other government organizations are responsible for managing rivers adjacent to the Bogong High Plains that contained substantial willow populations capable of contributing seed to the peatlands. Parks Victoria established partnerships with agencies and interested community groups to facilitate a landscape-scale approach to management of the invasion of willow into the alpine peatlands.

Parks Victoria and partners commenced managing willow on and around the high plains in early 2004 but were seeking assistance from researchers to establish the level of resources to allocate to control as well as how to allocate these resources to best effect—specifically, where in the landscape the resources should best be allocated, how often control should be undertaken in any particular location, and what types of resources (e.g., contractors vs. volunteers) should be utilized.

While the main mitigation strategy available was to remove willow individuals, it was unclear to Parks Victoria whether existing levels of control were sufficient. It was also unclear whether willow control effort should be focused exclusively on removing seedlings from peatlands or if some effort should be allocated to reducing the threat of continuing invasion (while peatland substrates remained incompletely vegetated) or future invasion (by removing populations that could act as seed sources following further fires). Removing willows from the peatlands was a priority, due to the significant impacts they were predicted to have on the peatlands if left untreated. However, Parks Victoria and their partner organizations were also investing substantial effort in removing potential source populations to minimize further invasion and with a view to reducing the threat in the long term. Preliminary modeling undertaken by Moore suggested that understanding the seed dispersal dynamics of willow was key to finding this balance, but uncertainty was high regarding both the population dynamics of the species and the effectiveness of control. Finally, there was also a need to account for the predicted increase in fire frequency due to climate change (Hennessy et al. 2005), which would likely increase the future threat of willow invasion. Indeed, a second major wildfire burned western and southern parts of the Bogong High Plains in late 2006, and further wildfires burned nearby alpine areas in 2009 and 2013.

In 2007, Moore established a research project in collaboration with Parks Victoria to assist in the management of willows across the Bogong High Plain (Moore et al. 2017). This chapter details a subproject focused on developing a medium- to long-term willow management strategy that accounted for population dynamics, dispersal, and stochastic fire events. Of particular interest was to identify whether control of mature (seed-source) populations was likely to be of long-term benefit and, if so, how far from the Bogong High Plains it was necessary to manage seed sources. The decision framing was developed in collaboration with Parks Victoria and partner agencies through a 2-day structured decision-making workshop (Moore and Runge 2012).

Decision Analysis

We approached the problem as a single-objective resource allocation problem and used a stochastic simulation model to evaluate the benefit of different alternatives over long time frames (200 years). There was substantial uncertainty with respect to population dynamics, wildfire frequency, and the effectiveness of management (willow control). This uncertainty was accounted for using sensitivity analysis and value of information analysis.

Structured Decision-Making Workshop

A structured decision-making workshop was held in 2008 with 8 staff from Parks Victoria and partner agencies. The aim of the workshop was to assist managers to develop a long-term management strategy for willow on the Bogong High Plains. The problem context, fundamental objective, alternatives, and consequences were identified and outlined during the workshop. This information provided the basis for the development of the simulation model, the evaluation of the alternatives, and the identification of an optimal resource allocation, which was carried out after the workshop.

Objectives

The fundamental objective was identified as protecting the alpine peatlands on the Bogong High Plains through a reduction in the occurrence of willow. Alpine peatlands are subject to a wide range of threats including other invasive plants, invasive animals such as feral horses and sambar deer, past grazing by cattle (which ceased in 2006), changed hydrological regimes caused by roads, aqueducts, and other infrastructure, and climate change. It was considered outside of scope to consider all these threats. Willow was identified as the most serious threat amenable to management as part of a previous risk assessment framework in place at Parks Victoria. High densities of willow were expected to be detrimental to peatland integrity and function through negative impacts on peatland-dependent flora via competition for light, water, and nutrients, increased water loss through willow transpiration, reduced oxygen availability in peatlands, impacts associated with an autumn leaf litter pulse (the native vegetation is evergreen), deeper rooting of willows, and subsequent impacts on peatland-dependent fauna, including endangered species. In addition, the willow threat was relatively new and hence was seen as a threat that should be addressed before significant damage occurred.

A second objective was to minimize the resources (e.g., cost or effort) required to achieve the fundamental objective. Because budgets are generally set through external processes that are constrained by factors not included in the analysis (e.g., available funding programs and government priorities), we included cost as an external constraint and evaluated how the optimal decision changed for different budgets. This also provided a way to evaluate how the size of the budget related to the expected reduction in willow, which provided a basis for identifying and seeking the necessary resources.

Performance Measure

To define a performance measure for our fundamental objective, we (1) developed a measure of peatland condition for individual peatlands, (2) aggregated this across the Bogong High Plains, and then (3) evaluated this over time to produce a performance measure that described the status of the region's peatlands with respect to willow across the time frame of the analysis (200 years).

Individual peatlands form discrete entities in the landscape and are treated as a management unit by Parks Victoria and in the analysis. That is, a peatland has a particular condition and is either managed or not. However, the peatlands also vary substantially in size (from 0.5 to >50 ha), so we aggregated the performance measure over the peatlands on an area basis so that small peatlands did not have disproportionate weight in the analysis.

The impacts of willow on peatland condition was considered to vary with both willow age (categorized as mature, juvenile, or seedling) and density. Hence, individual peatland condition was scored at any particular time as

- *Satisfactory:* no willow or no juvenile or mature willow and <100 seedlings per ha.
- *Occupied:* juvenile, but no mature, willow present or ≥100 seedlings per ha.
- *Mature:* at least 1 mature willow present.

The measure of performance in any given year was calculated by summing the area of all peatland classified as being in satisfactory condition.

Finally, we defined a measure of peatland status that is integrated over time to account for system dynamics. A threshold was chosen to describe the overall condition of peatlands on the Bogong High Plains. At any given time, the peatlands overall would be in satisfactory condition if at least 90% of total peatland area were in satisfactory condition. We then measured performance over time as the number of years that the peatlands were in satisfactory condition. This threshold approach ensured that each year contributed equally to the performance measure but sacrificed resolution by creating a binary classification of condition in each year. Subsequent analysis showed that the pattern of resource allocation was not affected by the threshold value. Capturing variation in condition through time was important for this analysis because the condition of the peatlands did not increase or decrease consistently but rather varied through time in response to a stochastic process (wildfire). Recording condition only at the end of the simulation might be misleading if a wildfire happened to occur just before the simulation endpoint.

Alternatives

Four broad dimensions of willow management were identified: location of control, who will undertake control, control method, and control intensity at a given location. It emerged from the workshop that the location of control and the associated cost and effectiveness of control (incorporating different methods) in different locations were key considerations for this analysis. The resource allocation problem was to identify an optimal proportion of effort allocated to each location while incorporating variations in cost and effectiveness of management at different locations.

LOCATION OF CONTROL

There were 2 broad categories of willow management based on location: removing willow (predominantly seedling and juvenile) from peatlands or controlling willow in other locations that may be acting as seed sources for the peatland populations. We classified areas of potential seed source populations into 3 categories, which were defined by a combination of proximity to peatlands and the terrain. Terrain affects accessibility and hence the cost or effort required to undertake willow control. The 3 source population categories were

- *BHP reaches*: waterways, other than peatlands, on the Bogong High Plains;

- *High reaches*: upper reaches of streams and rivers characterized by steep slopes (>20 degrees) adjacent to the Bogong High Plains; and
- *Low reaches*: lower reaches of streams and rivers on gentler slopes or more distant from the Bogong High Plains.

Thus, we identified 4 willow control zones: peatlands, BHP reaches, high reaches, and low reaches.

We created a set of discrete alternatives applying 6 proportional levels of budget (0, 0.2, 0.4, 0.6, 0.8, 1) to each of the 4 management zones and considering all combinations of partitioning a budget between the zones. The 56 resulting alternatives represented all possible combinations of allocating the entire budget to the 4 management zones. We also included a do-nothing alternative (no resources allocated to any management zone) to give a total of 57 budget partitioning alternatives.

OTHER DIMENSIONS OF WILLOW MANAGEMENT

Willow control had been undertaken by a number of different weed control companies as well as volunteers. However, volunteers had completed a relatively small proportion of previous control work, there was limited scope to increase volunteer effort, and volunteers could be safely deployed only in a few areas. Hence, the analysis focused exclusively on where the weed control companies should be deployed.

Willow control was carried out using a search and control approach where people moved through the landscape controlling all individuals found. Previous work had shown that not all individuals were detected by this method and that detection rate depended on search effort (Giljohann et al. 2011). The specifics of how individuals were treated depended on their size. Seedlings and juveniles (predominantly in the peatlands and BHP reaches) were controlled by cutting the stem of each individual with secateurs at the base, painting the stem with herbicide, and relocating cut foliage to dry ground where it could not resprout (cut and paint). Individuals too large for this method (predominantly in the high and low reaches) were poisoned by inserting herbicide into cuts made in the stems and left standing (frill and fill). Control could only be undertaken in summer and autumn due to access limitations, difficulties in locating willows when not in leaf, and the need for plants to be actively growing for herbicide to be effective.

Parallel work (Giljohann et al. 2011) had shown that there was likely to be a trade-off between control intensity at any specific location and the expected number of repeat visits at the site, with the number of repeat visits decreasing as intensity at a given location increased. To simplify the analysis, we did not explicitly incorporate this trade-off. Instead, we assumed that the chosen management action would be repeated annually for the entire period of the analysis and that the mean observed intensity of management to date was applied. Subsequent analysis (Moore and Runge 2012) showed that the optimal solution was unaffected by the assumption of continuous management.

Consequences

We built a system model focused on describing how peatland condition across the Bogong High Plains changed over time with parameters estimated by experts, from the literature, and from some field work. Alternatives were evaluated using simulation.

SYSTEM MODEL

To describe the dynamics of the system, we built a state and transition model of how wildfire, seed dispersal, and management influenced peatland condition. We modeled the system using a set of discrete equations that described the changes from state to state in each time step (1 year). The model contained 6 state variables—the amount of peatland area in satisfactory, occupied, and mature condition as well as the area of BHP, low, and high reaches occupied by mature (seed-producing) willows (fig. 12.1). The area of peatlands in satisfactory condition was the variable used to evaluate the status of the peatlands in any given year (peatlands had satisfactory status if

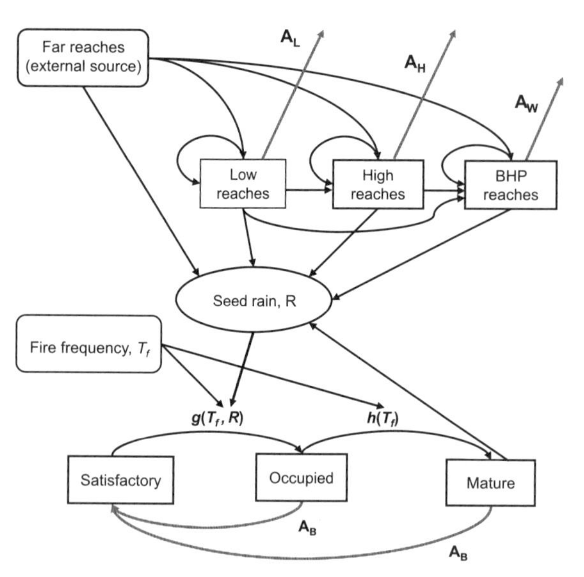

Figure 12.1. An outline of the willow spread model used to identify the optimal allocation of effort between the 4 candidate management areas (Peatlands, BHP stream reaches, high river reaches, low river reaches). The square boxes refer to the proportion of area occupied by mature willow in the low, high, and BHP reaches and the proportion of peatland area that contain no or few willow seedlings (satisfactory), juvenile *S. cinerea* individuals (occupied), and seed-producing individuals (mature—calculate by subtraction). We assume there is a small external source of seed that cannot be eliminated. Seed from the external source, reaches, and peatlands with mature individuals all contribute to a single seed rain which is assumed to be uniformly distributed across the peatlands. The rate of transition from the satisfactory state to the occupied state, $g(T_f,R)$, depends on the magnitude of the seed rain and the amount of bare ground (needed for seed germination). The amount of bare ground depends on fire frequency and the vegetation recovery rate. The rate of maturation $h(T_f)$ is determined by the mean time to maturity. We assume that fire effectively resets willow maturation. The management actions considered are to decrease the area occupied by mature willow in the low reaches (A_L), high reaches (A_H), BHP reaches (A_W), or peatlands (A_B), with control effort in the peatlands split between occupied and mature states in proportion to their prevalence. The effectiveness and cost of control varies between the 4 management zones. *Adapted from Moore and Runge 2012.*

>90% were in satisfactory condition). In the absence of management, peatlands in satisfactory condition transition to occupied and mature states over time as plants germinate and grow. Mature willows in peatlands and in the reaches contribute seed to a global seed rain that then increases the rate of transition from a satisfactory to an occupied state. The amount of seed contributed depends on the seed dispersal parameters. Because the objective was focused on condition of peatlands and the area of the reaches far exceeded the area of peatlands, we only modeled seed dispersal toward the peatlands and ignored spread in the opposite direction. The key wildfire parameter was fire frequency, modeled as a stochastic event. Wildfire increases bare ground, which facilitates willow establishment. Hence, wildfires increase the rate of transition from the satisfactory to the occupied state. The effect of wildfires on establishment decreases with time as vegetation cover increases after a fire; the rate of recovery of vegetation cover is the key parameter describing this process. Wildfires also decrease the rate of transition from the occupied to mature state, as maturation is delayed by the loss of above-ground biomass after a fire. Willow control directly reduces the size of the infestation in the relevant management zone. The degree to which a given resource allocation reduces population size depends on the control effectiveness parameters. Overall, the model has 30 parameters and 7 initial conditions which need to be set prior to running the model. Further details, including the equations, can be found in Moore and Runge (2012).

PARAMETER ESTIMATES

A digital elevation model (20 m grid) was used to identify watercourses based on topography, with reaches defined as being within 10 m of a watercourse. The reaches were then split into low reaches, high reaches, and BHP reaches based on their location. We assumed that all reaches were suitable for willow occupation. Peatland area was derived from an alpine peatlands GIS layer provided by Arn Tolsma (Arthur Rylah Institute for Environmental Research).

Parameters describing willow dynamics, dispersal and control effectiveness and costs, and fire frequency were estimated from a combination of the literature, measurements, and expert opinion (see "Supplementary Information," Moore and Runge 2012 for further details).

SIMULATIONS

We used exhaustive search to identify the optimal solution. We simulated the performance over 200 years for every alternative for each of the 10,000 sets of parameter values and initial conditions that were chosen from prespecified distributions that reflected the uncertainty associated with each parameter and initial condition. We used a long time horizon in order to evaluate the effect of different fire frequencies (5- to 100-year average) on the long-term dynamics of the system. We repeated these simulations for 13 budget levels (expressed as work days, based on an average work-day cost), ranging from 50 to 3,000 work days per year. We recorded the expected performance (number of years more than 90% of total peatland area was in a satisfactory state) for each combination of parameters (10,000), budget partitioning alternative (57), and budget level (13). We used sensitivity and value of information analyses to identify which parameters were most important for driving the decision. The modelling and analysis were undertaken using MATLAB 7.0.

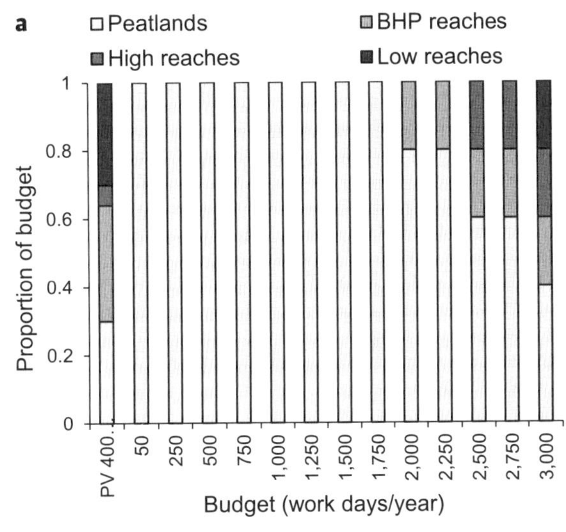

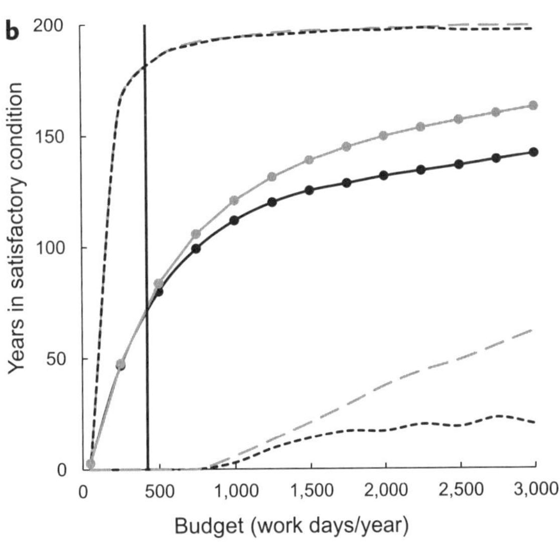

Figure 12.2. (a) The optimal allocation of effort between the 4 candidate management areas across a range of budgets, with the column at the far right labeled PV 400 indicating Parks Victoria's actual allocation over the 4-year period (2004–2008) prior to the development of the model (mean effort was 400 work days/year). (b) Expected performance of optimal action given best estimates for the parameters (black line, dashed lines 95% CI) and expected performance if we had perfect information and knew the parameter values prior to allocating effort (gray line, dashed gray lines 95% CI). The vertical line indicates the resources allocated to control in 2008–2009 by Parks Victoria. *Adapted from Moore and Runge 2012.*

Analyzing Trade-Offs

OPTIMAL STRATEGY AND EFFECTIVENESS OF MANAGEMENT

The optimal allocation was surprisingly insensitive: All resources should be allocated to controlling willows within the peatlands, unless the budget was at least 2,000 work days per year (fig. 12.2a). It is only when budgets were large (more than 5 times the contemporary Parks Victoria budget) that any resources should be allocated to management of seed sources outside of the peatlands, indicating that substantial resources were needed before the seed rain could be meaningfully reduced. The optimal allocation contrasted with Parks Victoria's resource allocation at the time. While the optimal solution was to allocate all resources to the peatlands, Parks Victoria were splitting their effort between the peatlands and the reaches, an approach that would only be optimal when budgets were very large. The relatively small amount of effort allocated by Parks Victoria to the high reaches reflected the high cost and logistical difficulty of accessing these remote and very steep (slope >20%) areas.

We also calculated the expected performance across a range of budgets, assuming the optimal allocation for each budget (fig. 12.2b). As available resources increased, so did the expected performance. However, the model predicted a pattern of diminishing returns; as budgets increased, additional investment in control resulted in only small improvements in expected performance once budgets reached 1,500 work days per year. The broad 90% quantiles illustrated that there was a lot of variability in the outcome, reflecting the high level of uncertainty about how the system was working (i.e., uncertain parameter values)

EXPECTED VALUE OF PERFECT INFORMATION

We calculated the expected value of perfect information (EVPI) to assess the overall robustness of the optimal resource allocation to uncertainty (Runge et al. 2011; Yokota and Thompson 2004). EVPI measures the increase in expected performance if we could resolve all uncertainty prior to making the management decision. A positive EVPI indicates that resolving uncertainty prior to making the decision would improve the performance of the management strategy; that is, the optimal strategy is different for different values of the parameters.

Calculation of EVPI for willow management indicated that when budgets were low, resolving uncertainty in the model parameters did not increase expected performance (fig. 12.2b). This indicated that the optimal allocation was the same irrespective of the parameter values. In this case, learning more about the system was of limited value to the decision maker, as they could not change their decisions in response to the new knowledge. The only options available were to change the resource allocation for willow control at different locations. When budgets were small, it was best to focus resources in the peatlands, regardless of parameters (e.g., fire frequency, seed dispersal characteristics, or willow growth rates), suggesting that in all possible scenarios all resources were required to deal with current invasion rather than trying to manage to reduce further invasion. As budgets increased, value of information also increased, indicating that reducing uncertainty (e.g., knowing the specific fire frequency, seed dispersal distance, or willow growth rate) would improve the expected management performance. However, even when budgets were high, the improvement in expected performance (difference between the black and gray lines in fig. 12.2b) was modest, with a maximum predicted increase in performance of 10% when budgets were very high (2,500–3,000 work days per year).

SENSITIVITY ANALYSIS

Although the value of information was low, the effectiveness of management (measured by expected performance) varied substantially (large confidence intervals), indicating that different model parameter values had a large impact on whether the peatlands could be effectively protected from willow infestation. We undertook a sensitivity analysis to identify which parameters had the biggest influence on expected performance. We did a linear regression of expected performance against each parameter and initial condition for all combinations of budget and allocation alternatives. We recorded the R^2 statistic for each budget and allocation combination as our measure of importance. In this case, the R^2 value indicates the proportion of variation in expected performance that is explained by the parameter. Parameters with a high R^2 are therefore highly correlated with expected performance. For each parameter and budget, we calculated an R^2 value for each of the 57 possible budget partitioning alternatives.

Overall, R^2 values were small, indicating that no single parameter explained the majority of the variation in performance (number of years with at least 90% of peatland area in satisfactory condition). For each budget level, we identified model parameters that had R^2 values of at least 0.05 for at least 1 alternative. The analysis showed that the expected performance was most sensitive to changes in fire frequency, with R^2 values approximately 3 times the size of any other parameter, and management more likely to succeed when fires were infrequent. Expected performance showed lesser sensitivity to the peatland revegetation rate after fire and the parameters that describe dispersal of seed from the reaches to the peatlands. When budgets were high, the germination and seedling survival rates of willows in peatlands were important, whereas when budgets were low the cost of controlling mature willow in peatlands and the initial size of the infestation were important.

EXPECTED VALUE OF PARTIAL INFORMATION

We calculated the partial EVPI to compare the importance of uncertainty about different key parameters, such as fire frequency or seed dispersal distance (Yokota and Thompson 2004). Partial EVPI measures the increase in expected performance if a specific part of the uncertainty is resolved (e.g., fire frequency) and so can help to target learning effort toward parameters that will most reduce uncertainty about the best allocation strategy. To do this, we chose the 8 parameters with the highest R^2 values from the sensitivity analysis and calculated the contribution on expected performance of knowing the value of each of these parameters prior to identifying the optimal resource allocation. We predicted the expected performance by running the simulation for 2 values (25th and 75th percentiles of the parameter distribution) for each of the 8 focal parameters in all 256 (2^8) possible combinations, with the remaining 29 parameters and initial conditions chosen from their distributions. We repeated the 256 simulations 1,000 times to reduce the effect of specific values for the 29 non-focal parameters. We then calculated how knowing the value of each of the 8 focal parameters before we made our allocation decision affected the expected performance of each management strategy. We also considered all pairwise interactions between the 8 focal parameters by calculating the contribution of knowing the values of 2 parameters in combination.

The analysis revealed that uncertainty regarding the contribution that willow populations in different reaches (management zones) made to the seed rain in peatlands was the most important factor affecting decision making. Knowing the dispersal capability of willow in 1 of the reaches increased performance by 2–3.5% if budgets were large (more than 2,000 work days per year). The remaining parameters examined (fire frequency, peatland recovery rate, cost of controlling mature willows, and willow growth parameters) had very small partial EVPI values (predominantly $<0.1\%$), indicating that learning about these parameters would not alter the allocation strategy. There was little evidence of interactions between the different parameters: Knowing the value of 2 parameters together rarely provided more insight than knowing both separately.

Decision Implementation

The decision analysis showed that Parks Victoria's status quo allocation of willow control resources and relatively modest budget were unlikely to best meet the stated objective of reducing willow populations to protect alpine peatlands. Parks Victoria therefore shifted focus over the subsequent 2–3 years so that all willow control was concentrated in peatlands. As of 2015–16, all peatlands had been treated at least once since the 2003 wildfire, and priority peatlands had been treated a second time.

The decision analysis process also influenced data collection and monitoring. Following the decision analysis, regular monitoring of peatland condition (with respect to willow) was initiated. GPS monitoring was integrated with willow control in order to document control effort (in person-days) along with the number of individual willows of each age class controlled in each peatland.

Discussion

The structured decision-making workshop clarified the willow management objective for Parks Victoria and its partner agencies. The subsequent decision analysis prompted them to shift control effort to improve the likelihood of achieving this objective. This transition from managing peatlands and reaches to just peatlands took a number of years as existing programs wound down. At the time of the structured decision-making workshop it was envisaged that an adaptive management approach (Runge 2011) would be used, monitoring management effectiveness and system responses to refine the model and optimal resource allocations over time. However, it turned out that a formal adaptive management approach did not appear to be warranted, as there appeared to be little value to the decision maker in reducing uncertainty regarding the important parameters (Moore and Runge 2012). Variation in management performance was mainly driven by a process (wildfire) which was neither predictable (frequency and intensity) nor readily manageable (wildfires severe enough to burn onto the Bogong High Plains are influenced substantially more by weather, fuels, and topography than by suppression efforts). This highlights the usefulness of undertaking value of information analysis in addition to sensitivity analysis. The fact that the value of information was relatively low (especially at the level of resources then being applied) meant that management recommendations could be made with considerable confidence even though there was substantial uncertainty in the efficacy of management and, indeed, the probability of a successful outcome.

From a researcher perspective, this experience highlighted how important it is to account for and analyze uncertainty as part of resource allocation decisions. Developing tools for effective and accessible uncertainty analysis is an important part of identifying reliable solutions. In this case, we were fortunate that our decision was reasonably robust to uncertainty, providing useful guidance despite imprecise knowledge. Given the high levels of uncertainty and stochasticity in environmental and ecological systems, optimizing can result in poor decisions if uncertainty is not considered. In addition, the collaborative process of decision framing by researchers and decision makers at the 2-day structured decision workshop made communication of the model results more effective over the following months and years and probably contributed substantially to the implementation of the model findings (Moore et al. 2017).

Another factor contributing to the success of the process was institutional support. Parks Victoria invests considerable resources (in-kind and some cash) toward developing a scientifically rigorous evidence base for their management decisions. This institutional outlook meant that the staff involved had a strong commitment to participating in, understanding, and implementing the analysis, even though the structured decision-making process itself was relatively unfamiliar.

Key learnings from a management perspective were the need to clarify management objectives and the critical importance of gaining a shared understanding between researchers and managers of the problem, the control options available, and the inherent assumptions in both, all down to the finest detail, including data collection and data curation (Moore et al. 2017). Time should be allocated to develop these shared understandings and facilitate the implementation of decision analysis.

The strong scientific underpinning of the Bogong High Plains willow control program from 2008 onward supported managers to articulate clear and achievable objectives, report on the efficiency and effectiveness of progress toward achieving these, and implement a wide range of improvements across the program. It also helped Parks Victoria and its partner agencies to attract substantial funding, through a variety of sources, to maintain the program over a sustained period.

ACKNOWLEDGMENTS

The authors acknowledge the traditional owners of the Bogong High Plains, the Jaithmathang and Dhudhuroa people. They thank the staff of Parks Victoria, Department of Sustainability and Environment, Department of Primary Industries, North East Catchment Management Authority, and the Falls Creek Resort Management Board for participating in the workshop and Michael Runge for facilitating the workshop. Brendan Wintle instigated the collaboration, and Kate Giljohann helped with field work. The research was supported by the Applied Environmental Decision Analysis Hub of the Commonwealth Environment Research Facility, Parks Victoria, the Australian Centre of Excellence for Risk Analysis, and the Threatened Species Recovery Hub of the National Environmental Science Program. Data collected were approved as part of Research Permit 10004458.

LITERATURE CITED

Carr GW, Bedgood SE, Muir AM, Peake PE. 1994. *Distribution and Management of Willows* (Salix) *in the Australian Alps National Parks*. Melbourne, Australia: Ecology Australia.

Cremer K. 2003. Introduced willows can become invasive pests in Australia. *Biodiversity and Conservation* 4:17–24.

Cremer K, van Kraayenoord C, Parker N, Streatfield S. 1995. Willows spreading by seed: implications for Australian river management. *Australian Journal of Soil and Water Conservation* 8:18–27.

Department of the Environment, Water, Heritage and the Arts. 2009. Alpine Sphagnum Bogs and Associated Fens in Community and Species Profile and Threats Database. Canberra, Australia: Department of the Environment, Water, Heritage and the Arts. www.environment.gov.au/sprat.

Giljohann KM, Hauser CE, Williams NSG, Moore JL. 2011. Optimizing invasive species control across space: willow invasion management in the Australian Alps. *Journal of Applied Ecology* 48:1286–1294.

Hennessy K, Lucas C, Nicholls N, Bathols J, Suppiah R, Ricketts J. 2005. *Climate Change Impacts on Fire-Weather in South-East Australia*. Aspendale, Australia: CSIRO.

Holland-Clift S, Davies J. 2007. *Willows National Management Guide: Current Management and Control Options for Willows* (Salix *spp.*) *in Australia*. Geelong, Australia: Victorian Department of Primary Industries.

Karrenberg S, Suter M. 2003. Phenotypic trade-offs in the sexual reproduction of Salicaceae from flood plains. *American Journal of Botany* 90:749–754.

McDougall KL. 1982. *Alpine Vegetation of the Bogong High Plains*. Melbourne, Australia: Victorian Ministry for Conservation.

McDougall KL, Morgan JW, Walsh NG, Williams RJ. 2005. Plant invasions in treeless vegetation of the Australian Alps. *Perspectives in Plant Ecology Evolution and Systematics* 7:159–171.

McDougall KL, Walsh NG. 2007. Treeless vegetation of the Australian Alps. *Cunninghamia* 10:1–57.

Moore JL, Pascoe C, Thomas E, Keatley M. 2017. Implementing decision analysis tools for invasive species management. Pages 125–155 in Bunnefeld N, Nicholson E, Milner-Gulland EJ, eds. *Decision-Making in Conservation and Natural Resource Management*. Cambridge, UK: Cambridge University Press.

Moore JL, Runge MC. 2012. Combining structured decision making and value-of-information analyses to identify robust management strategies. *Conservation Biology* 26:810–820.

Runge MC. 2011. An introduction to adaptive management for threatened and endangered species. *Journal of Fish and Wildlife Management* 2:220–233.

Runge MC, Converse SJ, Lyons JE. 2011. Which uncertainty? Using expert elicitation and expected value of information to design an adaptive program. *Biological Conservation* 144:1214–1223.

Yokota F, Thompson KM. 2004. Value of information literature analysis: a review of applications in health risk management. *Medical Decision Making* 24:287–298.

Zylstra P. 2006. *Fire History of the Australian Alps: Prehistory to 2003*. Canberra, Australia: Department of the Environment and Water Resources.

PART IV ADDRESSING RISK

13 Introduction to Risk Analysis

Michael C. Runge and
Sarah J. Converse

Many decisions are made in the face of uncertainty that either cannot or will not be reduced, and the challenge to the decision maker is how to manage the risk imposed by that uncertainty. This chapter will introduce the field of risk analysis, focusing on both the scientific tasks (estimating the probabilities and magnitudes of possible outcomes) and the policy-relevant value judgments needed (understanding the risk tolerances of the decision makers and stakeholders). The 3 case studies that follow demonstrate a range of approaches to risk management in natural-resource settings.

Introduction

Uncertainty is a hallmark challenge of natural resource management. Natural systems are complex, the humans who interact with them equally so, and our knowledge about the potential effects of management actions is never as complete as we would like. The next 3 parts of this book all address some aspect of this challenge.

It is valuable at the outset to distinguish 3 kinds of uncertainty; each plagues decision making in different ways, and the tools to address them differ (Regan et al. 2002). First, *linguistic uncertainty* arises from ambiguity, imprecision, or misunderstanding imbedded in the language used to describe natural systems and their management. Such uncertainty is most pernicious when it goes unnoticed. For example, a species might be described as "highly fecund," and this might have some bearing on its management. But experts hearing this term might have different impressions based on their background: a large mammal that has twins every other year might be described as highly fecund; a sea turtle that lays hundreds of eggs might be described as highly fecund; but an insect may lay thousands or millions of eggs before it is considered highly fecund. Indeed, even experts studying the same animal may interpret the phrase differently. Linguistic uncertainty can be reduced through awareness, structured discussion, and deliberate clarification and operational definition of terms and concepts.

Second, *aleatory uncertainty* (from *aleator*, the Latin word for "gambler") arises from the inherent variability in some systems and uncontrollability of some events. Environmental and demographic stochasticity fall in this category: while we can sometimes describe the frequency with which rain occurs in July, or with which adult manatees (*Trichechus manatus*) are hit by boats, the July rainfall in a specific year or precisely which manatees will be hit by boats next year are random events.

Third, *epistemic uncertainty* arises from the limits of our knowledge. The important difference between aleatory and epistemic uncertainty is that the latter is, at least theoretically, reducible. When epistemic uncertainty is theoretically, but not practically, reducible, it is often treated as if it were aleatory uncertainty. Epistemic uncertainty and the value of reducing it are the topics of part 5.

How do we make decisions in the face of aleatory uncertainty, or in the face of epistemic uncertainty that is impractical to reduce? That's the focus of risk analysis and management. A decision maker choosing whether to bring the last remaining individuals of a species into captivity faces 2 critical uncertainties: whether the species will go extinct in the wild and whether it will go extinct if brought into captivity. Either choice of action exposes the decision maker to risk—the chance of an undesirable outcome.

The scientific task for this type of decision problem is to characterize and estimate the uncertainty. Ideally, during the consequence analysis phase, distributions are developed to describe the probability of the possible outcomes associated with particular actions, thereby reflecting the range of possible outcomes and their likelihoods. In such analyses, it is valuable to distinguish the irreducible (aleatory) and reducible (epistemic) components of the uncertainty.

The value judgment at the heart of this type of decision problem is the risk tolerance of the decision maker. In the face of uncertainty, a decision maker needs to weigh the consequences of the uncertain outcomes to arrive at a preferred action. Especially in cases where some of those outcomes are undesirable (e.g., extinction), a decision maker faces a challenging policy task: how to reflect their willingness, or lack thereof, to accept risk. A risk-averse decision maker might forego a higher expected outcome in order to reduce the likelihood of a poor outcome; a risk-seeking decision maker might be willing to accept a higher chance of a poor outcome if it means a greater chance of an excellent outcome. These choices reflect the values of the decision maker, in the context of a particular decision, and may also consider the viewpoints of stakeholders. Helping decision makers and stakeholders articulate their risk tolerances in a way that can be combined with the scientific analysis is the primary policy-related challenge in problems where aleatory uncertainty plays a critical role.

We encounter risk decisions every day, in our personal and professional lives, but our innate cognitive processes do not always serve us well (Kahneman 2011; Slovic 1987). When the consequences are important, we can consult the extensive set of tools designed to help us structure, understand, and effectively navigate decisions involving risk.

Tools for Addressing Risk Problems

Games of chance and decisions about investment feature aleatory uncertainty and have occupied the attention of philosophers, economists, and mathematicians for centuries. In the late 20th century, engineers developed and applied methods for probabilistic risk analysis in a wide variety of contexts, including nuclear engineering (Rasmussen 1981); other fields, like public health, have followed suit (Cox 2007). Thus, the set of tools for undertaking risk analysis and management is well developed and has been well explored in the conservation and natural resource management setting (Burgman 2005). The tools can be grouped into 4 categories: tools for displaying the structure of risk problems; tools for estimating uncertainty; tools for integrating estimates of uncertainty with value judgments about risk tolerance to recommend the best action; and tools for eliciting risk tolerance.

The first set of tools is used to structure risk problems. The structure of a risk decision has the same elements as any other decision (Runge and Bean, chapter 1, fig. 1.1), with the addition of sources of uncertainty. As a result of uncertainty, any action has multiple possible outcomes, and part of the structuring is simply to identify those possible outcomes. Often, it is helpful just to visualize the uncertainty and its potential effect on the decision. A *fault tree* (fig. 13.1) can be used to investigate the causal chain of events that might lead to an undesired outcome (Burgman 2005). Like influence diagrams, fault trees

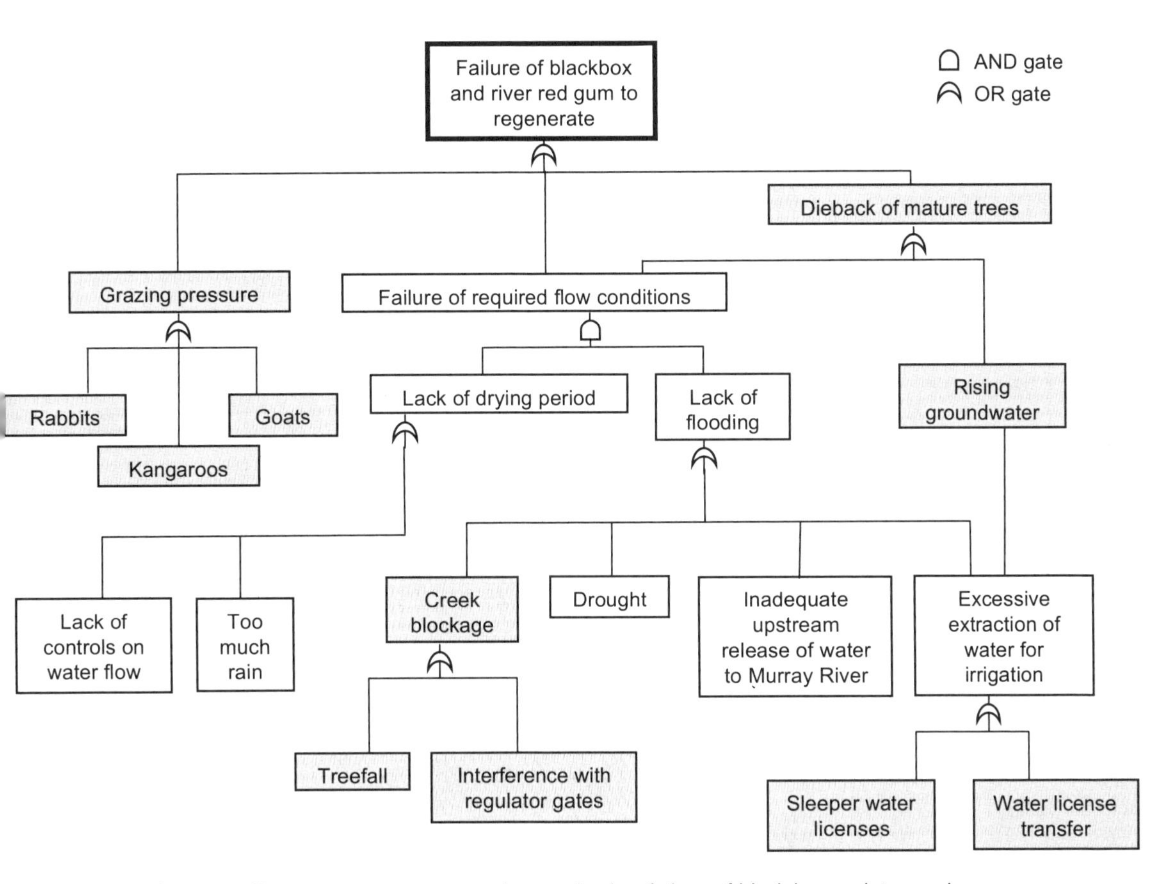

Figure 13.1. A fault tree illustrating various events that can lead to failure of black box and river red gum trees to regenerate. Each box represents a separate type of event, and the hierarchical structure shows how those events combine to lead to the top-level failure. "OR gates," such as that immediately below the top-level failure box, indicate that any 1 of the events below is sufficient to lead to the failure. "AND gates" indicate that all of the lower-level events must occur for the failure to occur. The shaded boxes were added in a second iteration, demonstrating the value of prototyping. *Example from Hattah-Kulkyne National Park, Victoria, Australia, Carey et al. 2005.*

can help motivate potential interventions, but they can also be used to identify the sources of uncertainty that need consideration. A *decision tree* (fig. 13.2) can be used to associate the potential outcomes with different management actions. Ultimately, quantitative methods can be used to "solve" a decision tree, identifying the choice of action that achieves the highest expected outcome in the face of uncertainty (Clemen 1996), but initially, the decision tree is a convenient way to see the structure of the decision.

The second set of tools is used to estimate uncertainty. As noted above, the primary scientific task in a risk analysis is to articulate the possible outcomes and estimate their likelihood. When there are data to support the development of predictive models, the estimation of the probability of various outcomes can occur through a variety of modeling methods. *Statistical models* are used to estimate the parameters of a model, and to characterize their uncertainty, based on empirical data. Forecasting models, like *Bayesian belief nets* (Marcot et al. 2006) or other simulation methods, can then propagate the uncertainty in the parameters through to uncertainty about the outcomes associated with particular actions. *Integrated population models* are now routinely used to combine these 2 steps for predicting population outcomes (Kéry and Schaub 2011). *Expert judgment methods* can be used to elicit and synthesize the knowledge

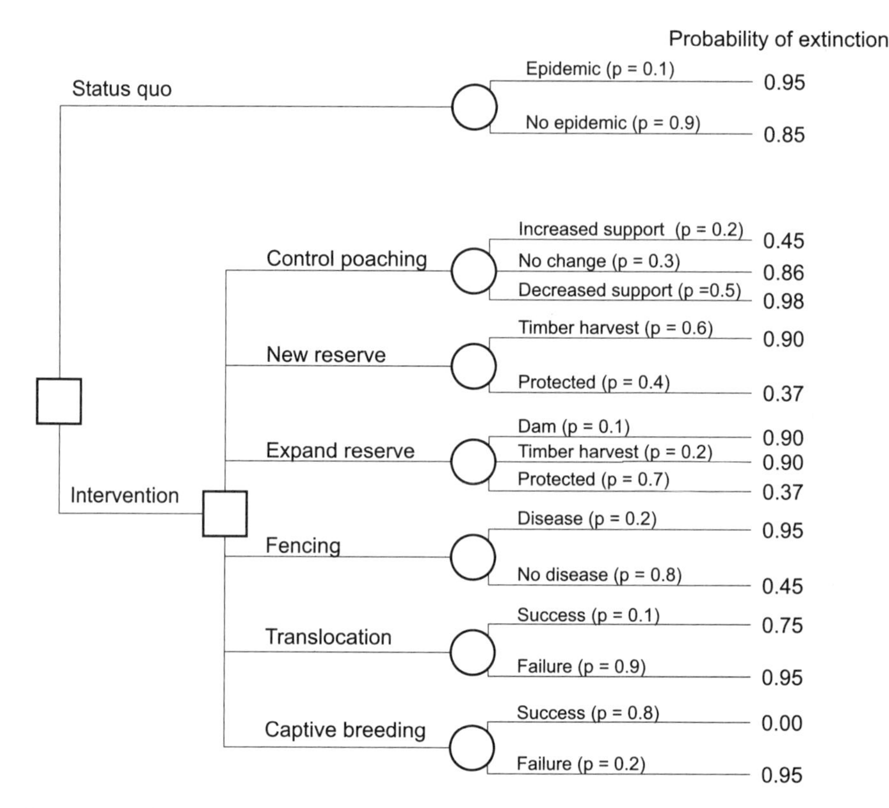

Figure 13.2. Decision tree for management of the Sumatran rhinoceros (Maguire et al. 1987). Squares represent decision nodes: the first decision is whether to have any intervention at all. If there is to be an intervention, the second decision indicates 6 different approaches. Circles represent uncertain events, with each branch indicating a possible outcome of that event. For example, following the status quo decision there is uncertainty about whether there will be an epidemic, and an expert has estimated that probability at 0.1. The value at the end of each decision-event combination is the outcome of interest; here, the probability of extinction over the ensuing 30 years.

of experts and to defensibly make predictions of the outcomes associated with alternative actions, while carefully accounting for uncertainty (Hanea et al. 2017; Speirs-Bridge et al. 2010). The point of both empirical and expert-based methods is to acknowledge, account for, and estimate the uncertainty in the predicted outcomes associated with each management action.

To make a decision in the face of uncertainty, a decision maker needs more than just an analysis of the likelihood of various outcomes associated with each alternative action; they also need to understand their own tolerance for risk, and possibly also the tolerances for risk of stakeholders. The third set of tools consists of approaches for making choices given risk, and these tools must capture different value judgments with regard to the trade-offs among outcomes. A common technique is to use the *expected value* (the weighted average) of the outcomes; a decision maker who chooses this approach is described as risk neutral, because a gain of, say, 10 units on the scale of the performance metric has the same absolute value as a loss of 10 units. A wide variety of other techniques can be used to reflect the values of risk-averse or risk-seeking decision makers. A risk-averse decision maker cares a great deal about guarding against bad outcomes. In that case, a *maximin criterion* might be used, in which they select the action with the best worst outcome across sources of uncertainty (a risk-seeking decision maker could use a maximax criterion, in which they would select the best best outcome.) *Robust satisficing* is an approach where a decision maker defines a minimum level of performance they consider to be satisfactory and then chooses the option that ensures at least that level of performance over the greatest range of uncertainty (Ben-Haim 2006).

The most flexible and well-studied approach for managing risk is *expected utility theory* (von Neumann and Morgenstern 1944); this approach develops a utility function that maps outcomes onto a utility scale that reflects the decision maker's or stakeholder's risk tolerance (see also Converse, chapter 5, this volume). As an example, consider a fish hatchery manager who is choosing among some actions that may affect the production of the facility. A risk-averse manager would guard against risk by assigning very low utility to decreases in production while being somewhat indifferent to increases in production (fig. 13.3). By contrast, a risk-seeking manager might be indifferent to decreases in production yet place a high utility on increases. To evaluate each alternative, the expected utility (the weighted average utility across uncertain outcomes) is calculated; the alternative with the highest expected utility best balances the risks as expressed by the decision maker's utility function.

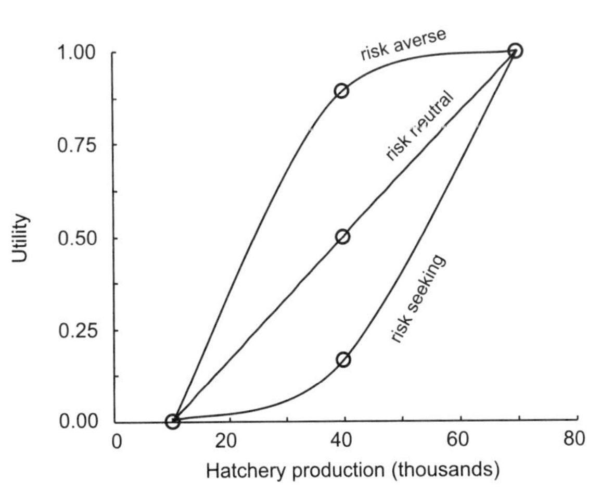

Figure 13.3. Hypothetical utility functions reflecting 3 different tolerances for risk associated with fish hatchery production.

Fourth, there are tools for eliciting a utility function from a decision maker or stakeholder, including *preference comparison* methods, *probability equivalence* methods, and *certainty equivalence* methods, among others (Farquhar 1984). All of these methods present a number of hypothetical gambles, expressed in terms of the performance metric for the decision in question, and ask the decision maker to choose a preferred action in each of the simplified cases. An analysis of the collection of responses is used to estimate the utility function. A well-constructed elicitation can focus in on the salient characteristics of the decision maker's or stakeholder's risk tolerance by using comparisons that are cognitively simpler to solve than the full decision at hand.

Case Studies

The next 3 chapters present case studies in which the risk posed by irreducible uncertainty was at least one of the impediments to a decision and showcase a number of the decision analytical tools available to help decision makers understand and manage risk.

In chapter 14, Sarah Sells and colleagues discuss their work with Montana Fish, Wildlife, and Parks, the state agency that manages wildlife in Montana, to develop management strategies for bighorn sheep (*Ovis canadensis*). At issue is the risk of epizootic pneumonia, a disease that can cause sharp declines in abundance and long periods of low reproduction. Management actions that might reduce the risk of disease have effects on other objectives, so in addition to considering the uncertainty in the consequences to bighorn sheep populations, the authors had to consider trade-offs among multiple objectives. Their approach involved embedding a decision tree in a multi-criteria decision analysis (see also Converse, chapter 5, this volume), developing a predictive model for the risk of disease, and creating a decision tool that allowed managers of individual herds to customize the analysis to their circumstances. The utility curves were represented with a power function, allowing the managers to explore different risk tolerances by varying a single parameter. Herd managers throughout the state are actively using the decision tool to manage bighorn sheep.

In chapter 15, Jean Cochrane and colleagues present a risk analysis concerning mitigation strategies for take of golden eagles (*Aquila chrysaetos*). Golden eagles are sometimes killed at wind energy generation facilities. To receive an incidental take permit from the US Fish and Wildlife Service, wind energy companies must undertake compensatory mitigation to offset the take. Novel mitigation methods pose a challenge, however, because their effectiveness is uncertain. The authors analyzed methods to abate lead poisoning in eagles as a compensatory mitigation strategy. They used formal methods of expert judgment to develop a probability distribution for the uncertain effect of this mitigation and certainty equivalence methods to elicit a utility curve to reflect the agency's risk tolerance. This work serves as a template for how other novel mitigation methods could be assessed in the future.

In chapter 16, Stefano Canessa and colleagues describe a decision analysis concerning supplementary feeding methods for the Mauritius olive white-eye (*Zosterops chloronothos*). For this critically endangered passerine, risk aversion had stymied efforts to test novel supplemental feeding methods, but as the population of birds increased, the status quo method was exhausting constrained resources. The authors used expert elicitation and empirical results from some limited field trials to develop estimates of the population growth rate under a number of alternative feeding methods. They then used several methods to account for the decision makers' risk tolerance, including a minimax criterion, stochastic dominance, and utility functions. They found that an alternative to the status quo significantly reduced costs without a very large risk of decreasing the population growth rate. The structured analysis allowed the managers to carefully consider their risk attitude and gave them confidence to implement this new feeding method.

Open Questions and Challenging Issues in Risk Analysis

With such a long history, the field of risk analysis has confronted many of its most persistent challenges, and abundant and diverse tools are available to decision makers faced with risk. In our experience, these tools are underutilized in natural resource management. The challenges in application arise more from a need for more awareness and training than from a need for more technical methods. Although decision makers in business, finance, and management are often exposed to the concepts of risk analysis in their training, decision makers in natural resource management typically are not. Training in the concepts of risk analysis may be valuable for current and future natural resource management decision makers.

Quantitative methods of risk analysis rely on probabilistic interpretations of uncertainty. At times, we have found it difficult to employ these methods when the ecologists and decision makers struggle with mathematical representations of probability (Hogarth 1975; Mosteller and Youtz 1990). Cognitive scientists have explored a variety of methods for conveying probabilistic understanding (Fischhoff et al. 1993), and it would be valuable to translate these insights into a wider set of tools for visualizing, evaluating, and communicating risk.

Unlike in finance or medicine, in many natural resource management settings, there are few data to develop empirical estimates of uncertainty, so we have to rely on expert judgment (Burgman 2015). The literature on expert judgment in conservation and natural resource management is expanding quickly (Martin et al. 2012), but training in and widespread acceptance of these methods could be improved.

LITERATURE CITED

Ben-Haim Y. 2006. *Info-Gap Decision Theory*. 2nd ed. Oxford, UK: Academic Press.

Burgman MA. 2005. *Risks and Decisions for Conservation and Environmental Management*. Cambridge, UK: Cambridge University Press.

———. 2015. *Trusting Judgements: How to Get the Best out of Experts*. Cambridge, UK: Cambridge University Press.

Carey JM, Burgman MA, Miller C, Chee YE. 2005. An application of qualitative risk assessment in park management. *Australasian Journal of Environmental Management* 12:6–15.

Clemen RT. 1996. *Making Hard Decisions: An Introduction to Decision Analysis*. Pacific Grove, CA: Duxbury Press.

Cox LA, Jr. 2007. Health risk analysis for risk-management decision-making. Pages 325–350 in Edwards W, Miles RFJ, von Winterfeldt D, eds. *Advances in Decision Analysis: From Foundations to Applications*. Cambridge, UK: Cambridge University Press.

Farquhar PH. 1984. Utility assessment methods. *Management Science* 30:1283–1300.

Fischhoff B, Bostrom A, Quadrel MJ. 1993. Risk perception and communication. *Annual Review of Public Health* 14:183–203.

Hanea A, McBride M, Burgman M, Wintle B, Fidler F, Flander L, Twardy C, Manning B, Mascaro S. 2017. I nvestigate D iscuss E stimate A ggregate for structured expert judgement. *International Journal of Forecasting* 33:267–279.

Hogarth RM. 1975. Cognitive processes and the assessment of subjective probability distributions. *Journal of the American Statistical Association* 70:271–289.

Kahneman D. 2011. *Thinking, Fast and Slow*. New York: Farrar, Straus and Giroux.

Kéry M, Schaub M. 2011. *Bayesian Population Analysis Using WinBUGS: A Hierarchical Perspective*. Waltham, MA: Academic Press.

Maguire LA, Seal US, Brussard PF. 1987. Managing critically endangered species: the Sumatran rhino as a case study. Pages 141–158 in Soulé ME, ed. *Viable Populations for Conservation*. Cambridge, UK: Cambridge University Press.

Marcot BG, Steventon JD, Sutherland GD, McCann RK. 2006. Guidelines for developing and updating Bayesian belief networks applied to ecological modeling and conservation. *Canadian Journal of Forest Research* 36:3063–3074.

Martin TG, Burgman MA, Fidler F, Kuhnert PM, Low-Choy S, McBride M, Mengersen K. 2012. Eliciting expert knowledge in conservation science. *Conservation Biology* 26:29–38.

Mosteller F, Youtz C. 1990. Quantifying probabilistic expressions. *Statistical Science* 5:2–12.

Rasmussen NC. 1981. The application of probabilistic risk assessment techniques to energy technologies. *Annual Review of Energy* 6:123–138.

Regan HM, Colyvan M, Burgman MA. 2002. A taxonomy and treatment of uncertainty for ecology and conservation biology. *Ecological Applications* 12:618–628.

Slovic P. 1987. Perception of risk. *Science* 236:280–285.

Speirs-Bridge A, Fidler F, McBride MF, Flander L, Cumming G, Burgman MA. 2010. Reducing overconfidence in the interval judgments of experts. *Risk Analysis* 30:512–523.

von Neumann J, Morgenstern O. 1944. *Theory of Games and Economic Behavior*. Princeton, NJ: Princeton University Press.

14 Addressing Disease Risk to Develop a Health Program for Bighorn Sheep in Montana

Sarah N. Sells,
Michael S. Mitchell,
and Justin A. Gude

Montana Fish, Wildlife and Parks (MFWP) is concerned that current management practices are responding inadequately to the increasing frequency and consequences of pneumonia die-offs (epizootics) in wild herds of bighorn sheep. Pneumonia epizootics threaten the persistence of herds and recovery of the species, causing rapid population declines and prolonged periods of poor recruitment. Generally, wildlife managers are poorly prepared to prevent disease outbreaks, relying instead on reactive "crisis management." MFWP sought to develop a more proactive, overarching strategy to assist herd managers in determining when it would be beneficial and cost-effective to conduct management to reduce the risk of an epizootic, before it takes place. The system needed to be broadly consistent across the state yet support flexibility for local decisions based on site-specific knowledge about conditions and risks. This case study illustrates how we developed a general framework that integrates a decision tree predicting site-specific probability of a pneumonia epizootic into a multi-criteria decision analysis to help guide local decisions about proactive disease management. We formalized this decision-making process in a spreadsheet model that has been made available to managers of bighorn herds in Montana. To illustrate an application of this decision tool, we also present the results of an analysis conducted by the manager of a herd at high risk of an epizootic.

Problem Background

Pneumonia is a critical challenge for managing herds of bighorn sheep (*Ovis canadensis*) and is considered a key factor limiting recovery of herds across much of their range (Cassirer et al. 2013; Plowright et al. 2013; Wehausen et al. 2011). Pneumonia die-offs (epizootics) are often characterized by high mortality that spreads quickly across all ages of a herd. Mortality rates typically range from 25 to 50% or more, including cases of 90–100% mortality (Sells et al. 2015). Records available for 43 of the 52 herds in Montana showed that between 1979 and 2013, a minimum of 22 epizootics of 25–100% mortality occurred in 18 of the herds (Sells et al. 2015). Pathogens responsible for the disease are associated with exposure to domestic sheep and goats (Besser et al. 2012a, 2012b, 2013; Miller et al. 2011) and may become endemic in a herd of bighorn sheep following exposure (Plowright et al. 2013). Although surviving adults may acquire a degree of immunity, lambs often experience low survival due to an apparent waning of immunity shortly after weaning. Coupled with the increased threat of other stochastic events in small populations, the long-term implication may be an inability to recover or even outright extirpation.

Decision Maker and Their Authority

Montana Fish, Wildlife and Parks (MFWP) manages most bighorn herds in the state. Many decisions for how to manage each herd are made at the local level, by the area biologist and manager directly responsible for that herd working in concert with private and federal landowners and local stakeholders. Decision authority for major management actions (e.g., translocations, harvest quotas) rests with the Montana Fish and Wildlife Commission. A consistent statewide approach is needed for deciding how to manage risk of epizootics in each herd. Flexibility in decision making, however, is also crucial for implementing unique, herd-specific decisions based on the biological uniqueness, decision timing, and management contexts of each herd.

Historically, management of pneumonia has been largely reactive after an epizootic, such as culling or augmenting herds by translocating sheep. A reactive approach, however, is often costly and can be ineffective. For example, after an epizootic in southwest Montana, managers augmented what remained of the herd with nearly 100 animals, yet it had still failed to recover 3 decades later (MFWP 2010). As an alternative to reactive management, a proactive approach trying to prevent epizootics from occurring in the first place may be less costly and more effective (Mitchell et al. 2013; Sells et al. 2016). For example, proactive management could include preventing exposure to or spread of pathogens among herds. It could also focus on actions that may reduce pathogen transmission within a herd, such as reducing the herd's density.

Risk of Epizootics

A key impediment to proactive management of pneumonia in bighorn sheep is understanding risk of future epizootics and how different factors might influence that risk. Epizootics are probabilistic events, with 2 possible outcomes given any management decision: an epizootic will occur within a certain timeframe, or it will not. To help predict the probability of future epizootics, Sells et al. (2015) developed an empirical model that identified 4 risk factors based on historical data. The first 3 risk factors were based on conditions existing within a herd's area of high risk (the herd distribution plus a 14.5-km buffer): (1) percentage of private land (private land), (2) use of domestic sheep or goats to control weeds (weed control), and (3) whether a herd or a neighboring herd had a pneumonia epizootic since 1979 (neighbor risk). These factors were expected to represent risk of pathogen exposure from domestic sheep or goats on hobby or commercial farms, domestic sheep or goats used for weed control operations, and infected wild bighorn sheep, respectively. The final risk factor was herd density (density, defined as low, medium, or high relative to each herd's historical densities from 1979 to 2013). Density was expected to represent risk of increased pathogen transmission rates within relatively dense herds.

Decision Structure

Managing pneumonia in bighorn sheep presents a risk problem compounded by multiple objectives. Generally, multi-objective problems can be compartmentalized by identifying fundamental objectives, developing alternatives, estimating how well each alternative will meet each objective to identify necessary trade-offs, and determining how each alternative provides overall utility at meeting the full set of objectives. In this case, however, another challenge is uncertainty about when and where epizootics will occur. Each alternative is likely to have a different effect on the probability of each outcome (epizootic or no epizootic), and each outcome may have a different effect on the ability to meet objectives. For example, an epizootic could necessitate a costly crisis management response; avoiding an epizootic eliminates this cost. Similarly, whereas an epizootic may reduce hunting opportunity for years, avoiding an epizootic would likely maintain hunting opportunity. Selection of an optimal alternative requires knowing the probability that an epizootic will occur, because the accuracy of estimated consequences depends on accounting for this risk. As such, this risk must be incorporated explicitly into the decision analysis.

To incorporate the risk of a pneumonia epizootic into the decision analysis, we used a decision tree. This entailed calculating the probability of (1) an epizootic and (2) no epizootic occurring for each alternative, along with predicting the consequences of each alternative for each objective under each scenario. We then integrated the probabilities and predicted consequences to estimate each alternative's expected value for each objective. As a result, our decision analysis accounts for the effect of uncertainty in when and where epizootics will occur.

Decision Analysis

Our approach represents efforts of 2 working groups, as presented previously (Mitchell et al. 2013; Sells 2014; Sells et al. 2016). An original working group of biologists and managers from MFWP and the Montana Cooperative Wildlife Research Unit (MTCWRU) met in 2010 to formalize a problem statement and fundamental objectives and to develop the prototype of a risk model based on expert opinion (Mitchell et al. 2013). Following these efforts, MTCWRU and MFWP collaborated to develop the aforementioned empirical risk model (Sells et al. 2015) and to initiate work on a decision tool based on structured decision making (SDM) that would be useful for herds statewide. In 2014, we held a workshop with a second working group of representative biologists and managers from MFWP who are the decision makers for individual herds of bighorn sheep (Sells et al. 2016). The group prepared groundwork for the decision tool by reviewing and expanding on the original working group's decision analysis steps and evaluating decisions for representative herds. This allowed us to ascertain whether each step would be comprehensible and efficacious for future decision makers once we formalized the decision tool. The decision tool now in use for herds statewide is based on the following components.

Objectives

The working group identified 6 fundamental objectives and measurable attributes for managing bighorn sheep (Mitchell et al. 2013; Sells et al. 2016):

1. Maximize probability of herd persistence (measured in terms of probability of pneumonia, modified by the decision maker's risk attitude towards this probability).
2. Minimize operating costs (i.e., cost in US dollars of activities associated with managing bighorn sheep).
3. Minimize personnel costs (i.e., cost in person-days associated with managing bighorn sheep).
4. Minimize crisis response costs (i.e., cost in US dollars of responding to an epizootic).
5. Maximize public satisfaction in terms of viewing opportunity (measured as relatively low, medium, or high viewing opportunity compared to current viewing opportunity).
6. Maximize public satisfaction in terms of hunting opportunity (measured as number of hunting licenses).

Objectives 1, 5, and 6 are interconnected (i.e., there cannot be viewing or hunting opportunity without herd persistence). The cost-related objectives (objectives 2–4) were separated because each cost is incurred from different budgets in MFWP. The importance of minimizing these costs may vary by herd, and each alternative or possible outcome may affect some costs but not others.

Dollars, person-days, and licenses are natural measures obtainable from MFWP records. The measurable attribute for herd persistence factors in risk attitude using a utility function (details below). The group chose a constructed scale for viewing opportunity because no relevant data are available on a natural scale.

Alternatives

Each herd in Montana has different risk levels of a pneumonia epizootic affected by different combinations of risk factors. Furthermore, management objectives and contexts (e.g., availability of funding) vary among herds, and the timing and duration of potential decisions must be flexible to accommodate the local situation. As such, alternatives need to be herd-specific to realistically address risk of pneumonia and management objectives for a herd.

To assist decision makers in developing herd-specific alternatives, the working group laid out possible actions to reduce risk of pneumonia (table 14.1). A decision maker can combine various sets of actions to develop competing alternatives.

Consequences and Trade-Offs

The working groups based the decision analysis on the simple multi-attribute rating technique (SMART; Edwards and Barron 1994; Goodwin and Wright 2004) combined with a decision tree for the probability of each potential outcome (epizootic or no epizootic). Decision makers first predict how actions in an alternative will affect risk factors in the risk model (fig. 14.1). Probability of each outcome is then calculated for each alternative as Pr(Epizootic$_t$) and Pr(No epizootic$_t$), where t is the number of years over which the prediction is made. Decision makers next use knowledge of the herd to predict the consequence of each alternative on each objective over t years.

Next, the decision maker accounts for risk attitude toward the likelihood of an epizootic. A utility function calculates Utility$_{\text{Risk}}$ for each alternative based on risk attitude and Pr(Epizootic$_t$):

$$\text{Utility}_{\text{Risk}} = 1 - \text{Pr}(\text{Epizootic}_t)^r$$

where r represents risk attitude and ranges 0.25–4.00 (fig. 14.2; Sells et al. 2016). If the decision maker

Table 14.1. Example management actions that could address risk factors for pneumonia epizootics based on expert opinion of MFWP biologists and managers

	Alternatives to address risk factors			
	Private land	Weed control	Neighbor risk	Density
Least aggressive actions	Do nothing	Do nothing	Do nothing	Do nothing
↓	Public education grazing systems / livestock replacement	Public education	Manage for young ram season	Harvest ewes and young rams, ranging from a large number of licenses to unlimited
↓	Conservation easements/ fee title purchases	Remove wandering bighorn	Create lethal removal zones around herd	Address range health by expanding or improving habitat
↓	Covenants/zoning	Create standards for fencing and herders	Cull herd	Trap and transplant to relocate away from herd
↓	Remove or haze wandering bighorn	Change timing of grazing using domestic sheep and goats		Trap and transplant to relocate within herd to expand range
Most aggressive actions	Remove wandering domestics	Replace domestic sheep and goats with biocontrols or herbicides		

Sources: Sells et al. 2015, 2016.

Note: Actions are ordered from least to most aggressive. Decision makers can combine any set of these or other actions to create an alternative they would consider for managing the herd; e.g., a "public education and increased harvest" alternative could include actions from the second row.

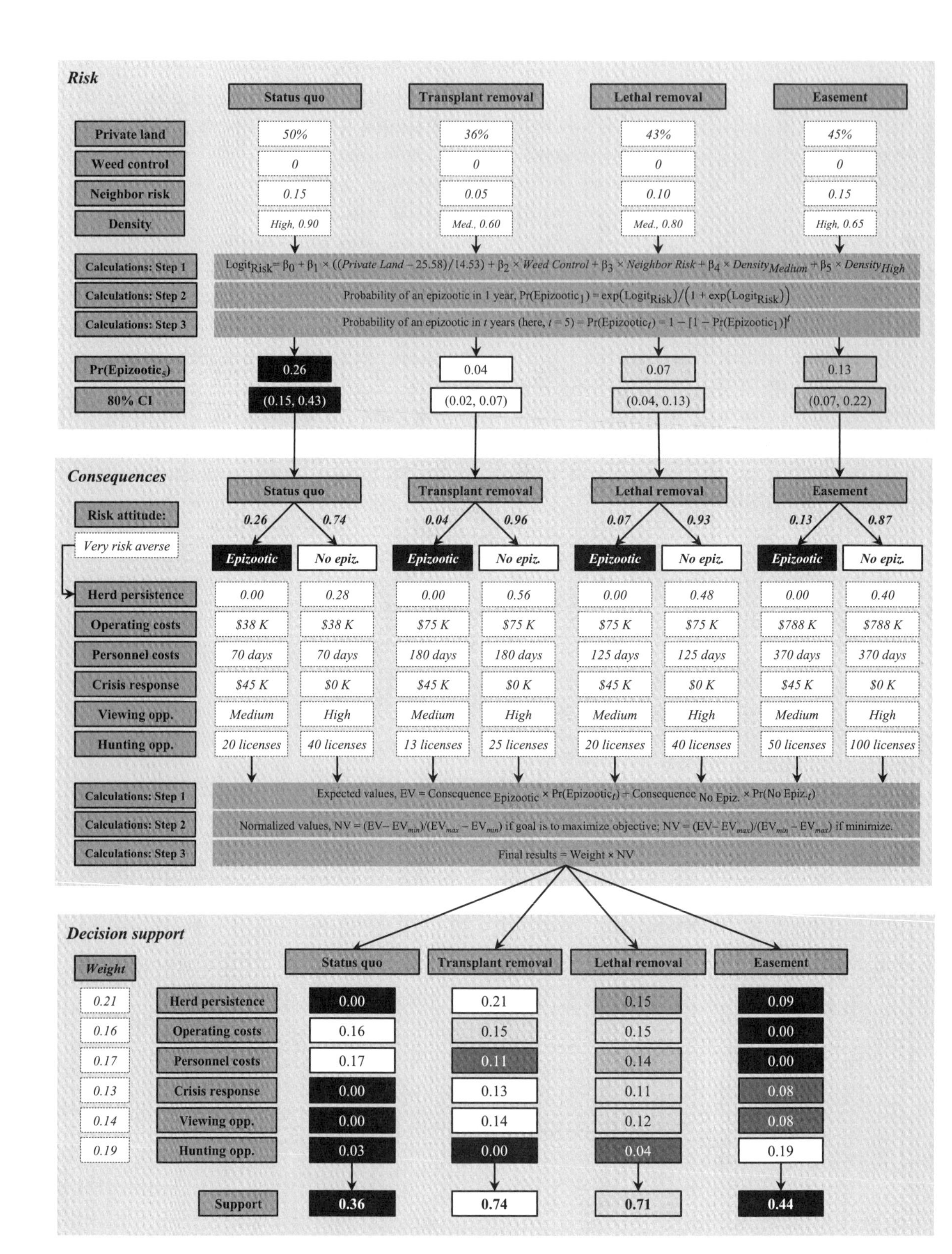
Risk
Status quo
Transplant removal
Lethal removal
Easement
Private land
Weed control
Neighbor risk
Density
50%
36%
43%
45%
0
0
0
0
0.15
0.05
0.10
0.15
High, 0.90
Med., 0.60
Med., 0.80
High, 0.65
Calculations: Step 1
Logit_Risk = β0 + β1 × ((Private Land – 25.58)/14.53) + β2 × Weed Control + β3 × Neighbor Risk + β4 × Density_Medium + β5 × Density_High
Calculations: Step 2
Probability of an epizootic in 1 year, Pr(Epizootic_1) = exp(Logit_Risk)/(1 + exp(Logit_Risk))
Calculations: Step 3
Probability of an epizootic in t years (here, t = 5) = Pr(Epizootic_t) = 1 – [1 – Pr(Epizootic_1)]^t
Pr(Epizootic_5)
0.26
0.04
0.07
0.13
80% CI
(0.15, 0.43)
(0.02, 0.07)
(0.04, 0.13)
(0.07, 0.22)
Consequences
Status quo
Transplant removal
Lethal removal
Easement
Risk attitude:
Very risk averse
0.26
0.74
0.04
0.96
0.07
0.93
0.13
0.87
Epizootic
No epiz.
Herd persistence
Operating costs
Personnel costs
Crisis response
Viewing opp.
Hunting opp.
0.00
0.28
0.00
0.56
0.00
0.48
0.00
0.40
$38 K
$38 K
$75 K
$75 K
$75 K
$75 K
$788 K
$788 K
70 days
70 days
180 days
180 days
125 days
125 days
370 days
370 days
$45 K
$0 K
$45 K
$0 K
$45 K
$0 K
$45 K
$0 K
Medium
High
Medium
High
Medium
High
Medium
High
20 licenses
40 licenses
13 licenses
25 licenses
20 licenses
40 licenses
50 licenses
100 licenses
Calculations: Step 1
Expected values, EV = Consequence Epizootic × Pr(Epizootic_t) + Consequence No Epiz. × Pr(No Epiz._t)
Calculations: Step 2
Normalized values, NV = (EV – EV_min)/(EV_max – EV_min) if goal is to maximize objective; NV = (EV – EV_max)/(EV_min – EV_max) if minimize.
Calculations: Step 3
Final results = Weight × NV
Decision support
Weight
Status quo
Transplant removal
Lethal removal
Easement
0.21
Herd persistence
0.00
0.21
0.15
0.09
0.16
Operating costs
0.16
0.15
0.15
0.00
0.17
Personnel costs
0.17
0.11
0.14
0.00
0.13
Crisis response
0.00
0.13
0.11
0.08
0.14
Viewing opp.
0.00
0.14
0.12
0.08
0.19
Hunting opp.
0.03
0.00
0.04
0.19
Support
0.36
0.74
0.71
0.44

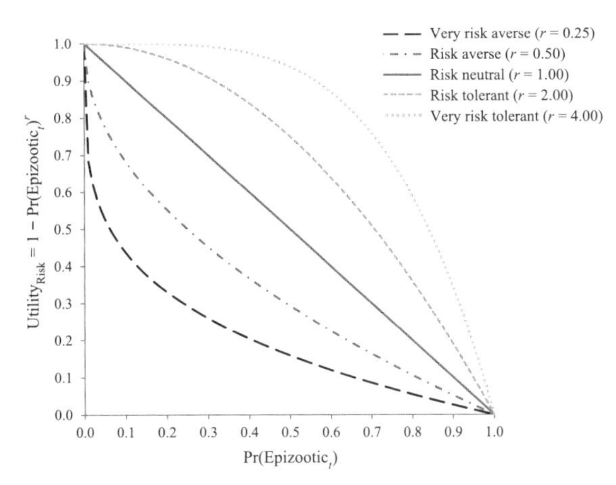

Figure 14.2. The decision maker defines risk attitude and calculates $\text{Utility}_{\text{Risk}}$ for alternatives based on which curve most closely reflects how he or she values Pr(Epizootic$_t$) as it ranges from low to high (Sells et al. 2016). A very risk-averse decision maker will be satisfied only with alternatives at very low Pr(Epizootic$_t$); a very risk-tolerant decision maker will be more equally satisfied with alternatives that range from low to relatively higher Pr(Epizootic$_t$).

is very risk averse ($r = 0.25$), $\text{Utility}_{\text{Risk}}$ is only high at very low Pr(Epizootic$_t$), whereas if they are very risk tolerant ($r = 4.00$) $\text{Utility}_{\text{Risk}}$ is high even at higher Pr(Epizootic$_t$). The decision maker defines risk attitude based on which curve most closely reflects how he or she values Pr(Epizootic$_t$) as it ranges from low to high. $\text{Utility}_{\text{Risk}}$ for each alternative's Pr(Epizootic$_t$) is entered as the consequence for the objective related to herd persistence.

Finally, the decision maker accounts for the relative importance of objectives (w_i). Although the 6 fundamental objectives are universal for MFWP, the relative importance of meeting each objective is herd-specific, based on what the decision maker for the herd values most. The decision maker therefore estimates w_i for each objective (fig. 14.1).

Decision Solution

In the last step, consequences are transformed to make them directly comparable and to reveal trade-offs. First, each pair of consequences on the decision tree is transformed to expected value (EV) to account for uncertainty (Behn and Vaupel 1982):

$$\text{EV} = \text{Consequence}_{\text{Epizootic}} \times \Pr(\text{Epizootic}_t) + \text{Consequence}_{\text{No epizootic}} \times \Pr(\text{No epizootic}_t)$$

EVs for each objective are then normalized and multiplied by the objective's w_i to calculate scores (fig. 14.1). The scores make trade-offs among alternatives explicit and allow decision makers to identify a decision solution. Within an objective, alternatives with higher scores do better at fulfilling the objective. The sum of each alternative's scores is its overall predicted performance toward meeting the full set of objectives, given the importance of each objective.

Sensitivity of Decision Analysis Results

After evaluating a herd, sensitivity analyses allow the decision maker to identify whether decision support is sensitive to any single input in the decision analysis. Different components of the analysis can be varied over the range of possible inputs to determine if decision support changes. For example, lower and upper credible intervals (similar to confidence intervals; e.g., fig. 14.1) quantify uncertainty in predictions

Figure 14.1. Decision analysis for the Petty Creek herd. The decision maker entered predictions (dashed boxes) for risk (i.e., effect of each alternative on each risk factor), consequences (for objectives if an epizootic occurs or not), risk attitude, and weights (relative importance for meeting objectives). These predictions are used to calculate Pr(Epizootic) and associated credible intervals (CI; Sells et al. 2015, 2016), expected values, normalized values, and final results. Shading of final results indicates the alternative performing best (white) or worst (black) on each objective. Overall support was highest for transplant removal and lethal removal. (Expected values are calculated for each pair of consequences except for herd persistence, which is based on a utility function incorporating Pr(Epizootic). Normalized values are calculated using an objective's set of expected values.)

from the risk model and can be used within the decision analysis in place of the mean estimates, in turn, to determine sensitivity to uncertainty in predicted risk. The decision maker can also consider uncertainty in predicted effects of actions on risk factors, adjust these inputs to re-predict risk, and determine if decision support changes. Each alternate risk attitude can also be used within the decision analysis to determine sensitivity to risk attitude. Additionally, weights for each objective can be varied, in turn, over the range of potential weights (0–1) to identify the weight at which decision support would switch to a different alternative.

In each of these sensitivity analyses, if decision support changes substantially, the decision maker can further consider his or her selection for that component. If uncertain about the selected input, the decision maker could use an average of the various inputs, use the input of another expert familiar with the herd, or average the input of multiple experts.

Another means for addressing sensitivity is a value-of-information analysis. This technique measures the importance of uncertainty to the decision maker (see Smith, chapter 17, this volume).

Application

To illustrate the application of a decision analysis, the manager for a herd west of Missoula, Montana, identified the optimal management approach to minimizing risk of pneumonia epizootics (Sells et al. 2016). The herd, Petty Creek, numbered at least 125 animals in 2014 and was not known to have experienced an epizootic. However, numerous epizootics had occurred among nearby herds in the preceding 5 years.

To carry out the decision analysis for Petty Creek, the herd manager first developed 4 alternatives:

1. *Status quo:* focused on public education about disease risk from domestic sheep and goats to bighorn sheep. Additional actions included removing wandering domestic sheep and goats, aerial surveys of the population, and harvest management to achieve density goals.
2. *Transplant removal:* focused on reducing density by removing bighorn sheep for transplant elsewhere. Additional actions included those from the status quo alternative, along with removing or hazing wandering bighorn sheep.
3. *Lethal removal:* focused on maintaining separation between bighorn sheep and domestic sheep and goats through lethal removal zones for bighorn sheep wandering outside of the regular herd distribution, plus status quo actions.
4. *Easement:* focused on conservation easements and fee title purchases to reduce risk from domestic sheep or goats on hobby farms, improving range health, and status quo actions.

The herd manager specified a 5-year timeframe for implementing the selected alternative, after which the herd would be re-evaluated to determine an alternative to implement next.

The herd manager next predicted consequences for Petty Creek (fig. 14.1). The manager used expert opinion about the herd and the risk model to predict risk of pneumonia for each alternative; risk of at least 1 epizootic over the next 5 years ranged from 0.26 at the status quo to 0.04 for the transplant removal alternative. Next, the manager predicted consequences for each objective if each alternative were implemented. The manager also determined that she was very risk averse when managing this herd, given the large number of recent epizootics nearby. Finally, the manager determined weights for each objective. These ranged from 0.21 for maximize persistence (most important) to 0.13 for minimize crisis response costs (least important; fig. 14.1). Summed weights for biological and social objectives totaled 0.54 versus 0.46 for cost-related objectives, which the manager deemed a realistic weighting given public interest in this herd and public preference for minimizing costs of wildlife management.

The alternatives with most support for Petty Creek were transplant removal (0.74 overall support) and lethal removal (0.71; fig. 14.1). Easement and status quo scored relatively low (0.44 and 0.36, respectively). The herd manager thus focused on trade-offs among transplant removal and lethal removal. Compared to other alternatives, transplant removal did best on half of the objectives (persistence, crisis response costs, and viewing opportunity), with trade-offs of lower performance on hunting opportunity and personnel costs. Lethal removal arguably provided a better balance among all objectives by scoring neither best nor worst on any objectives. It had scores somewhat higher than transplant removal on hunting opportunity and personnel costs, with trade-offs of slightly lower scores on the 3 objectives on which transplant removal did best. Based on these results, the manager found strong support for implementing either the transplant removal or lethal removal alternative.

After completing the decision analysis, the herd manager determined sensitivity of the decision support to the various model inputs. Decision support was unaffected by uncertainty in predicted risk of pneumonia, indicating this uncertainty did not solely influence the decision. Decision support also was not sensitive to risk attitude. Scores and overall support for both top-scoring alternatives remained nearly identical at any risk attitude, with lethal removal gaining slightly higher overall support than transplant removal at more risk-tolerant attitudes; both alternatives remained optimal. The only other appreciable change was a slight increase in overall support for easements under a very risk-tolerant attitude (0.56 versus 0.44 at very risk averse, still too low to be considered). Decision support was not sensitive to weights on objectives, either. The only instance where decision support changed was if weight on hunting opportunity reached 0.5, in which case the easements alternative became the decision with most support. The manager determined that this weight on hunting opportunity was not realistic. Based on these results, the manager was confident that decision support for transplant removal or lethal removal was not sensitive to any single input tested.

Decision Implementation

Developing a health program for herds statewide required the decision analysis be accessible to all biologists and managers. As such, we developed a spreadsheet-based tool that negates the need for expertise in SDM. The decision analysis only requires the decision maker to design alternatives and predict consequences, and the tool provides instructions for these steps. The remaining steps of the analysis are fully automated (e.g., all calculations are embedded as formulas). Results are provided in tables and graphs, and sensitivity analyses run automatically. Any biologist or manager can therefore use the tool to analyze decisions regarding how to manage a herd at any time, in any place, and for any duration of management cycle.

After developing the decision tool, we held workshops and disseminated the tool to all biologists and managers in Montana. Although use of our tool within MFWP is not mandatory, agency leadership has asked herd managers to explain how their recommended actions would affect disease risk, prompting use of the tool by local biologists and managers. To date, use of the tool has provided justification for 3 translocations out of herds (to reduce local herd density), 1 within-herd translocation (to expand the herd range, thereby decreasing local herd density), 1 conservation easement (to reduce risk from potential interactions with domestic sheep and goats on private lands), and 1 effort to depopulate a herd followed by translocations to start a new herd.

Discussion

Our decision tool provides an important advance for managing bighorn sheep by formally integrating risk into decision making for a complex, multi-objective issue. Previously, uncertainty in risk of future epizootics had been a primary impediment to making decisions for how to proactively manage herds of bighorn sheep in Montana. Although when or where an epizootic will occur in the future can never be known with certainty, a risk model can help predict the probability of such an event. As the first model available to predict the probability of pneumonia epizootics based on historical data, the Sells et al. (2015) risk model provides a key advance for decisions related to managing the risk of pneumonia in bighorn sheep. The model has strong empirical support and is reliable for making decisions (Sells et al. 2015). Integrating the risk model with the decision analysis via a decision tree–based approach makes use of the only available science that is tailor-made to this decision context and provides a rigorous method to formally deal with risk and incorporate both science- and value-based judgments. The model provides a formal method to modify the predicted risk of pneumonia by implementing new actions or alternatives, and the decision tree approach can help decision makers account for risk attitude toward the probability of undesirable outcomes. Results from each decision analysis for any herd in Montana therefore more accurately capture the ability of alternatives to meet objectives, given the probability of a pneumonia epizootic occurring.

In general, a decision tree–based approach can be useful for addressing complex, multi-objective problems compounded by uncertainty in the potential outcomes. Such an approach is important if alternatives are likely to have different effects on the probabilities of potential outcomes. Some alternatives may be relatively riskier or entail higher uncertainty. As a result, the ability of different alternatives to meet objectives must be estimated in a probabilistic sense. This provides more realistic estimates of consequences, accounting for uncertainty, which is important for identifying an optimal decision. Were these considerations ignored, a different alternative may be identified as the best course of action. Such a solution, however, is unlikely to capture the true range of possible consequences of alternatives and thus is less likely to be truly optimal.

Decision makers could incorporate a decision tree–based approach for problems of this type through various means. Probability of potential outcomes could be estimated by using an existing empirical model, developing a new empirical model, or using expert opinion. For example, an empirical model for risk of pneumonia was not yet available to the first working group in 2010. To capture uncertainty, that working group instead developed an expert opinion–based model to estimate the risk of pneumonia for each alternative (Mitchell et al. 2013). This model captured risk factors the group felt were most likely to be important for pneumonia in bighorn sheep. The model thus provided transparent predictions of risk and a means of factoring expert opinion and uncertainty into the decision analysis. This expert opinion–based model also provided numerous a priori hypotheses for potential risk factors for pneumonia and demonstrated the utility of a risk model, justifying the later development of an empirical model (Sells et al. 2015). The usefulness of a formal, empirical risk model for pneumonia was thus identified through this initial expert opinion–based model.

Incorporating risk attitude can also be important for addressing problems compounded by uncertainty in the potential outcomes. Risk does not mean the same thing to everyone, and therefore the absolute probability of an undesirable event (e.g., a pneumonia epizootic) does not evoke the same response in all decision makers. As such, the notion of risk in making decisions is inherently value-based. Our decision tool allows for explicit incorporation of this value judgment via a utility function. This has been useful to decision makers in MFWP who can better account for their attitudes toward risk of pneumonia epizootics in different herds. For example, a decision maker was relatively risk tolerant toward a herd with a recent epizootic because the decision maker felt that little could be done to help prevent recurring pneumonia outbreaks in the immediate future (Sells et al. 2016). Meanwhile, a decision maker may be highly risk averse regarding one of the few healthy herds in a wide geographical area (e.g., Petty Creek). Using the utility function, decision makers capture these risk attitudes toward herd persistence.

Incorporating risk attitude may also be valuable when risk attitude by subordinate decision makers (e.g., herd managers) differs from that of ultimate decision makers (e.g., regional managers or the Montana Fish and Wildlife Commission). Decision makers can ascertain and demonstrate how different risk attitudes affect the results, which can be useful for communicating with other biologists, managers, stakeholders, or politicians. For example, herd managers can evaluate potential risk, optimal management approaches, and the influence of risk attitude on decisions to transplant animals to existing herds or new areas. In several recent decisions, the commission has differed in risk attitude from herd managers, who felt it was optimal to transplant bighorn sheep from certain herds (to reduce the disease risk) to herds out of state. The herd managers were unable to find transplant sites with low enough risk to move the animals into within Montana, and they were averse to the risk of starting a new herd that might die off or cause a die-off in a nearby herd. The commission decided not to send animals out of state, however, wanting the agency instead to take the chance in establishing new herds in Montana. The commission reasoned that existing herds and the overall number of bighorn sheep are already stagnant at best, so that even if a new herd dies off, little is lost. The commission was more risk tolerant than MFWP biologists in making decisions in these cases.

Complex problems involving risk and numerous management decisions could be improved by developing a linked decision analysis. Management decisions are not made in isolation, and in this case, each decision affects others. For example, resources are finite, so a decision to allocate time and funds to one herd may reduce resources available to another herd. Additionally, how a herd is managed may affect the risk of pneumonia and resulting probability of persistence for other herds as well as the longer-term risk of pneumonia epizootics for that herd. Thus, an optimal decision for one herd may not be optimal in the context of managing multiple herds or for managing a particular herd over the longer term. In using our decision tool for a single herd, managers implicitly but informally consider how actions taken on that herd will affect others. Our decision analysis would likely be improved, however, if these considerations were made explicit via a linked decision-making process.

A linked decision-making process could be constructed as a hierarchical decision analysis that optimizes decisions across multiple herds. Decision makers would first need to identify fundamental objectives for managing multiple herds. For example, objectives may potentially include maximizing the number of herds persisting in a metapopulation, maximizing efficiency in allocation of time and money, etc. The decision analysis would also require an optimization function to identify decisions across a set of herds while also considering optimal decisions at the herd level. This may require prioritizing herds (e.g., native versus reintroduced, or by hunting or viewing value, etc.). Such a prioritization process could be accomplished via weighting the importance of each herd. Implementation of a linked decision analysis within MFWP, however, could be challenging due to the decentralized nature of decision making in the agency, which is what prompted our original approach to herd-level decisions. Close collaboration among local herd managers would be needed to develop and cultivate shared ownership in a linked decision tool, and the tool would need to be developed in an enterprise fashion for use by multiple, spatially disparate decision makers. Though such a tool would likely be complicated to develop and use, it may provide a valuable approach for considering trade-offs between decisions at different scales.

ACKNOWLEDGMENTS

The authors thank J. B. Grand, D. Ohlson, and G. Long for their feedback on earlier versions of this chapter. They thank N. J. Anderson, V. L. Edwards, P. R. Krausman, J. J. Nowak, P. M. Lukacs, and J. M. Ramsey for their assistance and expertise. They thank the US Geological Survey, the US Fish and Wildlife Service, and the staff of the National Conservation Training Center for organizing and implementing the 2010 workshop. They thank the 2010 workshop participants, N. J. Anderson, J. F. Cochrane, V. L. Edwards, C. N. Gower, E. R. Irwin, J. M. Ramsey, M. G. Sullivan, M. J. Thompson, and T. Walshe, and the 2014 workshop participants, N. J. Anderson, H. R. Burt, J. A. Cunningham, S. A. Hemmer, B. N. Lonner, J. M. Ramsey, B. A. Sterling, S. T. Stewart, and M. J. Thompson. In developing the risk model, for their assistance and expertise, the authors thank the biologists and employees of Montana Fish, Wildlife and Parks, USFWS, Bureau of Land Management, US Forest Service, USGS, National Park Service, Confederated Salish and Kootenai Tribes, Chippewa Cree Tribe, Montana Conservation Science Institute, British Columbia Fish and Wildlife Branch, Idaho Fish and Game, and Wyoming Game and Fish. Funding was provided by the general sale of hunting and fishing licenses in Montana, the annual auction sale of bighorn sheep hunting licenses in Montana, matching Pittman-Robertson grants to MFWP, the Montana Cooperative Wildlife Research Unit, and the USGS Cooperative Research Units.

LITERATURE CITED

Behn RD, Vaupel JW. 1982. *Quick Analysis for Busy Decision Makers.* New York: Basic Books.

Besser TE, Cassirer EF, Highland MA, Wolff P, Justice-Allen A, Mansfield K, Davis MA, Foreyt W. 2013. Bighorn sheep pneumonia: sorting out the cause of a polymicrobial disease. *Preventative Veterinary Medicine* 108:85–93.

Besser TE, Cassirer EF, Yamada C, Potter KA, Herndon C, Foreyt WJ, Knowles DP, Srikumaran S. 2012a. Survival of bighorn sheep (*Ovis canadensis*) commingled with domestic sheep (*Ovis aries*) in the absence of *Mycoplasma ovipneumoniae*. *Journal of Wildlife Diseases* 48:168–172.

Besser TE, Highland MA, Baker K, Cassirer EF, Anderson NJ, Ramsey JM, Mansfield K, Bruning DL, Wolff P, Smith JB, Jenks JA. 2012b. Causes of pneumonia epizootics among bighorn sheep, western United States, 2008–2010. *Emerging Infectious Diseases* 18:406–414.

Cassirer EF, Plowright RK, Manlove KR, Cross PC, Dobson AP, Potter KA, Hudson PJ. 2013. Spatio-temporal dynamics of pneumonia in bighorn sheep. *Journal of Animal Ecology* 82:518–528.

Edwards W, Barron FH. 1994. SMARTS and SMARTER: improved simple methods for multi-attribute utility measurement. *Organizational Behavior and Human Decision Processes* 60:306–325.

Goodwin P, Wright G. 2004. *Decision Analysis for Management Judgement.* 3rd ed. West Sussex, UK: John Wiley and Sons.

Miller DS, Weiser GC, Aune K, Roeder B, Atkinson M, Anderson N, Roffe TJ, Keating KA, Chapman PL, Kimberling C, Rhyan J, Clarke PR. 2011. Shared bacterial and viral respiratory agents in bighorn sheep (*Ovis canadensis*), domestic sheep (*Ovis aries*), and goats (*Capra hircus*) in Montana. *Veterinary Medicine International.* www.hindawi.com/journals/vmi/2011/162520/.

Mitchell MS, Gude JA, Anderson NJ, Ramsey JM, Thompson MJ, Sullivan MG, Edwards VL, Gower CN, Cochrane JF, Irwin ER, Walshe T. 2013. Using structured decision making to manage disease risk for Montana wildlife. *Wildlife Society Bulletin* 37:107–114.

Montana Fish, Wildlife and Parks [MFWP]. 2010. *Montana Bighorn Sheep Conservation Strategy.* Helena, MT: Montana Fish, Wildlife and Parks. http://fwp.mt.gov/fwpDoc.html?id=39746.

Plowright RK, Manlove K, Cassirer EF, Cross PC, Besser TE, Hudson PJ. 2013. Use of exposure history to identify patterns of immunity to pneumonia in bighorn sheep (*Ovis canadensis*). *PLOS ONE* 8(4):e61919.

Sells SN. 2014. Proactive management of pneumonia epizootics in bighorn sheep in Montana. PhD thesis. Missoula, MT: University of Montana.

Sells SN, Mitchell MS, Edwards VL, Gude JA, Anderson NJ. 2016. Structured decision making for managing pneumonia epizootics in bighorn sheep. *Journal of Wildlife Management* 80:957–969.

Sells SN, Mitchell MS, Nowak JJ, Lukacs PM, Anderson NJ, Ramsey JM, Gude JA, Krausman PR. 2015. Modeling risk of pneumonia epizootics in bighorn sheep. *Journal of Wildlife Management* 79:195–210.

Wehausen JD, Kelley ST, Ramey RR II. 2011. Domestic sheep, bighorn sheep, and respiratory disease: a review of the experimental evidence. *California Fish and Game* 97:7–24.

15 Hedging against Uncertainty When Granting Permits for Mitigation

Jean Fitts Cochrane, Taber D. Allison, and Eric V. Lonsdorf

The US Fish and Wildlife Service (Service) must decide how much credit to grant for mitigation actions, such as methods designed to compensate for incidental taking of eagles by wind energy facilities or other activities requiring federal permits. When established mitigation approaches are insufficient or unavailable for avoiding or offsetting losses, conservation goals may still be achievable through experimental implementation of novel mitigation methods. The uncertainty in outcomes, or risks of not meeting conservation targets, must be analyzed thoroughly and addressed explicitly in the decision analysis. We used simulation modeling and a decision model with utility, a quantitative expression of the agency's risk tolerance, to demonstrate how the Service can evaluate a plan to voluntarily abate lead poisoning of golden eagles in central Wyoming. This example illustrates how to characterize and respond to uncertainty in a regulatory decision.

Problem Background

More than half of annual mortality in North American golden eagle (*Aquila chrysaetos*) populations is due to anthropogenic sources, especially poisoning, shooting, electrocution, and collisions (Millsap et al. 2016). The Bald and Golden Eagle Protection Act of 1940 (Eagle Act) and implementing regulations prohibit such "take" of golden and bald eagles (*Haliaeetus leucocephalus*), except when incidental to otherwise lawful activity and compatible with the preservation of eagles (USFWS 2016). The potential for eagles to collide with wind turbines has affected development of new wind power facilities in regions of the United States where high wind energy potential overlaps golden eagle range, prompting investigation into novel mitigation approaches.

Permits may be authorized for incidental taking under the Eagle Act, provided the take is residual after application of all reasonable avoidance and minimization measures and would not affect eagle population status (USFWS 2013, 2016). For golden eagles, the US Fish and Wildlife Service (Service) has determined that no permits will be issued that result in a net increase in mortality (USFWS 2016). Thus, any incidental take of golden eagles by wind facilities or other permitted activities, such as electric utility distribution, mining, road construction, and animal damage control operations, must be fully offset with compensatory mitigation that either decreases a pre-existing mortality factor or increases carrying capacity in the affected eagle management unit. The challenge is finding methods to achieve these rigorous mitigation requirements. The only method employed in permits to date, retrofitting power poles to eliminate electrocution of eagles (USFWS 2016), will be insufficient to offset eagle mortality as wind energy expands to meet growing demand for renewable energy.

Since 2011, the American Wind Wildlife Institute (AWWI), a partnership of wind industry and conservation organizations, has collaborated with the Service, eagle scientists, and other stakeholders to address the pressing need for alternative compensation methods. The overarching goals of the effort have been eagle conservation and mitigating climate change via renewable energy development. This case study describes work completed by AWWI in consultation with its partners and the Service on abating a well-documented source of anthropogenic eagle mortality: lead poisoning due to ingestion of spent game hunting ammunition (e.g., Haig et al. 2014). Abatement depends on hunters voluntarily using non-lead ammunition, which is widely available but more expensive than lead ammunition and not used commonly for big game hunting at present.

The Service specifies that novel mitigation methods may be approved if mortality reductions are predicted with a "credible, scientifically peer-reviewed model" (USFWS 2016). Cochrane et al. (2015) developed a lead abatement model for golden eagles, which we employed to estimate the effects of management actions (consequences) in this case study. The model uses empirical data and expert judgment inputs to simulate lead poisoning rates with and without mitigation. Because the estimates incorporate substantial scientific uncertainty about underlying biological processes and mitigation effects, the model produces a probability distribution of eagles saved by any mitigation proposal. Before the Service can grant a permit for incidental take compensated with lead abatement, managers must decide how much mitigation credit to award for these uncertain predictions, articulating how much they want to discount estimated reductions in eagle mortality (i.e., hedge against uncertainty). In turn, permit applicants need some assurance that sufficient mitigation credits would be generated to warrant their investment before proceeding with a mitigation program.

Our case study simulated a lead abatement program on private land that is managed as a conservation mitigation bank. We assumed hunter access would be controlled and all hunters would be provided with non-lead ammunition or a subsidy, thus all lead poisoning due to gut pile exposure would be eliminated within the reserve area. As sensitivity analysis, we also evaluated voluntary or partial adoption of non-lead ammunition with potential for more eagle credits over time on a much larger area of public land, but with implementation uncertainty. Voluntary abatement programs typically employ hunter education in addition to distributing free or reduced cost non-lead ammunition, encouraging big game hunters to use the new ammunition voluntarily. During lead abatement programs in parts of Arizona and Wyoming, 24–83% of big game hunters voluntarily switched to non-lead ammunition over time (e.g., Bedrosian et al. 2012; Epps 2014; Sieg et al. 2009).

We simulated lead impacts under conditions in the region surrounding Casper in Natrona County, Wyoming. We assumed the mitigation bank was on a reserve covering 2,000 square km with approximately 83 total eagles and potential availability of 6.46 big game gut piles per golden eagle based on big game harvest and eagle densities in the region (Nielson et al. 2014; Ryan Nielson, Western EcoSystems Technology, Inc., personal communication, 2014; Wyoming Game and Fish Department 2013).

Decision Maker and Their Authority

The director of the Service has authority to decide whether to issue incidental take permits under the Eagle Act (USFWS 2016). We demonstrate how Service managers, to whom this authority is delegated, could use risk or utility assessment to decide how many eagle offset credits will accrue from lead abatement actions, once they determine that a proposed mitigation program meets Eagle Act standards. Wind energy companies also make critical decisions in developing their mitigation proposals and permit applications, combining concern for economic trade-offs with clearing regulatory hurdles. This case study focuses on the Service's decision about risk and mitigation credit.

Ecological Background

Golden eagles frequently feed on carrion including big game gut piles left by hunters (Kochert et al. 2002; Legagneux et al. 2014; Watson 2010), which commonly contain fragmented lead ammunition (Hunt et al. 2006, 2009; Warner et al. 2014). As eagles ingest spent lead, the maximum concentration of lead in the blood increases until they suffer from lead poisoning and the likelihood of mortality increases (e.g., Cruz-Martinez et al. 2012). Lead toxicosis accounted for at least 2.7% of 97 dead radio-tagged golden eagles recovered between 1997 and 2013 in North America. Sublethal blood lead concentrations may contribute to mortality from other proximate sources (e.g., Hunt 2012; Millsap et al. 2016 and personal communication). Hunt et al. (2006, 2009) and Bedrosian et al. (2012) found that blood lead levels in eagles were highly correlated with the number of big game animals hunted with lead ammunition. On average, 90% of shot large game are retrieved and field-gutted by hunters (Fuller 1990, Nixon et al. 2001), leaving gut piles with lead fragments for eagles and other species to scavenge.

Decision Structure

This case study typifies decisions where the outcome of a management action is markedly uncertain. The solution requires 2 interlinked analyses: risk analysis describing and quantifying the potential mitigation results, and risk management deciding how to respond to the uncertainty when granting mitigation credits. When laws and policies do not provide specific direction, managers must decide how protective to be based upon their interpretation of legislative intent and the public's attitudes about risk. We illustrate how agency decision makers can build a decision model of risk management preferences as a utility function (Farquhar 1984) that assigns mitigation credits based on uncertain or probabilistic consequences.

Decision Analysis

Objectives

The Service's objective in permitting decisions under the Eagle Act is no net loss of golden eagles over the permit period, or over 5-year intervals for permits up to 30 years in length (USFWS 2016). We refined this objective statement as attaining a sufficiently high likelihood that permitted compensatory mitigation results in no net loss of eagles, given uncertainty about how much eagle mortality will decline with mitigation actions. Compensation sufficiency is ultimately a subjective or values-driven choice (Gardner et al. 2013). The Service employs a quantitative approach to defining take limits and the uncertainty in those limits because it is "explicit, allows less room for subjective interpretation, and can be consistently implemented throughout the country and across the types of activities that require permits" (USFWS 2016, 91497).

Alternatives

The alternatives in our case study are simply the amount of mitigation credit to permit for a program of lead abatement when the estimated reduction in eagle mortality is uncertain.

Consequences

We developed a simulation model to estimate how golden eagles ingest spent lead and die from that exposure, and thus how eagle mortality declines when hunters use non-lead ammunition. Before using probabilistic estimates from a simulation model in agency decisions, responsible managers must be thoroughly familiar with the model's design, assumptions, and calculations, and with the process for eliciting expert judgments used in modeling. Cochrane et al. (2015, including online supplements) provide full documentation for the lead abatement model. Here, we summarize the approach to illustrate key elements employed to deal transparently and rigorously with uncertainty, particularly structured elicitation of expert judgments.

The lead model estimates eagle deaths due only to a particular set of influential variables (fig. 15.1). It simulates stochastically how many scavenging golden eagles in a geographic region consume spent ammunition and accumulate lead in their blood and how many die from acute lead exposure during a month of big-game hunting season. The overarching relationship is that as the density of gut piles containing spent lead ammunition increases, the likelihood of eagle exposure and acute mortality increases proportionally, accounting for variation in eagle and gut pile densities.

Expert judgment served 2 critical and distinct roles in this realm of incomplete understanding and data gaps. The first role was provided by a group of experts who advised the authors on all aspects of model development from initial concepts to the final simulations. The second role for expert judgment was providing quantitative estimates for specific model parameters where data from empirical research were limited or absent. At various stages of the project we engaged with 16 carefully selected experts in (1) eagle behavior, ecology, and management, (2) raptor lead poisoning, (3) quantitative skills and modeling, (4) regulatory requirements and mitigation planning, and (5) field conditions in different regions of the Western United States.

We developed the model in stages beginning with a conceptual model at an initial workshop with experts. Subsequent stages were preliminary deterministic and stochastic models, structured elicitation of expert judgments for select parameter values, and prototype simulations and sensitivity analysis. Throughout we employed best practices of expert elicitation to help maximize critical thinking and information sharing, while minimizing biases such as anchoring and group think (e.g., Drescher et al. 2013; Martin et al. 2012; McBride and Burgman 2012). An iterative build-review-revise process was repeated over multiple months before we settled on all the functional relationships to include in a fully computational model. The model was implemented in Matlab (version R2010a, MathWorks) with stochastic sampling from the distributions we assigned to the input variables and their functional relationships (5,000 iterations for each scenario).

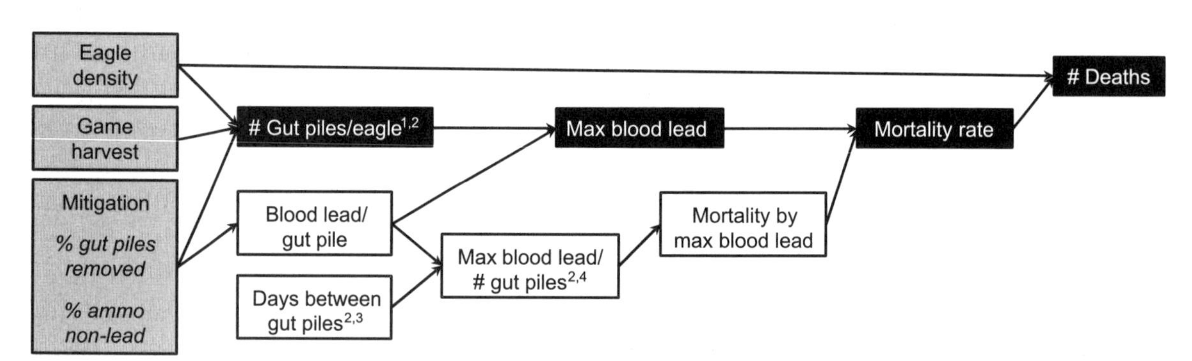

Figure 15.1. Diagram of the golden eagle lead abatement model illustrating the cause-to-effect relationships (directional arrows) between input and output parameters (boxes). The predictive variables or inputs (light shaded boxes) are set for each scenario modeled. The subsequent parameters or response variables result from the modeling steps; of these, the dark boxes are location-specific responses (dependent upon the gut piles available per eagle). Four additional model inputs are indicated by superscript numbers where they influence a response parameter: the percentage of shot animals that hunters retrieve and gut in the field (game recovery rate) (1), maximum number of gut piles scavenged per month (2), minimum days' lag between gut piles scavenged (3), and daily blood lead decay rate derived from the blood concentration half-life (4). The model output is a probability distribution of golden eagle deaths produced from repeated stochastic simulations.

For 3 parameters where we lacked empirical data, we elicited expert judgments more formally to develop functional distributions for the model's cause-effect relationships. A group of 4 experts who were highly experienced in golden eagle behavior provided estimates for eagle scavenging rate (the average expected number of gut piles scavenged per eagle in association with specific levels of eagle and gut pile densities). A separate group of 4 experts in lead poisoning in raptors provided estimates for 2 parameters addressing lead toxicity (blood lead level increase per scavenge, and mortality per maximum blood lead level). We used a modified Delphi approach for the elicitation (Runge et al. 2011) and followed Speirs-Bridge et al.'s (2010) 4-point method. The questioning sequence was minimum, maximum, and most likely estimates for a parameter, and confidence that the "true" answer was within the stated min-max bounds. In addition, we elicited a discrete probability distribution for the incremental increase in blood lead per gut pile scavenged. For probability estimates, we employed direct elicitation while encouraging the experts to think about probabilities as frequencies or proportions of events they could envision from their experiences (e.g., indirect elicitation, Burgman 2005). We developed distributions for the 3 elicited parameters by looking at the patterns among the elicited values and subjectively applying functional relationships that best matched plots of elicited values (Cochrane et al. 2015).

For the mitigation bank program, we used simulation results from Cochrane et al. (2015, Natrona County example) to estimate how many eagles may be saved by eliminating all lead poisoning in a 2,000-square-km area in central Wyoming. The probability distribution of 5,000 simulations (fig. 15.2, gray curve) illustrates both the wide range (0.4–18.1) and clustering of results near the median 2.0 eagles saved (80% credible interval 0.9–5.9). Only the top 10% of simulations estimated more than 6 eagles saved per year (fig. 15.3). For sensitivity analysis we increased the area affected to 10,000 square km and reduced the percentage of non-lead ammunition adoption to 10–30%, which resulted in somewhat increased range of 0–25 eagles saved per year.

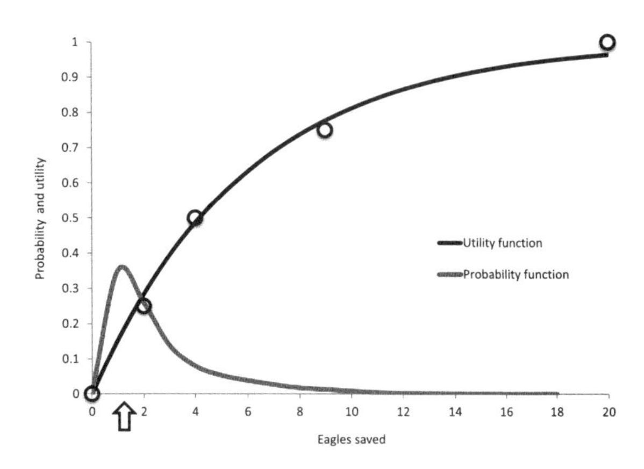

Figure 15.2. Estimated probability distribution function (gray line) and exponential utility function (black line) for the number of eagles saved in the mitigation bank example. The utility function was fit to the elicited utilities (circles) by setting the risk tolerance parameter, R, at 6 in the equation $U(x) = 1 - e^{-x/R}$ (Clemen and Reilly 2001), where x is the number of eagles saved by mitigation.

Decision Solution

The next step was using the probability distribution of eagles saved by the mitigation bank to decide how much credit to grant for the proposed mitigation action. A simple solution would have been to grant credit for the median prediction from the stochastic model runs: 2.0 eagle credits per year on average. This assumes, however, that decision makers could accept a 50% chance that fewer eagles would be saved from lead poisoning, down to 1 or less. The Service has rejected this "liberal" (risk neutral) approach for predicting eagle mortality and compensation (USFWS 2016). Instead, the Service prefers to err toward more conservative estimates, at least until more data are gathered through monitoring and research.

A rigorous decision model for managing risks considers the extent of uncertainty unique to the decision, or typical for a type of decision, before determining the appropriate level of risk tolerance (Burgman 2005). The Service has adopted a quantitative

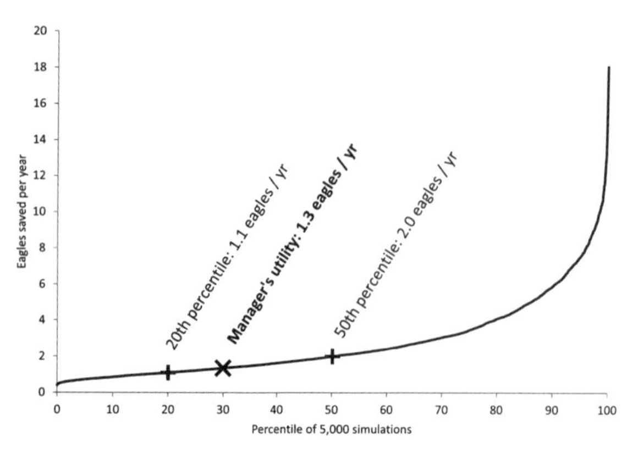

Figure 15.3. Distribution of eagles saved per year from 5,000 simulations of the mitigation bank program. Marks on the graph show the simulation equivalent to the elicited total utility or risk preference compared with other decision criteria mentioned in the text: the 20th percentile and 50th percentile or median simulation.

standard for predicting incidental eagle mortality that incorporates uncertainty into their models and uses the upper 80th quantile or upper credible limit as the level of take to regulate "in favor of being overprotective of eagles" (USFWS 2016, 91497; Millsap et al. 2016). A parallel risk standard for compensatory mitigation decisions would grant credits for the 20th percentile of the simulated credible interval of eagles saved. However, if an 80th-quantile take estimate and 20th-percentile mitigation credit were employed simultaneously in a regulatory decision, the combined effect could overrepresent the Service's risk aversion.

Given that the Service has not implemented a risk standard for mitigation crediting, we demonstrate how to elicit a risk tolerance or decision model for lead abatement based on the utility of a full probability distribution of simulated mitigation results. Risk attitude may be represented as a utility function that translates quantitative outcomes such as eagles saved to relative utility (scaled 0–1 for least to most satisfaction; Farquhar 1984). When the potential outcomes from mitigation are multiplied (weighted) by the utility scale, the resulting product approximates how much the decision makers value the outcome, given the uncertainty. For lead abatement, utility would represent the amount of credit (number of eagles saved) the agency is willing to grant for uncertain mitigation results.

Fortunately, well-known methods allowed us to elicit utility and model risk management choices. We used the lottery method with certainty equivalents (Clemen and Reilly 2001; Farquhar 1984) for the elicitation. Since our case study was not an actual decision, we conducted the elicitation with a non-agency team member who was familiar with the permitting process. In a series of questions, the subject was asked which of 2 hypothetical choices they preferred as the quantitative outcomes of a mitigation project. These choices were always between a hypothetical scenario with an uncertain outcome—0.5 probabilities each of a low and high number of eagles saved (called a reference lottery)—and a scenario with a certain outcome with an intermediate number of eagles saved. For example, the assessor asked: "would you prefer a 50:50 chance of saving either 0 or 10 eagles, or saving 5 eagles with certainty?" Once the subject responded, the question was repeated with a new offer for the certain value (higher certain value if the subject chose the gamble, lower certain value if they chose the certainty). The questions continued until the subject did not express a preference. The certain outcome where the subject was indifferent between that value and the lottery was their "certainty equivalent" to the uncertain results, which we inferred to be the utility the subject assigns to the gamble among uncertain results (Goodwin and Wright 2004). We assumed utility increases monotonically from the lowest to the greatest possible outcome, to which we assigned the 0 and 1 utility (relative preference dependent on the actual range on offer). The certain equivalent to the lottery between the lowest and greatest potential number of eagles saved (assigned utilities 0 and 1) represented the 0.5 utility. The questioning process was repeated to elicit the 0.25 and 0.75 utilities, producing 5 points on the utility function for this subject and decision context.

In the sensitivity analysis, this subject expressed generally consistent risk aversion responses as we varied the range of eagles potentially saved between lows of 0 to 5 and highs of 3 to 25. Thus, we graphed the results across the range of 0–20 eagles saved and fit an exponential utility function (Clemen and Reilly 2001), $U(x) = 1 - e^{-x/R}$, with $R = 6$ to visually align the curve with the 5 elicited utility points (fig. 15.2). The result was an upward-sloping or concave utility curve representing marked risk aversion (Goodwin and Wright 2004). The subject described their greatest concern as avoiding any outcome with no or close to no eagles saved, consistent with a risk-averse utility function.

We weighted the outcome of each stochastic simulation in the mitigation bank example by the respective utility for that number of eagles saved to calculate a total utility or mitigation credit of 1.3 eagles per year. This utility represents approximately the 30th percentile of the credible interval for number of eagles estimated to be saved by the mitigation bank program per year. It represents a risk-averse decision because it would hedge against uncertainty by granting less credit than the median or "expected" outcome. Figure 15.3 illustrates how this credit falls between the 20th and 50th (median) percentiles mentioned previously as potential decision criteria (1.1 and 2.0 eagles per year in this example). The 20th percentile represents an even more precautionary decision, or low credit award, while the 50th percentile reflects neutrality toward risk.

Discussion

Compensatory mitigation actions that offset eagle take must be "scientifically credible and verifiable" (USFWS 2013) to be considered for permitting approval. The lead mortality estimates from our model seemed reasonable and consistent with prior knowledge of eagle lead poisoning rates (Cochrane et al. 2015). Unbiased estimates of golden eagle mortality rates and causes derived from telemetry studies (Millsap et al. 2016) are forthcoming and will be used to update the model.

In our demonstration, we assumed non-lead ammunition would be required for hunting access in the mitigation bank area. However, if ammunition control were incomplete or if the lead abatement program were implemented on open hunting lands, it would be difficult to predict in advance the percentage of hunters who would adopt non-lead ammunition (Epps 2014). Rather than attempting to predict adoption rates in advance with a detailed model (perhaps with social and economic parameters), monitoring during implementation could provide direct measures of hunters' ammunition use before and after the program to determine final mitigation credits.

Because of all the uncertainties and limitations in the modeling process, the estimates it produces should be treated as hypotheses about cause-effect relationships, and lead abatement actions should be implemented experimentally. A mitigation bank is well suited for experimental adaptive management, with a goal to reduce uncertainty about lead mortality and abatement processes while accumulating mitigation credits. With appropriate experimental design and monitoring, the experts would be able to update their beliefs about eagle scavenging, lead poisoning and abatement relationships, in addition to updating the model directly with empirical data.

Permit decisions need not wait for experimental results or model updating, given the tools available for eliciting and implementing risk management standards. The lottery method for eliciting risk tolerance is supported by theory and detailed guidance (Farquhar 1984). Although utility elicitation is prone to inconsistencies because the judgments are so subjective (Goodwin and Wright 2004), utility functions can be approximated roughly and still provide satisfactory decision models (Clemen and Reilly 2001). The exponential utility function is frequently employed to represent risk attitudes (Clemen and Reilly 2001), and it represented our subject's elicitation values well. However, this curve may not match the Service's preferences for mitigation crediting. For policy development, the elicitation exercise should be conducted with multiple managers in a modified Delphi setting to produce an overarching standard reflecting agency interpretation of Eagle Act mandates.

A decision model based on a set percentile of credible results, such as the Service's 80 to 20 risk allocation for mortality, also considers uncertainty and is fairly straightforward to implement. In our mitigation bank example, the total utility value was equivalent to the 30th percentile of all results, but in other lead abatement scenarios we simulated the same utility curve equated with percentiles up to the 40th. Utility elicitation across a wide selection of potential permit decisions, including when take was predicted by the 80th quantile, might help identify a robust percentile value for general use.

After this case study was completed, the Service published an updated final Eagle Rule in December 2016 that proposes a risk management model for compensatory mitigation for the first time. The rule states that compensatory mitigation must be designed to offset eagle take at a 1.2 to 1 mitigation ratio (USFWS 2016). When determining credits for action taken, this ratio converts to 0.833 (1/1.2) credits per estimated eagle saved. This standard can be applied to any point estimate of eagles saved without detailed quantitative modeling. The disadvantage of a fixed multiplier, however, is neglecting to consider the extent of uncertainty in estimated compensation benefits. In our example, the ratio results in 1.7 credits for a median estimate of 2.0 eagles saved, which is roughly the 41st percentile of our simulation results. If the range of outcomes were very different around the same median, however, the credits would not change (still 1.7) but would equate with a far different treatment of uncertainty. In other words, fixed mitigation ratios provide a deceptively consistent decision model, potentially applying very different risk preferences across cases. The ratio approach is not as robust as the Service's 80th percentile standard for predicting fatalities and not necessarily "overprotective of eagles" (USFWS 2016, 91497).

The final rule states the Service may require further adjustments to the mitigation ratios based on uncertainty in the effectiveness of a particular mitigation practice. Moreover, the "Service will develop guidance for different types of compensatory mitigation projects for eagles . . . [that] include methods and standards for determining credits (i.e., how much of the type of mitigation is needed to offset one eagle) and mitigation ratios based on uncertainty" (USFWS 2016, 91505). The Service could establish a compensation utility function to apply to permit applications where variance in potential mortality has been modeled. At minimum, eliciting utilities from agency decision makers will inform policy improvements.

ACKNOWLEDGMENTS

Carol Sanders-Reed was a core team member and ace programmer for this project. The authors thank the experts in eagle biology, lead toxicology, and mitigation assessment who provided their time and expertise in assisting with the development and parameterization of the model: B Bedrosian, P Bloom, M Collopy, C Franson, G Hunt, T Katzner, T Kelly, M Kochert, B Millsap, B Murphy, L New, P Redig, B Rideout, and L Wilkinson. Ryan Nielson and WEST, Inc, provided eagle density estimates. The National Renewable Energy Laboratory and US Department of Energy, and the wind industry and conservation partners of the American Wind Wildlife Institute, provided financial support for this project.

LITERATURE CITED

Bedrosian B, Craighead D, Crandall R. 2012. Lead exposure in Bald Eagles from big game hunting, the continental implications and successful mitigation efforts. *PLOS ONE* 7(12):e51978. doi:10.1371/journal.pone.0051978.

Burgman MA. 2005. *Risks and Decisions for Conservation and Environmental Management*. Cambridge, UK: Cambridge University Press.

Clemen RT, Reilly T. 2001. *Making Hard Decisions with DecisionTools*. Pacific Grove, CA: Duxbury Press.

Cochrane JF, Lonsdorf E, Allison TD, Sanders-Reed CA. 2015. Modeling with uncertain science: estimating mitigation credits from abating lead poisoning in Golden Eagles. *Ecological Applications* 25:1518–1533.

Cruz-Martinez L, Redig PT, Deen J. 2012. Lead from spent ammunition: a source of exposure and poisoning in bald eagles. *Human-Wildlife Interactions* 6:94–104.

Drescher M, Perera AH, Johnson CJ, Buse LJ, Drew CA, Burgman MA. 2013. Toward rigorous use of expert knowledge in ecological research. *Ecosphere* 4(7):83.

Epps CW. 2014. Considering the switch: challenges of transitioning to non-lead hunting ammunition. *The Condor* 116:429–434.

Farquhar PH. 1984. Utility assessment methods. *Management Science* 30:1283–1300.

Fuller TK. 1990. Dynamics of a declining white-tailed deer population in north-central Minnesota. *Wildlife Monograph* 110.

Gardner TA, von Hase A, Brownlie S, Ekstrom JMM, Pilgrim JD, Savy CE, Stephens RTT, Treweek J, Ussher GT, Ward G, Ten Kate K. 2013. Biodiversity offsets and the challenge of achieving no net loss. *Conservation Biology* 27(6):1254–1264.

Goodwin P, Wright G. 2004. *Decision Analysis for Management Judgment*. 3rd ed. Chichester, UK: John Wiley and Sons.

Haig S, D'Elia J, Eagles-Smith C, Fair JM, Gervais J, Herring G, Rivers JW, Schulz JH. 2014. The persistent problem of lead poisoning in birds from ammunition and fishing tackle. *The Condor* 116:408–428.

Hunt WG. 2012. Implications of sublethal lead exposure in avian scavengers. *Journal of Raptor Research* 46:389–393.

Hunt WG, Burnham W, Parish C, Burnham K, Mutch B, Oaks JL. 2006. Bullet fragments in deer remains: implications for lead exposure in avian scavengers. *Wildlife Society Bulletin* 34:167–170.

Hunt WG, Parish CN, Orr K, Aguilar RF. 2009. Lead poisoning and the reintroduction of the California Condor in Northern Arizona. *Journal of Avian Medicine and Surgery* 23:145–150.

Kochert MN, Steenhof K, Mcintyre CL, Craig EH. 2002. Golden Eagle (*Aquila chrysaetos*). In Poole E, ed. *The Birds of North America Online*. Ithaca, NY: Cornell Lab of Ornithology.

Legagneux P, Suffice P, Messier JS, Lelievre F, Tremblay JA, Maisonneuve C, Saint-Louis R, Bêty J. 2014. High risk of lead contamination for scavengers in an area with high moose hunting success. *PLOS ONE* 9(11):e111546. doi:10.1371/journal.pone.0111546.

Marcot BG, Steventon JD, Sutherland GD, McCann RK. 2006. Guidelines for developing and updating Bayesian belief networks applied to ecological modeling and conservation. *Canadian Journal of Forest Research* 36:3063–3074.

Martin TG, Burgman M, Fidler F, Kuhnert PM, Low Choy S, McBride M, Mengersen K. 2012. Eliciting expert knowledge in conservation science. *Conservation Biology* 26:29–38.

McBride MF, Burgman MA. 2012. What is expert knowledge, how is such knowledge gathered, and how do we use it to address questions in landscape ecology? Pages 11–38 in Perera AH, Drew CA, Johnson CJ, eds. *Expert Knowledge and Its Application in Landscape Ecology*. New York: Springer.

Millsap BA, Bjerre ER, Otto MC, Zimmerman GS, Zimpfer NL. 2016. *Bald and Golden Eagles: Population Demographics and Estimation of Sustainable Take in the United States, 2016 Update*. Washington, DC: US Fish and Wildlife Service, Division of Migratory Bird Management.

Nielson R, McManus L, Rintz T, McDonald LL, Murphy RK, Howe WH, Good RE. 2014. Monitoring abundance of golden eagles in the western United States. *Journal of Wildlife Management* 78:721–730.

Nixon CM, Hansen LP, Brewer PA, Chelsvig JE, Esker TL, Etter D, Sullivan JB, Koerkenmeier RG, Mankin PC. 2001. Survival of white-tailed deer in intensively farmed areas of Illinois. *Canadian Journal of Zoology* 79:581–588.

Runge MC, Converse SJ, Lyons JE. 2011. Which uncertainty? Using expert elicitation and expected value of information to design an adaptive program. *Biological Conservation* 144:1214–1223.

Sieg R, Sullivan KA, Parish CN. 2009. Voluntary lead reduction efforts within the northern Arizona range of the California Condor. Pages 341–349 in Watson RT, Fuller M, Pokras M, Hunt WG, eds. *Ingestion of Lead from Spent Ammunition: Implications for Wildlife and Humans*. Boise, ID: The Peregrine Fund.

Speirs-Bridge A, Fidler F, McBride M, Flander L, Cumming G, Burgman M. 2010. Reducing overconfidence in the interval judgments of experts. *Risk Analysis* 30:512–523.

US Fish and Wildlife Service (USFWS). 2013. *Eagle Conservation Plan Guidance. Module 1—Land-Based Wind Energy Version 2*. Arlington, VA: US Fish and Wildlife Service, Division of Migratory Bird Management.

———. 2014. *Final Environmental Assessment. Shiloh IV Wind Project Eagle Conservation Plan, California. June 2014*. Sacramento, CA: US Fish and Wildlife Service, Division of Migratory Bird Management.

———. 2015. *Draft Environmental Assessment. Alta East Wind Project Eagle Conservation Plan, California. October 2015*. Sacramento, CA: US Fish and Wildlife Service, Division of Migratory Bird Management.

———. 2016. Eagle permits; revisions to regulations for eagle incidental take and take of eagle nests. *Federal Register* 81(242):91494–91554.

Warner S, Britton E, Becker D, Coffey M. 2014. Bald eagle lead exposure in the Upper Midwest. *Journal of Fish and Wildlife Management* 5(2):208–216.

Watson J. 2010. *The Golden Eagle*. 2nd ed. London, UK: T and AD Poyser.

Wyoming Game and Fish Department. 2013. *2012 Annual Report of Big and Trophy Game Harvest*. Cheyenne, WY: Wyoming Game and Fish Department.

16 Dealing with Risk Attitudes in Supplementary Feeding of Mauritius Olive White-Eyes

Stefano Canessa, Christelle Ferrière, Nicolas Zuël, and John G. Ewen

Many populations of endangered species require ongoing, intensive management to be viable, and as populations grow, meeting resource constraints may become increasingly difficult. However, managers hesitate to adopt less demanding management options, fearing this may jeopardize growth or viability. As a result, decision makers who are averse to risk may be reluctant to change from the status quo. In this chapter, we illustrate how we solved such a problem in the management of Mauritius olive white-eyes, an endangered passerine that has received ongoing provision of supplementary food. We combined 2 decision-support tools to determine whether risk-averse decision makers with multiple objectives should change to a less intensive and cheaper feeding regime.

Problem Background

Conservation scientists believe that many critically endangered species can still be saved from extinction, but their recovery will necessitate hands-on interventions. However, choosing the best action to take is difficult because of uncertainty. The stakes of decisions are high, given the risk of extinction of the population or species (Black and Groombridge 2010). The resources, especially time, that managers can commit are usually limited (Martin et al. 2012b), and managers often allocate those resources with limited knowledge about the species, the causes of its rarity, and how effective alternative actions may be. Moreover, small and declining populations offer little opportunity to learn, for example because of reduced sample sizes (Canessa et al. 2016b).

Despite these challenges, successful examples worldwide show action can improve the chance of persistence of endangered species; examples include the provision of breeding sites (Norris and Mcculloch 2003), predator management (Jones et al. 2016), and supplementary feeding (Ewen et al. 2014). Ideally, the choice of action should be based on clearly stated a priori hypotheses; for example, food supplementation may be applied to compensate for a hypothesized lack of natural food resources (Ewen et al. 2014). This facilitates evidence-based choices, comparison of possible actions, and collection and use of monitoring data (Runge et al. 2011). However, in management of critically endangered species, the urgency of decline and lack of information may often force managers to apply a management strategy involving a battery of support actions, targeting a broad range of potentially limiting factors (Jones 2004; Jones and Merton 2012). In such a scenario, managers hope that one or more of the broad-spectrum actions will help populations recover, even though the proximate cause of recovery is not known. In theory, once populations have thus been secured, individual components of the overall strategy can be assessed and those not necessary for population viability removed. In practice, however, managers may then become reluctant to abandon support actions and jeopardize success; this resistance to change can then clash with the challenge of bearing the costs of increasing management as populations recover (Goble et al. 2012). Ultimately, uncertainty generates a situation where a change might provide benefits but may also lead to negative results: in other words, uncertainty (insufficient knowledge or stochasticity) leads to risk (possible positive or negative outcomes of making a decision in the face of that uncertainty). Decision analysis can help in such situations.

Ecological Background

The endemic Mauritius olive white-eye *Zosterops chloronothos* (hereafter olive white-eye) was historically widespread across Mauritius, but the combined pressures of habitat loss, competition from exotic bird species, and predation from exotic mammal species (particularly rats) has largely restricted the species to an area of approximately 25 square km within the Black River Gorges National Park (Maggs et al. 2015; Nichols et al. 2004). This remnant population is predicted to decline at an estimated 14% per annum without exotic rat species control (Maggs et al. 2015). Currently, the olive white-eye is the rarest of the remaining endemic land bird species of Mauritius. A recovery effort for olive white-eyes started in 2005 and has developed with 2 major components: the protection of the remnant population through rat control and a translocation to Ile aux Aigrettes, a small (27 ha) offshore island free of exotic rat species (Parnell et al. 1989). While the remnant population continues to decline despite the best efforts of the field teams, here we focus on management of the translocated population.

The translocated population was founded by 38 individuals that were released between 2006 and 2010. The population continues to grow, presumably aided by the intensive ongoing management on the island. Supplementary feeding stations have been established across the island, within each breeding pair's territory. At the time of this study, 15 feeding stations were available to pairs. Originally, each station was stocked with 3 types of food: Aves Nectar, a commercially available water-soluble powder deemed to provide a full dietary supplement to nectar-feeding birds; fresh grapes; and an insectivorous mix (Insectivorous feast, Birdcare Company) mixed with egg, carrot, and apple. Feeding was done twice daily (morning and midday) at all stations. The morning routine consisted of providing all the types of supplementary food mentioned above, and the midday routine consisted of changing the nectar. Twice-daily feeding was necessary because Aves Nectar fermented quickly in the warm weather on Ile aux Aigrettes.

The belief of managers that supplementary feeding was assisting with population recovery was justified by the observation that only one pair was known to have successfully fledged a chick without access to a feeding station. On the other hand, twice-daily feeding of the growing population was quickly becoming unsustainable in terms of material and staff costs, putting a strain on the limited resources available to management.

Decision Makers and Process Design

The desire of the decision makers (authors Zuël and Ferrière of the Mauritian Wildlife Foundation) to control management costs without jeopardizing population recovery triggered the decision of whether to change the feeding regime, which in turn was linked to the broader decision of how to supplement food for olive white-eyes. Zuël and Ferrière explored the literature on supplementary feeding and noticed the similarities between olive white-eye management and challenges faced by managers with supplementary feeding of hihi (*Notiomystis cincta*) in New Zealand (Chauvenet et al. 2012); decision support tools were used successfully in that case (Ewen et al. 2014). Zuël and Ferrière then engaged with coauthors Ewen and Canessa to provide advice on experimental designs and act as decision analysts. This process was achieved via frequent emails between all authors and reciprocal visits by Zuël and Ewen between Mauritius and the United Kingdom.

Zuël and Ferrière, in their role as decision makers, were, in principle, risk averse: that is, they prioritized avoiding any disruption of the ongoing population growth over budget savings. In the remainder of this chapter, we illustrate how different decision-analytic methods were combined to facilitate their decision, and we step back in time to walk through how we structured and solved the problem.

Decision Analysis

Objectives

For this problem, Zuël and Ferrière identified 2 fundamental objectives, one related to the conservation of the species and the other to management costs. As measurable attributes for these objectives, we used the growth rate of the population (λ), in the short term without consideration of density dependence, and the monthly expenditure in MUR rupees for food (per feeding station per month) and staff time required to carry out the feeding (staff salary for hours worked per month), respectively. We chose a per-station cost because, given the short-term analysis and the small scale of the island, as the population grows, increases in costs would largely be driven by increases in food, rather than by adding feeding stations or staff effort.

Alternative Actions

Next, we identified a set of alternative feeding strategies. Besides (1) the current ad libitum feeding regime (*status quo;* more food available than is consumed to ensure any level of demand is always fulfilled), we considered (2) implementing a dynamic feeding regime where supply more closely matched demand (*dynamic feeding*), (3) replacing Aves Nectar with a more heat-stable solution, 20% by mass of raw sugar crystals dissolved in water (*sugar water*), (4) using sugar water and also removing fresh egg, carrot, and apple from the insectivorous mix (*perishable removal*), and (5) providing no feeding at all (*no feeding*). The dynamic feeding strategy was based on recent recommendations by Maggs (2017) that the afternoon feed should be removed, as consumption appears greater for all 3 food types during the morning. In addition, food should be reduced in response to dominant pair breeding activity, including reducing or removing insectivorous mix when pairs are not breeding, reducing Aves Nectar when breeding pairs do not have fledglings, and reducing or removing fruit when pairs are not incubating and do not have fledglings.

Consequences

To help make a rational decision given these actions, we needed to make predictions about their expected outcomes in terms of Zuël and Ferrière's objectives (population growth and cost). These predictions must explicitly state uncertainty, because such uncertainty is what leads to risk (again, risk corresponds to the potential for positive or negative outcomes of a given decision in the face of uncertainty). In other words, the decision involves risk because we are not sure what will happen under any given alternative.

We used a combination of approaches to estimate population growth. For the alternatives *status quo* and *sugar water*, the growth rate was estimated using the results of a feeding experiment carried out over 2 years (September–August annual cycles in 2013/2014 and 2014/2015), where the 2 feeding regimes were compared as experimental treatments. All breeding pairs were assigned to one of 2 treatment groups for the duration of the study ($N=14$ pairs; 7 pairs in group 1 and 7 pairs in group 2). In the 2013/2014 year, group 1 pairs were fed Aves Nectar and then switched to sugar water in 2014/2015.

Group 2 pairs were fed the reciprocal, with sugar water provided in 2013/2014 and Aves Nectar provided in 2014/2015. Switching diets in different years provided a crossed design that allowed us to control for background environmental stochasticity. Since pairs remain faithful to their chosen breeding site across years, we were able to simply change the food provided in each territory, minimizing potential autocorrelation and spillover effects (pairs accessing a different treatment by using neighboring stations). Using model comparison based on information criteria, we derived model-averaged estimates of survival and fecundity (annual number of fledglings per female) for juveniles and adults under the 2 treatments. We then used those estimates to populate a 2-stage matrix and calculated the growth rate as the dominant eigenvalue (λ). To represent uncertainty, we rescaled the 95% confidence intervals of individual estimates obtained from model averaging to 100%, to obtain a maximum and a minimum estimate of growth rate under the *status quo* and *sugar water* alternatives (table 16.1).

Estimates of population growth under the remaining 3 alternatives (*dynamic feeding, perishable removal*, and *no feeding*) were obtained using expert elicitation. This was required because a full experiment comparing all actions was not possible given the small size of the population (sample sizes would have been too small to provide informative results) and because some alternatives (such as *no feeding*) were considered too risky to implement in practice before a preliminary risk analysis (the one we present here). In this case, the experts we consulted were involved in olive white-eye management (Zuël, Ferrière, Vikash Tatayah, and Carl Jones from the Mauritian Wildlife Foundation) and were conditioned prior to the elicitation through detailed discussion of experimental results, longer-term monitoring reports, and a thorough explanation of the elicitation process. The chosen number of 4 experts reflects the high specificity of our application, which required close knowledge of the species, and fits within the range of 3 to 7 recommended by Clemen and Winkler (1985) and Hora (2004). We recognized uncertainty by defining a full probability distribution of λ under each action. We used a modified Delphi protocol for formal expert elicitation (Martin et al. 2012a; McBride et al. 2012), whereby during 2 rounds of elicitation interspersed with discussions, Ewen asked each expert to define their most likely value, as well as minimum and maximum bounds to reflect their uncertainty. Observing the general agreement among experts, we averaged the individual estimates to obtain a unique set of values. Combining these elicited estimates with those obtained from the feeding experiment, we compiled a table of the predicted range of growth rates for each action (in table 16.1). Since estimated budgets can be based on values known with certainty (cost of material and staff salaries), we express those as point values without uncertainty, assuming possible deviations would be minor (table 16.1).

Table 16.1. Predicted population growth rates and management costs for alternative feeding regimes in a translocated population of Mauritius olive white-eyes

	Population growth rate (λ)			
Alternative actions	Minimum	Most likely	Maximum	Cost (MUR)
1. Status quo	1.60	1.83	2.20	6,232
2. Dynamic feeding	1.32	1.60	2.05	6,132
3. Sugar water	1.50	1.80	2.30	4,227
4. Perishable removal	1.15	1.50	1.98	3,667
5. No feeding	0.60	0.86	1.23	0

Note: Costs are expressed in Mauritian rupees per month and include food and staffing costs.

Trade-Offs and Optimization

Only 2 actions (*perishable removal* and *no feeding*) had some chance of resulting in a population decline ($\lambda < 1$), and only the latter was more likely than not to result in a decline (table 16.1), based on the estimates we compiled from a combination of empirical information and expert judgment. Also, even though in the worst case *sugar water* was predicted to perform worse than *status quo*, the expected growth rate was still positive, and there was a relatively minor difference between the 2 alternatives. *Status quo* and *dynamic feeding* were determined to have very similar costs, while *sugar water* and *perishable removal* were both cheaper, by 32% and 42% respectively, compared to *status quo*. To solve the decision problem, we needed to find the optimal solution across all the objectives of the decision makers: maximizing growth and minimizing costs, while minimizing risks to the current positive population trend.

The decision makers in this problem (Zuël and Ferrière) were risk-averse, since their priority was to avoid particularly bad outcomes. In this sense, any outcome leading to a population decline ($\lambda < 1$) would be a failure, regardless of its cost. For any non-negative outcome for growth ($\lambda \geq 1$), the decision would be based on a combination of growth and cost. We reflected these multiple objectives using the following function to represent utility (the satisfaction derived from a given outcome):

$$\begin{aligned} &\mathit{for}\ \lambda_a < 1,\ U_a = 0 \\ &\mathit{for}\ \lambda_a \geq 1,\ U_a = \lambda'_a + C'_a \end{aligned} \quad (1)$$

where λ' and C' are, respectively, growth rate and cost for action a, rescaled to a 0–1 interval to allow summation, where 0 and 1 are the worst- and best-case scenarios: respectively, applying the most expensive action without achieving growth (i.e., $\lambda' = 0$ and $C' = 0$ rescaled correspond respectively to $\lambda \leq 1$ and cost of *status quo* management on the original scale) and obtaining the maximum possible growth for no cost (i.e., $\lambda' = 1$ and $C' = 1$ rescaled correspond respectively to $\lambda = 2.3$ and $C = 0$ on the original scale; table 16.1).

To calculate the rescaled λ' and C', we took into account the attitudes of managers as they emerged from discussions. We rescaled cost from C to C' linearly between the worst- and best-case values, reflecting the fact that the money allocated to the white-eye project is only a component of the total Mauritian Wildlife Foundation budget. However, discussions about growth rate highlighted that decision makers did not simply seek to maximize this value. For example, choosing between 2 actions with outcomes $\lambda = 2.1$ and $\lambda = 2.2$ would be of lesser concern than choosing between 2 actions with outcomes $\lambda = 1.1$ and $\lambda = 1.2$, even though the absolute difference is the same. In other words, the utility of Ferrière and Zuël (the satisfaction derived from a given growth rate) was not linearly related to λ even when $\lambda > 1$. We expressed this by calculating the rescaled λ' for action a in eq. 1 as

$$\lambda'_a = \frac{1 - e^{\frac{-(\lambda_a - \lambda_{\min})}{k}}}{1 - e^{\frac{-(\lambda_{max} - \lambda_{min})}{k}}} \quad (2)$$

where $\lambda_{\min}$ and $\lambda_{\max}$ correspond to the worst and best outcomes respectively ($\lambda = 1$ and $\lambda = 2.3$), and k is an exponential constant determining the rate of diminishing returns (Kirkwood 1997; here, after visually comparing different utility curves, we concluded that a value of $k = 0.25$ best described the decision makers' attitudes).

Stochastic Dominance

While the utility function described above is straightforward to implement with unique values, in our case uncertainty surrounding our predictions was a significant part of the decision problem. If Zuël and Ferrière were risk-neutral (i.e., they did not consider uncertainty, but only the highest mean return), they could, for example, base their choice on the most likely outcome for each scenario. If they only aimed to maximize growth, they would choose *status quo*, since it is most likely to result in the highest λ (table 16.1). If they considered both growth and cost as equally important, resolving eq. 1 and 2 with the most likely values for λ, they would choose *perishable removal* because it maximizes aggregate utility. However, we knew the decision makers were risk averse in this case (i.e., they were sensitive not only to the mean return, but also to the possibility of negative outcomes), so using the most likely values would not account for their values adequately. Therefore, we carried out a further analytical step, comparing the distributions of the utility of the alternative actions using stochastic dominance (Levy 1998), a technique that allows decision makers to rank available actions considering the full range of uncertainty about the expected outcomes of actions, and making progressive assumptions about their own preferences. Although well known in economic theory, its application to conservation problems is uncommon (Benítez et al. 2006; Canessa et al. 2016a; Knoke et al. 2008; Yemshanov et al. 2012).

Figure 16.1 provides an illustrative example of stochastic dominance analysis. An action *X* stochastically dominates an action *Y* at the first order if, for every value of the utility function, *X* has a greater chance than *Y* to result in that outcome or in a better one. This can be verified visually by comparing the cumulative distribution functions of the predicted outcomes. In figure 16.1a, we present as an example the distribution of predicted outcomes for 3 actions X, Y, and Z. In figure 16.1b, we compare their cumulative distribution functions: for any value on the x-axis, curve X corresponds to a lower value on the y-axis than curve Z, so X dominates Z at the first order. First-order dominance only requires the assumption that greater utility is always better. Formally, this equates to the assumption that the utility function is non-decreasing and its first derivative is always positive.

However, in some cases action *Y* might have a greater chance than X of very positive outcomes, but also a greater chance of particularly negative ones; in other words, X is a riskier action (fig. 16.1a). In this case, there is no first-order dominance; second-order dominance can be verified instead, assuming managers are risk-averse (as in the olive white-eye example). The comparison is repeated using the integral of the cumulative distribution function for each action: a risk-averse decision maker would choose the action that is dominant at the second order (note

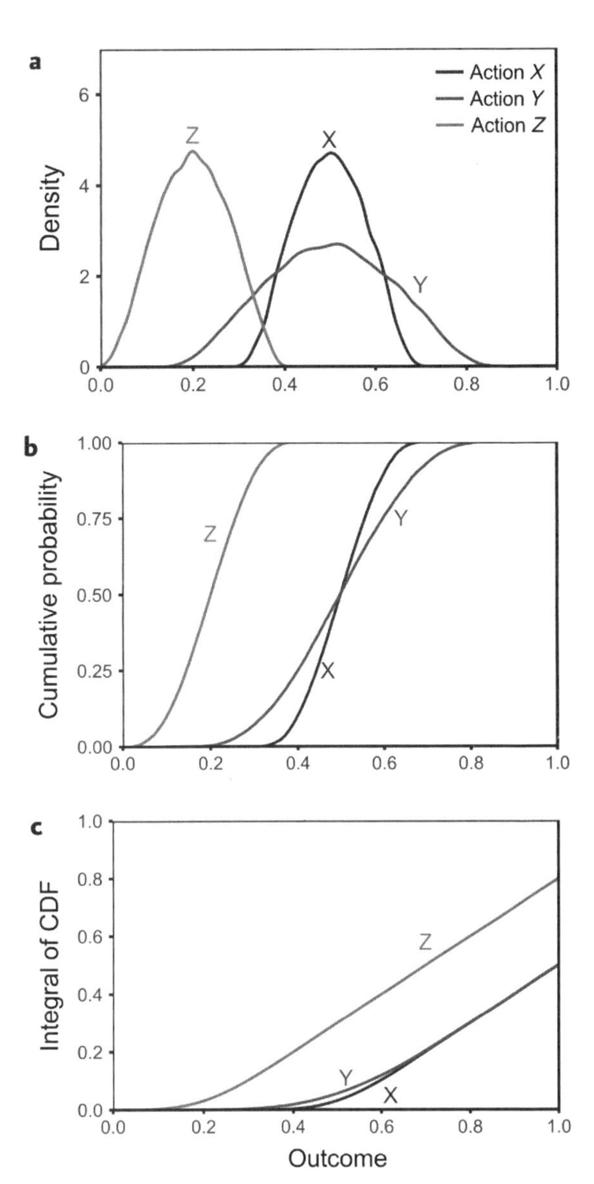

Figure 16.1. Example of stochastic dominance analysis. Plot a indicates the distribution of outcomes for 3 actions, *X*, *Y*, and *Z*. Actions *X* and *Y* have the same mean outcome; action *Y* has a wider distribution than *X*, meaning a greater chance of both worse and better outcomes. Plot b shows the cumulative distribution functions: actions *X* and *Y* always have a greater chance of providing better (greater) outcomes than *Z*, so they both dominate *Z* at the first order; however, their cumulative distributions cross, so they cannot be discriminated at the first order. Plot c shows the integrals of the CDFs: the curve for *X* is always at least equal to that of *Y*, and lower in at least one point, indicating that *X* dominates at the second order and thus is the best action for a risk-averse decision maker.

that actions that dominate at the first order also dominate at the second order). In figure 16.1b, the cumulative distribution functions for *X* and *Y* cross, so there is no dominance at the first order. Therefore, we can take the integrals of the cumulative distribution functions and compare them in figure 16.1c. Here, for any value on the x-axis, curve *X* corresponds to a lower value on the y-axis than curve *Y*, so *X* dominates *Y* at the second order. If the condition for second-order dominance is not met, higher orders can be explored, although they require more complicated assumptions, and differences between actions can be marginal in practice (Canessa et al. 2016a; McCarthy 2014).

In the case of olive white-eyes, both assumptions for first and second-order stochastic dominance were met: greater utility (defined using eq. 1) was always preferred, and the decision makers were risk averse. To analyze dominance, we first defined the entire distribution of results empirically. Since the worst- and best-case scenarios in table 16.1 represent 100% confidence intervals (minimum and maximum values, respectively), we assumed the most likely values as the mode and used them to fit a PERT distribution, a modified beta distribution that is especially suitable for expert-elicited values (Vose 1996). Drawing 10,000 random samples from each PERT distribution, we obtained an empirical distribution of possible outcomes for each action. To better discuss results with decision makers, we repeated the analysis with and without including management costs C' in eq. 2 (figs. 16.2a and 16.2b, respectively).

We assessed whether any action had first-order stochastic dominance over the others. When ignoring cost (assuming utility derives only from growth rate, rescaled to represent diminishing returns using eq. 1), *status quo* dominated all other actions at the first order (figs. 16.2a, b). In other words, maintaining the current feeding regime was the optimal choice for our decision makers regardless of their risk attitude. However, the decision makers did have cost as an objective: the more realistic utility function including costs showed a different optimal decision for our decision maker. Using this utility function, *status quo* was dominated at the first order by both *sugar water* and *perishable removal*, which in turn could not be discriminated at the first order (fig. 16.2e). The high ranking of *perishable removal* largely reflected its substantial economic savings: however, the uncertainty surrounding the predicted growth rate for this action meant that it was not the optimal choice for a risk-averse decision maker, and *sugar water* dominated at the second order (fig. 16.2f).

Decision Implementation

The Ile aux Aigrettes population of olive white-eyes has benefitted from strong growth. However, this success had resulted in increasing costs of supporting the growing population. From their initial position of risk aversion, the decision makers preferred to avoid changing the feeding regime if this reduced the chances of population recovery. However, the results of our experiment suggested that replacing an expensive and environmentally unstable food (Aves Nectar) with a cheap and more environmentally robust alternative (sugar water) had only a limited chance of reducing population growth. Risk analysis then showed that this marginal risk was more than compensated by substantial economic savings. The decision makers therefore made the choice to switch from Aves Nectar to sugar water. Prior to this analysis, discussions about changing management were mostly qualitative and difficult to rationally judge. Outside of a structured process, discussions were complicated by different poorly stated objectives and implicit weightings placed on the importance of these objectives. We found value in working through the structured-decision-making steps toward our final representation of utility and the recommended decision, as it yielded better understanding of the decision and how objective preferences, uncertainty, and the decision makers' attitude toward this uncertainty shaped the optimal choice. Importantly, this helped ensure that Ewen and Canessa clearly understood the decision problem that needed to be solved,

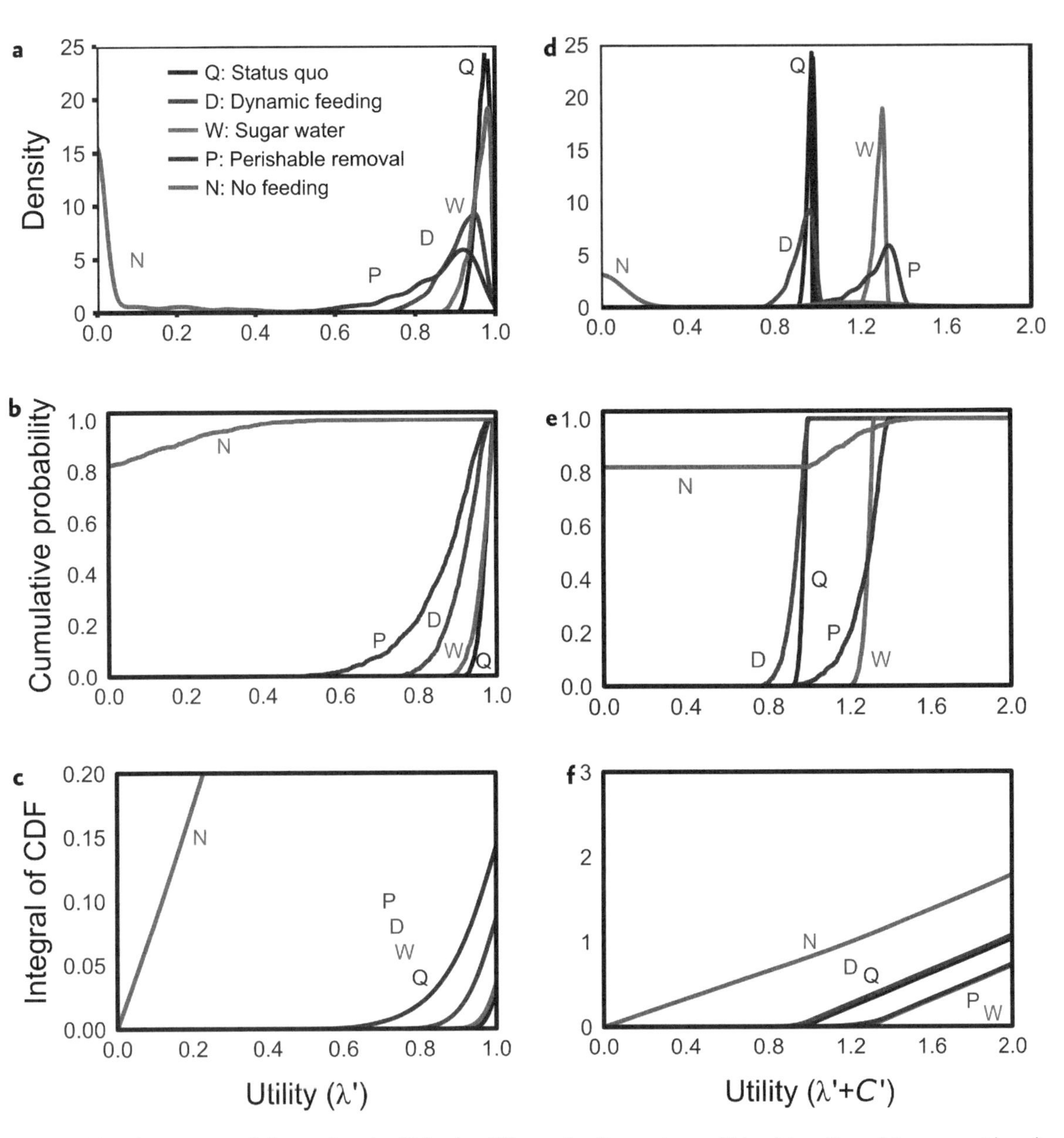

Figure 16.2. Comparison of the predicted utilities for different feeding regimes of Mauritius olive white-eyes, with and without consideration of management costs. Plots a–c and d–f indicate results when cost is respectively removed from and included in the calculation of utility (eq. 1-2). Plots a and d indicate the empirical distributions of utilities; plots b and e represent the cumulative distribution functions (CDFs) of the distributions in plots a and d, used to determine first-order stochastic dominance; plots c and f show the integrals of the CDFs in b and e, used to determine second-order stochastic dominance. Acronyms indicate different actions: status quo (Q), dynamic feeding (D), sugar water (W), perishable removal (P), and no feeding (N). Note all plots represent utility based on the rescaled growth rate and costs (x-axis; λ' and C' in Eq. 1-2).

that Zuël and Ferrière knew exactly how they arrived at their choice, and that a good choice was made.

Discussion

The challenge faced by managers of olive white-eyes on Ile aux Aigrettes is common to recovery programs worldwide, which struggle to maintain their capacity to support growing populations. A successful, but demanding management regime may lead to a breakdown in the quality of work done as staff struggle with resource constraints; however, changing to a less demanding regime can be perceived as a serious risk when uncertainty exists about its potential to maintain successful outcomes. Although such uncertainty is unlikely to be eliminated completely, we have illustrated how it can be dealt with by combining applied ecological research (in our case, an experimental test of feeding strategies), expert judgment, and formal methods for risk analysis.

First, making rational decisions requires making predictions about outcomes. In our case, experts believed from the beginning that *no feeding* and *dynamic feeding* of olive white-eyes were unlikely to ever perform better than other actions, so collecting additional data about them was not considered immediately necessary. This was justified because feeding olive white-eyes was strongly believed to contribute to population growth, and removing food was viewed as too likely to cause a population decline. Nevertheless, we found it useful to include the expected outcomes of removing feeding in our analysis, to better assess the absolute and relative differences of feeding strategies against this cheapest option. In addition, there was concern that *dynamic feeding* was too complicated to implement accurately such that mistakes in timing of delivery could be made (i.e., periods of high food demand and no supply). Conversely, experts were uncertain about the differences between *status quo* and *sugar water*, and this led to the initial decision to focus the experiment on *status quo* and *sugar water*, concentrating efforts, increasing sample sizes, and more effectively reducing uncertainty that was relevant to management.

Endangered species management often requires urgent decisions, and postponing direct management in order to learn is sometimes not feasible. For these reasons, expert judgment is often the first source of information for management decisions. Our elicitation of expert judgment was necessary given the longer time required to collect empirical data on all alternatives. Best-practice formal protocols for elicitation help protect against cognitive biases of experts (Burgman et al. 2011; Martin et al. 2012a). Once completed, our analysis suggested that *perishable removal* might provide substantial savings, but its effect on population growth remained too uncertain for a risk-averse decision maker. At the time of writing, an additional experiment was ongoing to reduce that uncertainty and clarify the potential of this action.

Second, the precautionary principle often advocated in conservation is an expression of risk aversion, where the priority is to avoid the risk of greater losses (Dickson and Cooney 2005; Myers 1993; Schwartz et al. 2009). However, for most management decisions risk aversion is not independent of the actual expected consequences, so it may be unrealistic to state it as a generality. Indeed, utility theory has long recognized that the willingness to gamble (risk attitude) changes with wealth (Arrow 1971; Pratt 1964), and this may equally apply to conservation problems. For example, managers may be less willing to experiment when expectations are low and the risk of losing populations greater, or because they fear negative judgment from peers on outcomes (Black et al. 2011). On the other hand, making large gains when expectations are already positive may be less important to managers. In our case, risk aversion decreased after the results of the feeding experiment suggested a positive population growth rate for both *status quo* and *sugar water* feeding, with only marginal differences.

Third, the initial problem of whether to change feeding regime was driven by the desire to reduce management cost. This is representative of how decision problems that involve some degree of risk are often driven by the presence of multiple, competing objectives: in the absence of a cost objective, there would have been no reason to consider changing the management strategy, as confirmed by the results of our analysis. Alternatively, the decision might have been triggered by other objectives such as moving toward a more "natural" condition, promoting public interaction facilitated by feeding stations (Walpole 2001), or reducing the impact on other species (Cortés-Avizanda et al. 2009). For a rational decision to be possible, these objectives must be recognized formally.

We have illustrated a combination of 2 methods for making decisions in the face of uncertainty and risk (utility functions and stochastic dominance). These methods can be effectively combined with applied ecological studies to inform management decisions. However, for this potential to be realized, it is necessary to recognize some key conditions:

- Decisions are triggered by objectives, and these must be clearly recognized in order to identify a rational decision.
- Risk is a result of uncertainty surrounding the possible outcomes of alternative actions. That uncertainty should be articulated explicitly.
- The optimal decision depends on risk attitude, but this can change depending on the expected outcomes. It may be misleading to assume a general precautionary principle, or risk aversion, before those expected outcomes have been evaluated.

In practice, ignoring the complexity of risky conservation decisions will not make them better; however, if such complexity is recognized, it becomes possible to deal with it using the appropriate methods provided by decision analysis.

ACKNOWLEDGMENTS

This study was only possible with substantial support and input from the Mauritian Wildlife Foundation Conservation Director Vikash Tatayah and Scientific Director Carl Jones. The authors thank the Mauritian Wildlife Foundation staff and volunteers for assistance with fieldwork; Alejandra Morán-Ordóñez, Alienor Chauvenet, and Doug Armstrong for helpful advice on data analysis and for commenting on early drafts of the manuscript; and the National Parks and Conservation Services for their collaboration. This project was funded by the Government of Mauritius, Hong Kong and Shanghai Banking Corporation and Chester Zoo. Stefano Canessa is supported by the Research Foundation Flanders (FWO16/PDO/019).

LITERATURE CITED

Arrow KJ. 1971. *Essays in the Theory of Risk-Bearing*. Chicago, IL: Markham.

Benítez PC, Kuosmanen T, Olschewski R, Van Kooten GC. 2006. Conservation payments under risk: a stochastic dominance approach. *American Journal of Agricultural Economics* 88:1–15.

Black SA, Groombridge JJ. 2010. Use of a business excellence model to improve conservation programs. *Conservation Biology* 24:1448–1458.

Black SA, Groombridge JJ, Jones CG. 2011. Leadership and conservation effectiveness: finding a better way to lead. *Conservation Letters* 4:329–339.

Burgman MA, McBride M, Ashton R, Speirs-Bridge A, Flander L, Wintle B, Fidler F, Rumpff L, Twardy C. 2011. Expert status and performance. *PLOS ONE* 6:e22998.

Canessa S, Ewen JG, West M, McCarthy MA, Walshe TV. 2016a. Stochastic dominance to account for uncertainty and risk in conservation decisions. *Conservation Letters* 9:260–266.

Canessa S, Guillera-Arroita G, Lahoz-Monfort JJ, Southwell DM, Armstrong DP, Chadès I, Lacy RC, Converse SJ. 2016b. Adaptive management for improving species conservation across the captive-wild spectrum. *Biological Conservation* 199:123–131.

Chauvenet ALM, Ewen JG, Armstrong DP, Coulson T, Blackburn TM, Adams L, Walker LK, Pettorelli N. 2012. Does supplemental feeding affect the viability of translocated populations? The example of the hihi. *Animal Conservation* 15:337–350.

Clemen RT, Winkler RL. 1985. Limits for the precision and value of information from dependent sources. *Operations Research* 33:427–442.

Cortés-Avizanda A, Carrete M, Serrano D, Donázar JA. 2009. Carcasses increase the probability of predation of ground-nesting birds: a caveat regarding the conservation value of vulture restaurants. *Animal Conservation* 12:85–88.

Dickson B, Cooney R. 2005. *Biodiversity and the Precautionary Principle: Risk and Uncertainty in Conservation and Sustainable Use*. London, UK: Earthscan.

Ewen JG, Walker L, Canessa S, Groombridge JJ. 2014. Improving supplementary feeding in species conservation. *Conservation Biology* 29:341–349.

Goble DD, Wiens JA, Scott JM, Male TD, Hall JA. 2012. Conservation-reliant species. *BioScience* 62:869–873.

Hora SC. 2004. Probability judgments for continuous quantities: linear combinations and calibration. *Management Science* 50:597–604.

Jones CG. 2004. Conservation management of endangered birds. Pages 269–301 in Sutherland WJ, Newton I, Green RE, eds. *Bird Ecology and Conservation*. Oxford, UK: Oxford University Press.

Jones CG, Merton DV. 2012. A tale of two islands: the rescue and recovery of endemic birds in New Zealand and Mauritius. Pages 165–222 in Ewen JG, Armstrong DP, Parker KA, Seddon PJ, eds. *Reintroduction Biology: Integrating Science and Management*. Chichester, UK: John Wiley and Sons.

Jones HP, Holmes ND, Butchart SH, Tershy BR, Kappes PJ, Corkery I, Aguirre-Muñoz A, Armstrong DP, Bonnaud E, Burbidge AA. 2016. Invasive mammal eradication on islands results in substantial conservation gains. *Proceedings of the National Academy of Sciences* 113:4033–4038.

Kirkwood CW. 1997. *Strategic Decision Making*. Belmont, CA: Duxbury.

Knoke T, Hildebrandt P, Klein D, Mujica R, Moog M, Mosandl R. 2008. Financial compensation and uncertainty: using mean-variance rule and stochastic dominance to derive conservation payments for secondary forests. *Canadian Journal of Forest Research* 38:3033–3046.

Levy H. 1998. *Stochastic Dominance: Investment Decision Making under Uncertainty*. The Netherlands: Kluwer Academic Publishers.

Maggs G. 2017. The ecology and conservation of wild and reintroduced populations of the critically endangered Mauritius olive white-eye *Zosterops chloronothos*. Doctoral thesis, University College London.

Maggs G, Nicoll M, Zuël N, White PJ, Winfield E, Poongavanan S, Tatayah V, Jones CG, Norris K. 2015. *Rattus* management is essential for population persistence in a critically endangered passerine: combining small-scale field experiments and population modelling. *Biological Conservation* 191:274–281.

Martin TG, Burgman MA, Fidler F, Kuhnert PM, Low-Choy S, McBride M, Mengersen K. 2012a. Eliciting expert knowledge in conservation science. *Conservation Biology* 26:29–38.

Martin TG, Nally S, Burbidge AA, Arnall S, Garnett ST, Hayward MW, Lumsden LF, Menkhorst P, McDonald-Madden E, Possingham HP. 2012b. Acting fast helps avoid extinction. *Conservation Letters* 5:274–280.

McBride MF, Garnett ST, Szabo JK, Burbidge AH, Butchart SH, Christidis L, Dutson G, Ford HA, Loyn RH, Watson DM. 2012. Structured elicitation of expert judgments for threatened species assessment: a case study on a continental scale using email. *Methods in Ecology and Evolution* 3:906–920.

McCarthy MA. 2014. Contending with uncertainty in conservation management decisions. *Annals of the New York Academy of Sciences* 1322:77–91.

Myers N. 1993. Biodiversity and the precautionary principle. *Ambio* 22:74–79.

Nichols R, Woolaver L, Jones CG. 2004. Continued decline and conservation needs of the endangered Mauritius olive white-eye *Zosterops chloronothos*. *Oryx* 38:291–296.

Norris K, Mcculloch N. 2003. Demographic models and the management of endangered species: a case study of the critically endangered Seychelles magpie robin. *Journal of Applied Ecology* 40:890–899.

Parnell JA, Cronk Q, Jackson PW, Strahm W. 1989. A study of the ecological history, vegetation and conservation management of Ile aux Aigrettes, Mauritius. *Journal of Tropical Ecology* 5:355–374.

Pratt J. 1964. Risk aversion in the small and in the large. *Econometrica: Journal of the Econometric Society* 30:122–136.

Runge MC, Converse SJ, Lyons JE. 2011. Which uncertainty? Using expert elicitation and expected value of information to design an adaptive program. *Biological Conservation* 144:1214–1223.

Schwartz MW, Hellmann JJ, McLachlan JS. 2009. The precautionary principle in managed relocation is misguided advice. *Trends in Ecology & Evolution* 24:474.

Vose D. 1996. *Quantitative Risk Analysis: A Guide to Monte Carlo Simulation Modelling*. Chichester, UK: John Wiley and Sons.

Walpole MJ. 2001. Feeding dragons in Komodo National Park: a tourism tool with conservation complications. *Animal Conservation* 4:67–73.

Yemshanov D, Koch FH, Lyons DB, Ducey M, Koehler K. 2012. A dominance-based approach to map risks of ecological invasions in the presence of severe uncertainty. *Diversity and Distributions* 18:33–46.

PART V ADDRESSING KNOWLEDGE GAPS

17 Introduction to Prediction and the Value of Information

David R. Smith

Predicting the consequences of alternative actions in terms of the objectives is central to decision making. Modeling in the broadest sense, from simple to complex and based on data or expert judgment, provides the essential toolkit for making decision-relevant predictions. Gaps in knowledge and the resulting uncertainty can make predictive modeling challenging. Gathering information to address knowledge gaps, thereby reducing uncertainty, can improve predictions. However, within a decision analysis, the value of gathering information depends on the extent that reduced uncertainty will improve the decision's outcome. Decision makers commonly confront the choice either to proceed directly to a decision in the face of uncertainty or to delay and attempt to reduce the uncertainty significantly before making the decision. Value of information analysis can help the decision maker choose intelligently. This chapter introduces the purpose, approaches, and tools for addressing knowledge gaps in a decision setting. The 3 case studies that follow illustrate some of the challenges and solutions encountered when addressing knowledge gaps.

Introduction

Decision making is necessarily forward-looking. Thus, the process of prediction is at the heart of decision making. By comparing the predicted consequences of alternative courses of actions, a decision maker can select the action that has the best chance of resulting in the preferred outcomes. Recall Ron Howard's 3-legged stool of decision essentials (Howard 2007): each leg relates to a question—what do you want, what can you do, and what do you know? That last question—what do you know?—links the actions to the prospective outcomes. "If you do not have any information linking what you do to what will happen in the future, then all alternatives serve equally well because you do not see how your actions will have any effect" (Howard 2007, 37).

While prediction is always at the heart of decision analyses, it is not always a hard task. For example, whether I need my raincoat depends on the daily weather forecast, which is reliable and easy to find. In other cases, however, prediction is the main impediment to solving a problem. For example, in chapter 18 Szymanski and Pruitt describe a case study where a predictive model was needed for effective conservation of Indiana bat (*Myotis sodalis*) in the face of the emergence of infectious disease (white-nose syndrome) and local impacts (e.g., wind energy projects). In the absence of those predictions, investment in conservation was ad hoc and of questionable effectiveness and efficiency. Knowledge gaps can make prediction challenging.

Rarely, if ever, do we lack knowledge completely, nor are we clairvoyant. Instead, we face gaps in knowledge, and before we can predict the consequences of alternative actions, we need to address those gaps. Scientists refer to knowledge gaps as epistemic uncertainty (Regan et al. 2002; see also Runge and Converse, chapter 13, this volume). However, a scientist's view of uncertainty can differ distinctly from a decision maker's perspective. To a scientist, the key uncertainties are those that suggest interesting alternative hypotheses. To a decision maker, the key uncertainties are those that affect decision choice; in some cases, it is possible to make a smart choice despite unresolved uncertainty. Addressing knowledge gaps in the context of decision analysis begins with figuring out which uncertainty is worth reducing (Maxwell et al. 2015; Runge et al. 2011) and then considers how best to reduce that uncertainty (McCarthy et al. 2004). For example, in chapter 19, Green and Bailey use a combination of modeling and sensitivity analysis to identify key uncertainties for vernal pool management, which then become the targets of focused research. In chapter 20, Converse describes the use of expert judgment to evaluate the effectiveness of management actions to recover whooping cranes (*Grus americana*) in the face of deep structural uncertainty.

Even Herculean efforts cannot fill some knowledge gaps, and some level of uncertainty will remain (Gregory et al. 2012). Irreducible uncertainty arises from both the unavoidable randomness of nature (aleatory uncertainty) and persistent gaps in knowledge (the epistemic uncertainty not reducible within the available time and effort). Failure to incorporate uncertainty into the decision analysis can cause overly optimistic expectations for decision outcomes and can cause a failure to realize that across the range of uncertainty, different courses of action would be best (Burgman et al. 2005). The basic approaches to coping with irreducible uncertainty in decision analysis are to proceed directly to a decision using risk analysis tools, such as decision trees (see part 4); to move to a decision but plan to reduce epistemic uncertainty while implementing recurrent actions coupled with follow-up monitoring through the process of adaptive management (see part 6; Failing et al. 2013); or to delay the decision and use the time to reduce epistemic uncertainty as much as possible prior to committing to a course of action (this part). Thus, addressing knowledge gaps (i.e., reducing epistemic uncertainty) is fundamental to the information problems discussed in this section and to adaptive management as discussed by Runge in chapter 21.

Delaying an action to address knowledge gaps can be a smart choice. For example, the endangered Shenandoah salamander (*Plethodon shenandoah*), with a range restricted to Shenandoah National Park, is known to compete with the red-backed salamander (*Plethodon cinereus*) and is subject to potential climate effects (Dallalio et al. 2017). Grant et al. (2014) evaluated ways that the National Park Service (NPS) could conserve the Shenandoah salamander, including active management to reduce competition and mitigate climate change effects. However, the institutional philosophy of the NPS strongly emphasizes minimal anthropogenic influence on the park ecosystem, and aggressive management can deviate from that philosophy. Thus, consideration of management options weighed conservation benefits against adherence to NPS philosophy. Given these considerations, Grant et al. (2014) compared acting now versus conducting further research prior to action and determined that the best option would be to continue research to test whether active management was necessary to conserve *P. shenandoah*. Importantly, Grant et al. (2014) treated further study before taking action as a distinct decision option and compared the consequences on all considerations against the other options to reach a recommendation.

Tools for Addressing Knowledge Gaps

In the broadest sense, the terms "predictive model" or simply "model" encompass the full range of tools used in the consequence analysis step of the PrOACT cycle (Runge and Bean, chapter 1, this volume) to predict the outcomes resulting from the alternative actions. Thus, models useful for decision analysis must accept alternative actions (among other influential factors) as input and produce performance metrics related to the fundamental objectives as output (Addison et al. 2013).

Models can range from conceptual to highly quantitative and from simple to complex and can include empirical evidence, expert judgment, or a combination of the two. For example, to decide when to start my commute, I rely on a prediction for drive time and whether I will arrive in time for a meeting. That prediction, which comes from an app on my phone, accounts for distance and speed adjusted by predicted traffic patterns. The phone app and its underlying algorithm is a predictive model. In another example, selection of a harvest rate to maximize sustainable yield relies on a prediction for harvest over time. The harvest prediction comes from a model that takes the harvest rate and current population size as inputs and produces harvest yield and future population size as outputs (Sweka et al. 2007).

Because a decision-relevant model fits a specific decision, a generic model might not be useful. For example, a red knot (*Calidris canutus*) population model that is useful for deciding management within stopover habitat in Delaware Bay (see McGowan et al., chapter 24, this volume) does not need to include survival rates specific to wintering grounds or nesting areas (McGowan et al. 2011). To build decision-relevant models, first include components essential to the decision and then use prototyping to find the minimal complexity needed to make a good choice (Garrard et al. 2017). An initial uncomplicated model (e.g., influence diagrams), developed collaboratively, can enhance communication among scientists and decision makers, contributing to a common understanding of the decision problem and building trust in the final predictive model (Addison et al. 2013; Howard 1988). While the goal is to develop a model no more complicated than necessary for the decision context (parsimony), additional model complexity might be required to maintain support among stakeholders who would not trust a simple model.

When parameterizing models, Addison et al. (2013) recommend reliance on data and empirical estimates when possible. However, it is common that the decision cannot or should not wait for new data or analyses. In that event, turn to expert judgment to estimate the parameters in models (Burgman 2016). Experts have knowledge and experience that can be useful in making predictions but are just as susceptible to cognitive biases as other people. Fortunately, formal and rigorous methods for elicitation have been established (Burgman 2016; Hemming et al. 2018). These protocols and procedures have been shown to prevent or reduce cognitive biases, such as anchoring, prediction overconfidence, and groupthink (Drescher et al. 2013; Speirs-Bridge et al. 2010).

An important question is how much time and effort should be spent to reduce uncertainty before committing to a course of action (Canessa et al. 2015; Maxwell et al. 2015; Runge et al. 2011; Williams et al. 2011). Fundamental to this question is the degree to which more information will improve the decision. High levels of uncertainty characterize natural resource management. However, the best management choice might not change across the range of uncertainty, in which case, there would be no benefit from resolving uncertainty. Also, delaying a decision imposes an opportunity cost; for example, waiting to act might allow habitat or population status to degrade further before improvements can take hold. Postponing a decision to learn more is a separate decision alternative to be evaluated against the option of deciding in the face of uncertainty. *Value of information* (VOI) analysis is a set of tools that compare the expected outcomes from 2 alternatives: (1) proceeding directly to a decision or (2) taking the time and effort to reduce the uncertainty prior to a decision (Canessa et al. 2015; Walters 1986; Williams et al. 2011). VOI analysis typically considers the benefits from eliminating all uncertainty (i.e., possessing perfect knowledge), which is called the *expected value of perfect information* (EVPI). Acknowledging the purpose of VOI analysis, Howard (2007, 39) refers to EVPI as the "value of clairvoyance," and Walters (1986, 195) regards EVPI as "bounding the importance of learning." A VOI analysis can also incorporate less-than-perfect knowledge by reducing some but not all uncertainty (i.e., the expected value of sample information; Canessa et al. 2015) or eliminating some but not all sources of uncertainty (i.e., the expected value of partial information; Runge et al. 2011).

An EVPI calculation can be viewed in the form of a decision tree with linked decisions (fig. 17.1). The tree has 2 main branches. Along the top branch, uncertainty is resolved first, and then the management option is selected. Along the bottom branch, the sequence is reversed; the management option is selected in the face of uncertainty, using the principles of risk analysis (see Runge and Converse, chapter 13, this volume). As an example, consider the biological control of an invasive pest where there is significant uncertainty about the effectiveness of one of the proposed control agents (adapted from Blomquist et al. 2010). Hemlock woolly adelgid (*Adelges tsugae*) is an invasive pest that causes high mortality in eastern and Carolina hemlock (*Tsuga canadensis* and *T. caroliniana*). Suppose that the options for biological control agents include predator beetles or insect-killing fungi, but not both. The effectiveness of fungi as a biological control is well established. However, the efficacy of predatory beetles is uncertain, and there is a 50:50 chance that beetles are highly effective or marginally effective. The VOI question is this: Should the effectiveness of beetles

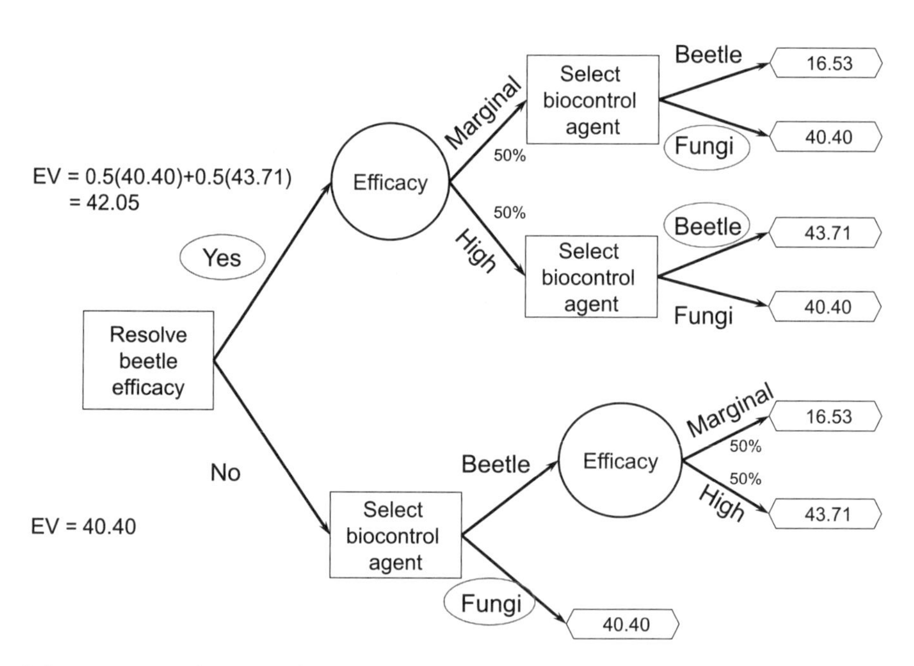

Figure 17.1. A decision tree to determine the value of resolving uncertainty before choosing a biological control agent for Hemlock wooly adelgid (adapted from Blomquist et al. 2010). Biological control agents include insect-killing fungi, with established efficacy, or a predatory beetle, whose effectiveness is not yet understood. The beetle effectiveness is hypothesized to be marginal or high with equal likelihood. From left to right, the initial decision is whether to resolve the efficacy before selecting the biological control or to choose the biological control agent in the face of the uncertain effectiveness. The predicted outcomes, on the far right, are the number of hemlock stands in 30 years. The expected value (EV) is the number of hemlock stands resulting from making the circled choices, averaged across the uncertainty where applicable. The ovals indicate the optimal choice for each decision node. The expected value of information (EVPI) is the difference in the EVs; in this case, EVPI = 42.05 – 40.40 = 1.65 stands of hemlock.

be established before the selection of a biological control agent? Suppose a model is available to predict the number of hemlock stands in 30 years under different biological control treatments. The EVPI is 1.65 stands of hemlock (fig. 17.1). Thus, in this example, some benefit results from reducing uncertainty. However, the cost of a research project aimed at resolving the uncertainty of beetle effectiveness should be no more than the monetary value of 1.65 stands of hemlock.

Case Studies

The chapters in this section illustrate some of the challenges and solutions encountered when addressing knowledge gaps within a decision analysis.

In chapter 18, Jennifer Szymanski and Lori Pruitt describe challenges inherent to the recovery of Indiana bats, a federally endangered species, owing to the emergence and rapid spread of white-nose syndrome, a highly infectious fungal disease. The lack of a predictive model hampered the identification of effective recovery actions. The model needed to predict population response to threats and conservation efforts at the appropriate temporal and spatial scales. The authors recognized that the model-building problem was nested within the analysis of endangered species conservation (fig. 18.1). Thus, they developed performance criteria for a request for proposals based on the objectives for how the conservation decision-making process would make use of the model. Information-gathering efforts, such as how best to build a predictive model, design a monitoring program, or design a research study, can be framed as decision problems nested within a more comprehensive decision context.

In chapter 19, Adam Green and Larissa Bailey present an analysis that focused on management of vernal pools for amphibian conservation within National Wildlife Refuges in the northeastern United States. An initial prototyping workshop framed the problem, developed objectives and alternatives, and identified gaps in knowledge needed to predict consequences. The authors then took a multiple-step approach to address the knowledge gaps. Their first step was to model the existing monitoring data to estimate parameters determining amphibian metapopulation dynamics. Their second step was to use the estimated parameters in a simulation study to identify the parameters most sensitive to hydroperiod. Their third step was to design an experiment, informed by the simulation study, to determine the effect of hydroperiod length on metapopulation dynamics. Their fourth step was to apply the new knowledge to predict metapopulation dynamics in response to management scenarios and to identify the vernal pool management that would optimize amphibian persistence while considering costs. The optimal strategy for manipulating or creating pools over time depended on characteristics of the study area, but the approach is adaptable to other conservation areas.

In chapter 20, Sarah Converse analyzes the VOI associated with resolving multiple hypotheses for poor reproduction in the eastern migratory population of the whooping crane. Multiple competing explanations for reproductive failure persisted despite concerted efforts to research the problem. Management of the eastern migratory population and the ultimate success of reintroduction was sensitive to that structural uncertainty because the effectiveness of management remedies depended on the cause of reproductive failure. An expert elicitation process was used to predict the reproductive success in response to management actions conditional on each hypothesis, which provided the basis for EVPI and partial EVPI (EVPXI) calculations. Results from the VOI analyses were used to rank research and monitoring priorities.

Open Questions and Challenging Issues in Addressing Knowledge Gaps

Knowledge has specific roles to play in structured decision making (SDM), among them to inform the consequence analysis step in the PrOACT cycle

(Runge and Bean, chapter 1, this volume). Because of the central role of predictive science, SDM features inquiry—in contrast to advocacy—as the core approach to decision making (Garvin and Roberto 2013). The scientists, who build the predictive tools, serve as honest brokers in the SDM process (Pielke 2007). However, for this all to work as intended, the decision makers need to accept and trust the scientific predictions (Addison et al. 2013; Doremus and Tarlock 2005). Addison et al. (2013) discuss the issues that erode trust in predictive modeling, and they also outline ways to build trust, which can be a slow and easily reversed process. As the adage goes, "trust is gained by the thimbleful and lost by the bucketful." Integrating the science and the decision-making processes, for example through shared governance (Burgman 2015) and collaborative model building (Langsdale et al. 2013), provides the opportunity for science to play a useful role in addressing knowledge gaps.

A central question in the administration of science and information gathering is how best to allocate limited research effort. When that scientific effort is in service of public decision making, then VOI considerations can be an essential tool (Runge et al. 2011). Strictly speaking, an efficient allocation of research funding would align with the monetary equivalent of the VOI associated with the uncertainty the research is aimed to reduce (Howard 2007). The information-gathering professions that strive to provide decision-relevant knowledge would benefit from more fully integrating VOI analyses into how they allocate funds to research and monitoring (Canessa et al. 2015; Williams et al. 2011).

LITERATURE CITED

Addison PFE, Rumpff L, Bau SS, Carey JM, Chee YE, Jarrad FC, McBride MF, Burgman MA. 2013. Practical solutions for making models indispensable in conservation decision-making. *Diversity and Distributions* 19:490–502.

Blomquist SM, Johnson TD, Smith DR, Call GP, Miller BN, Thurman WM, McFadden JE, Parkin MJ, Boomer GS. 2010. Structured decision-making and rapid prototyping to plan a management response to an invasive species. *Journal of Fish and Wildlife Management* 1:19–32.

Burgman MA. 2015. Governance for effective policy-relevant scientific research: the shared governance model. *Asia & the Pacific Policy Studies* 2:441–451.

———. 2016. *Trusting Judgement: How to Get the Best out of Experts*. Cambridge, UK: Cambridge University Press.

Burgman MA, Lindenmayer DB, Elith J. 2005. Managing landscapes for conservation under uncertainty. *Ecology* 86:2007–2017.

Canessa S, Guillera-Arroita G, Lahoz-Monfort JJ, Southwell DM, Armstrong DP, Chadès I, Lacy RC, Converse SJ. 2015. When do we need more data? A primer on calculating the value of information for applied ecologists. *Methods in Ecology and Evolution* 6:1219–1228.

Dallalio EA, Brand AB, Grant EHC. 2017. Climate-mediated competition in a high-elevation salamander community. *Journal of Herpetology* 51:190–196.

Doremus H, Tarlock DA. 2005. Science, judgment, and controversy in natural resource regulation. *Public Land & Resources Law Review* 26:1–38.

Drescher M, Perera AH, Johnson CJ, Buse LI, Drew CA, Burgman MA. 2013. Toward rigorous use of expert knowledge in ecological research. *Ecosphere* 4(7):1–26.

Failing L, Gregory R, Higgins P. 2013. Science, uncertainty, and values in ecological restoration: a case study in structured decision-making and adaptive management. *Restoration Ecology* 21:422–430.

Garrard GE, Rumpff L, Runge MC, Converse SJ. 2017. Rapid prototyping for decision structuring: an efficient approach to conservation decision analysis. Pages 46–64 in Bunnefeld N, Nicholson E, Milner-Gulland EJ, eds. *Decision-Making in Conservation and Natural Resource Management: Models for Interdisciplinary Approaches*. Cambridge, UK: Cambridge University Press.

Gavin DA, Roberto MA. 2013. What you don't know about making decisions. Pages 75–93 in *On Making Smart Decisions: HBR's 10 Must Reads*. Cambridge, MA: Harvard Business Review Press.

Grant EHC, Wofford JEB, Smith DR, Dennis J, Hawkins-Hoffman C, Schaberl J, Foley M, Bogle M. 2014. Management and monitoring of the endangered Shenandoah salamander under climate change. *Natural Resource Report* NPS/SHEN/NRR—2014/867. Fort Collins, CO: National Park Service. http://pubs.er.usgs.gov/publication/70169065.

Gregory R, Long G, Colligan M, Geiger JG, Laser M. 2012. When experts disagree (and better science won't help much): using structured deliberations to support endangered species recovery planning. *Journal of Environmental Management* 105:30–43.

Hemming V, Burgman MA, Hanea AM, McBride MF, Wintle BC. 2018. A practical guide to structured expert elicitation using the IDEA protocol. *Methods in Ecology and Evolution* 9(1):169–180.

Howard RA. 1988. Decision analysis: practice and promise. *Management Science* 34(6):679–695.

———. 2007. The foundations of decision analysis revisited. Pages 32–56 in Edwards W, Miles RFJ, von Winterfeldt D, eds. *Advances in Decision Analysis: From Foundations to Applications*. Cambridge, UK: Cambridge University Press.

Langsdale S, Beall A, Bourget E, Hagen E, Kudlas S, Palmer R, Tate D, Werick W. 2013. Collaborative modeling for decision support in water resources: principles and best practices. *Journal of the American Water Resources Association* 49:629–638.

Maxwell SL, Rhodes JR, Runge MC, Possingham HP, Ng CF, McDonald-Madden E. 2015. How much is new information worth? Evaluating the financial benefit of resolving management uncertainty. *Journal of Applied Ecology* 52:12–20.

McCarthy MA, Keith D, Tietjen J, Burgman MA, Maunder M, Master L, Brook BW, Mace G, Possingham HP, Medellin R, Andelman S, Regan H, Regan T, Ruckelshaus M. 2004. Comparing predictions of extinction risk using models and subjective judgement. *Acta Oecologica* 26:67–74.

McGowan CP, Smith DR, Sweka JA, Martin J, Nichols JD. 2011. Multi-species modeling for adaptive management of horseshoe crabs and red knots in the Delaware Bay. *Natural Resource Modeling* 24:117–156.

Pielke RA, Jr. 2007. *The Honest Broker: Making Sense of Science in Policy and Politics*. Cambridge, UK: Cambridge University Press.

Regan HM, Colyvan M, Burgman MA. 2002. A taxonomy and treatment of uncertainty for ecology and conservation biology. *Ecological Applications* 12:618–628.

Runge MC, Converse SJ, Lyons JE. 2011. Which uncertainty? Using expert elicitation and expected value of information to design an adaptive program. *Biological Conservation* 144:1214–1223.

Speirs-Bridge A, Fidler F, McBride M, Flander L, Cumming G, Burgman M. 2010. Reducing overconfidence in the interval judgments of experts. *Risk Analysis: An International Journal* 30:512–523.

Sweka JA, Smith DR, Millard MJ. 2007. An age-structured population model for horseshoe crabs in the Delaware Bay area to assess harvest and egg availability for shorebirds. *Estuaries and Coasts* 30:277–286.

Walters CJ. 1986. *Adaptive Management of Renewable Resources*. New York: Macmillan.

Williams BK, Eaton MJ, Breininger DR. 2011. Adaptive resource management and the value of information. *Ecological Modelling* 222:3429–3436.

18 Developing Performance Criteria for a Population Model for Indiana Bat Conservation

Jennifer A. Szymanski and Lori B. Pruitt

The Indiana bat (*Myotis sodalis*) is a federally endangered species found in 22 states in the eastern United States. The US Fish and Wildlife Service (USFWS) is responsible for developing and implementing recovery programs for species listed under the Endangered Species Act of 1973 (ESA). In meeting these regulatory obligations, the USFWS is routinely required to make decisions on how specific actions may affect recovery objectives. The emergence and spread of the infectious disease white-nose syndrome (WNS) poses a significant threat to the Indiana bat and underscores the importance of being able to predict both near- and far-term consequences of actions that may be implemented across the range of the species. The lack of a predictive population model for the effects of alternative actions on the species' viability has hampered the USFWS's ability to make consistent, scientifically robust decisions across the range of this species. Rather than issue a generic request for proposals (RFP) to develop such a model, we used a decision analytic framework to identify specific performance criteria for the model. This process led to a successful contract for model development. The model is currently used throughout the range of the species to evaluate impacts from proposed actions such as wind energy projects to recovery related conservation actions. The result is improved consistency, transparency, and rigor of USFWS' impact analyses.

Problem Background

The Indiana bat is a federally endangered species pursuant to the Endangered Species Act of 1973 (ESA 1973, as amended) and thus afforded the full protections of the ESA. The US Fish and Wildlife Service (USFWS), as the federal agency charged with implementation of the ESA, is tasked with making management decisions through several ESA-mandated programs. The primary programs include (1) interagency consultation with federal agencies under Section 7; (2) issuance of incidental take permits under Section 10(a)(1)(B) of the ESA; (3) issuance of scientific and recovery permits under Section 10(a)(1)(A); and (4) development and implementation of recovery actions. In carrying out these programs, the USFWS recognized the need for a predictive population model to evaluate the consequences of alternative actions on the survival and recovery of the species. This need was further accentuated in 2007 by the outbreak of white-nose syndrome (WNS), a highly infectious disease of hibernating bats (Blehert et al. 2009). Since its discovery, WNS has spread across the Indiana bat range (www.whitenosesyndrome

.org/resources/map), leaving in its wake devastated bat populations. The emergence of this novel threat increased the urgency for a population model that could forecast short- and long-term consequences of various alternative actions.

Typically, USFWS would have developed a simple RFP describing the intended purpose for the model and the required outputs. However, building a predictive model entails making a multitude of decisions about the structure of a model. As these model-specific decisions can influence the performance of the model, they should be explicitly assessed. Furthermore, USFWS needed the model to be scientifically rigorous to ensure the reliability of the model results and to engender credibility. They also wanted the model to be readily explainable to foster internal (USFWS biologists and managers) and external (project proponents and other stakeholders) support for the model results. As the objectives for the model are complex and may require trade-offs, USFWS elected to use a structured approach to discern the desired elements and functionality of the model. These performance criteria were used to develop a precise and objective-specific RFP for the development of a standardized, robust population model to predict consequences at multiple temporal and spatial scales. The model building effort was constrained by time (due to the urgency of the conservation problem) and funding.

Ecological Background

The Indiana bat is a temperate, insectivorous, migratory bat that hibernates colonially in caves and mines in the winter. In spring, reproductive females migrate and form maternity colonies in wooded areas where each female bears a single pup that is raised within the colony. Maternity roosts are typically behind the exfoliating bark of large, often dead trees. Both males and females return to hibernacula in late summer or early fall to mate and enter hibernation.

Indiana bats are exposed to varying stressors throughout the annual cycle including loss and degradation of forested roosting and foraging habitat, degradation of hibernacula, human disturbance during hibernation, and pesticides. The degree and severity of these threats vary across the geographic range of the species. The Indiana bat was originally listed as in danger of extinction under the Endangered Species Preservation Act of 1966 and is currently listed as endangered under the ESA. Long before standardized surveys were conducted for Indiana bats, bat biologists noted long-term declines in populations of Indiana bats (Greenhall 1973). From 1965 to 2001, there was a consistent decline, estimated at nearly 60% (USFWS unpublished data), in Indiana bat numbers across the species' range. From 2001 to 2007, biennial surveys suggested an increase of more than 25% in the rangewide population numbers, creating optimism that species recovery was underway.

The occurrence of WNS in the winter of 2006–2007 abruptly halted increasing trends in population numbers (Turner et al. 2011). From 2007 through 2015, the Indiana bat population rangewide has declined almost 20% (USFWS unpublished data; www.fws.gov/midwest/endangered/mammals/inba/index.html). WNS is now found throughout the range of the Indiana bat, and declines in the rangewide population are expected to continue. WNS is the most pervasive threat throughout the range of the Indiana bat; it has increased the sensitivity of the species to adverse impacts of other stressors (Thogmartin et al. 2013) and hence heightened the need and urgency for a population model. The rapid spread of WNS and ensuing precipitous decline prompted USFWS to pursue the development of a population model and was the primary impetus for the structured decision-making exercise described in this chapter.

Decision Maker and Their Authority

The decision maker for development of the population model was the field supervisor of the Bloomington, Indiana, field office, the national recovery lead for this species. USFWS biologists with Indiana bat recovery and consultation responsibilities across the species range counseled the decision maker on the utility of the model.

Decision Structure

The decision problem—to specify the performance criteria for development of a population model that predicts the consequences of management decisions—was inherently nested within a larger decision framework. In other words, the model-building problem is nested within the endangered species conservation problem (fig. 18.1). It is this ESA conservation decision context that we needed to understand to discern the required elements and functionality of the model. Thus, we looked to the decision contexts for which USFWS must make Indiana bat management decisions to decipher the specifications for the model. Although the model would be refined over time, we viewed developing the model as a multiple-objective, one-off decision.

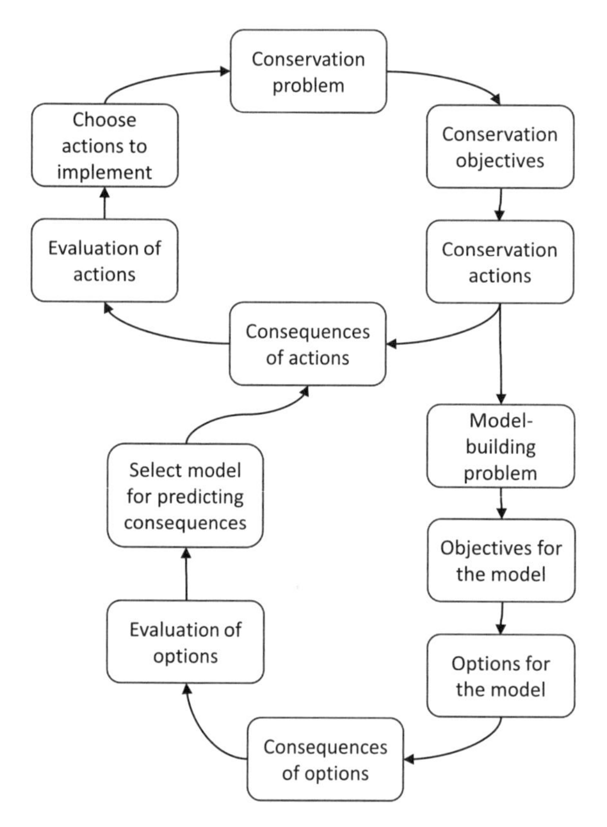

Figure 18.1. Schematic showing the nested decision structure where a model building problem is nested within an endangered species (ES) conservation problem. The process we describe in this chapter addresses the nested problem of building a population model to use as a tool for predicting consequences for ES decisions.

ESA Conservation Decision Context

To guide management decisions and implement effective conservation strategies in light of WNS, USFWS needed a population model that credibly predicted the effects of conservation-related and non–conservation-related actions. We developed decision structures for the types of decisions in which the model would be used: recovery planning, scientific take permitting, interagency consultation, and incidental take permitting decisions. We began by sketching out decision structures for case studies representing each decision context.

For all ESA programs, the overarching objective is to conserve listed species. How each program contributes to this overall goal varies, and while the details of decision problems are case-specific, the decision structures are consistent within program areas. Therefore, we used specific decision problems to identify the objectives and alternatives associated with each program area. The case studies evaluated included

1. Recovery planning: allocating funds among several potential recovery actions.
2. Section 7 interagency consultation: determining whether the US Federal Highways Administration's proposed action of providing funds to construct, maintain, and operate an interstate highway from Indianapolis to Evansville, Indiana, would appreciably reduce the likelihood of survival and recovery of the Indiana bat.
3. Scientific take permit issuance: determining whether to issue a scientific permit to collect little brown bats (*Myotis lucifugus*) from Great Scott Cave in Missouri, as part of a study of WNS. Indiana bats also use the cave for hiberna-

tion and thus were likely to be disturbed by the collection of little brown bats. USFWS would decide whether or not to issue the requested permit and, if applicable, specify conditions to minimize take to the extent practicable.

These case studies were used to understand the desired functionality and to discern the required elements of the population model.

ESA Conservation Decision Context Objectives

The fundamental objectives identified for the 4 program areas included: (1) minimize the impact of incidental take (e.g., harm or harassment); (2) maximize the probability of recovery; (3) minimize the probability of appreciably reducing the likelihood of survival and recovery (referred to as "jeopardy"); and (4) minimize the probability of rangewide extinction. These objectives are interrelated but include program-specific aspects. For example, although reducing the impact of incidental take is a means to achieving recovery, the ESA also specifically requires that the impact from incidental take be minimized to the maximum extent practicable. Similarly, minimizing the likelihood of jeopardy might appear to be a means to avoiding rangewide extinction, but the two are not often the same. Jeopardy entails more than simply avoiding extinction; it also requires that the likelihood of recovery is not appreciably reduced. Thus, these are unique program-specific fundamental objectives for which we need a population model to compare our management alternatives against.

There are policy elements of these objectives that are not yet resolved: Over what time frames should extinction and recovery be evaluated? What thresholds of risk are tolerable? Because of these policy uncertainties, the model needed to predict the cumulative probability of extinction over a range of time frames and a range of risk levels. Additionally, the model needed to predict the probability of extinction or recovery at multiple biological scales. For example, to assess the effect of a management decision on the probability of rangewide extinction, the population model must forecast extinction risk at the species level. However, to assess the effect of a management decision on the likelihood of recovery, the model needed to forecast the consequences of take of individuals at recovery unit and population scales. Recovery units are designations the USFWS may use to maintain the distribution of wide-ranging species that have multiple populations or varying ecological pressures in different parts of the range.

An assortment of questions arose from our case studies highlighting the need for a series of structural requirements in the model. For example, the behavior of Indiana bats gives rise to differences in structure among sex and age classes. Specifically, males and females are separate for much of the annual cycle and experience different energetic needs and ecological stressors (Bergeson 2017; Brack et al. 2002). Based on studies of other vespertilionids, for example, it is likely that first-year females have lower and more variable reproductive rates than older females (e.g., Pearson et al. 1952; Racey 1982; Schowalter and Gunson 1979) and in some cases more closely resemble males in behavior and thus energetic needs (USFWS 2007). As the types of actions typically reviewed by USFWS can take place at any time during the year, the model needed to account for these differences between sex and age classes. This was accomplished by tracking the population from the beginning of hibernation through summer just before parturition and by including 5 age and sex classes in the summer and 4 age and sex classes in winter. The resulting model structure and the state variables with their respective survival (s) and reproductive (p, b) parameters are shown in figure 18.2.

The case studies also illuminated the need to incorporate environmental stochasticity. Specifically, we noted the importance of capturing variability in the survival and reproductive rates as well as in the hibernacula-specific winter survival rates.

Similarly, we determined that it was important to account for the temporal differences among the 3

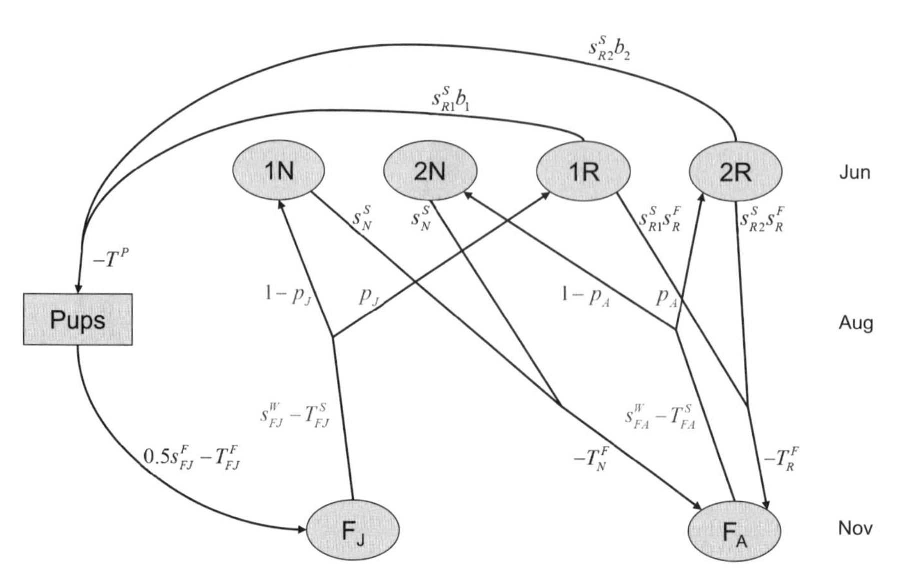

Figure 18.2. Schematic life-cycle diagram for the Indiana bat population model, female segment only and without spatial structure. The winter state variables (number of juvenile females, F_J; and number of adult females, F_A) are measured in November, just at the beginning of hibernation. The summer state variables (number of nonreproductive yearlings, 1N; number of nonreproductive older adults, 2N; number of reproductive yearlings, 1R; and number of reproductive older adults, 2R) are measured in June, just before parturition. Pups is an intermediate state variable, measured at volancy. The transitions from winter to summer include overwintering survival (s^W), spring take (T^S), and propensity to reproduce. The transitions from summer to winter include summer survival (s^S), fall survival (s^F), and fall take (T^F). The reproductive transitions include summer survival of reproductive females (s^S), reproductive rates (b), take of pups (T^P), fall juvenile survival (s^F_{FJ}), and fall take of juveniles (T^F_{FJ}). Note that lowercase transition rates (s, b, and p) are multipliers to the size of the population segment being affected; uppercase transition parameters (T) are absolute measures of take.

different types of take (T). Specifically, we defined spring take (TS) as take during late winter, spring migration, and pre-reproductive season (pre-volancy); fall take (TF) as take that occurs during fall migration through early winter; and pup take (TP) as take of pre-volant pups without coincident take of the mother. Although we differentiated among the sex classes, only female take will be modeled.

As we mentioned previously, the model must be able to predict consequences at various biological scales. Given the urgency for the model, we elected to focus on the recovery unit scale as our first prototype. We developed separate models for the 4 recovery units. Taken together, these models constitute a rangewide model, and thus, by focusing on the recovery unit scale, we could predict consequences at both the recovery unit and rangewide scales. However, we recognized that it was important to track some spatial heterogeneity at the scale of the hibernacula. Thus, the model was structured to allow specification of several (on the order of 2–10) major winter populations per recovery unit (fig. 18.3).

Last, the model will be used to investigate a single baseline scenario or compare multiple scenarios with or without proposed actions. For use of the model to investigate a single, baseline scenario, the user specifies a set of parameters (survival, take, and recruitment values) with the intent of describing biological status. However, often the model will be used to compare 2 or more scenarios, where one of the scenarios describes the baseline conditions and the other describes implementation of the proposed actions. Here, the user is interested specifically in the biological condition of the species with and without the proposed scenarios. To

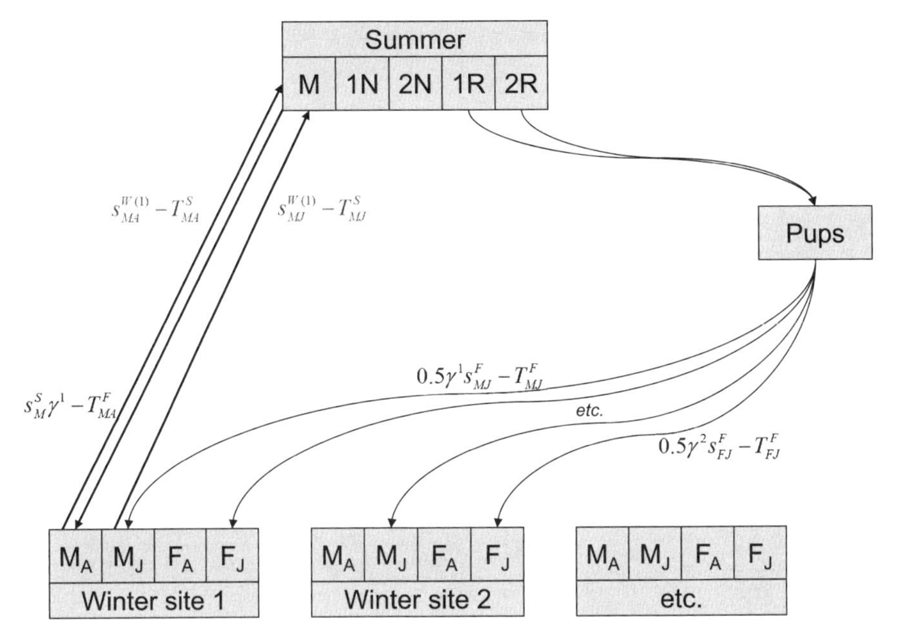

Figure 18.3. Schematic life-cycle diagram for the Indiana bat population model, including spatial structure and all sex and age classes. Note, the female transitions are not shown (see fig. 18.1), and not all of the spatial transitions are drawn. The desired model will be structured at the regional level, with a single regional summer population but several distinct winter populations with differing survival rates. The spatial structure is captured by an additional set of parameters (γ's) to govern the distribution of the summer population to the winter sites. Take of absolute numbers of animals can occur at 3 points: "spring" take (T^S) includes late winter take, take during spring migration, and prereproductive (pre-volancy) take on the breeding grounds; "fall" take (T^F) includes take during fall migration, swarming, and early winter; and pup take (T^P) includes take of prevolant pups without concurrent take of the mother.

make these comparisons, the model should be capable of running the scenarios in parallel, controlling all aspects of the model among the scenarios, while varying only the parameters in question.

ESA Conservation Decision Context Alternatives

The alternatives for interagency consultation and incidental take permitting do not vary among decisions; however, those for scientific take permitting and recovery planning are case-specific. All alternatives are expected to yield some portfolio of changes in demographic parameters. For example, an alternative for a permitting decision, such as "issue the research permit as requested," could specify a level of harassment during hibernation to predict a subsequent reduction in reproductive potential for some fraction of survivors. Similarly, an alternative for recovery planning, such as "implement a specific recovery action," might be designed to improve roosting habitat quality and predict an increase in reproductive success for some individuals. Thus, the model needed to accommodate different combinations and levels of the demographic parameters. Specifically, the user should be able to specify the following demographic parameters:

1. Lethal take of individuals at various points in the annual cycle.
2. Change in recruitment, positive or negative (either in propensity to reproduce, or in survivorship of pups to volancy or adulthood).
3. Change in survival, either positive or negative (winter survival, summer survival, or survival during migration).

Changes to these parameters may have a temporal component. For example, take might be expected to occur at a particular level for 5 years and then decrease to some lower level after that. Conversely, a recovery action might be expected to improve recruitment gradually over 10 years. Thus, the model needed to also account for temporal effects.

Understanding the decision structures, including the objectives and alternatives, of the specific decision contexts for which the model will be used allowed us to describe the desired functionality of the model and discern the required elements. These, in turn, became the fundamental objectives and performance criteria for developing the model.

Objectives for Predictive Model Building Decision

Based on agency-specific practices and the specific decision contexts in which the model will be used, we identified 3 fundamental objectives for developing a population model. Two are relatively simple and straightforward. The third objective addresses model functionality and thus is multifaceted. The 3 objectives were used to develop an objective-based RFP. The objectives included:

- *Maximize scientific rigor and integrity.* This objective entailed ensuring the model structure and parameterization comport with the biology of the species and are based on the best available information. This will ensure USFWS meets the mandate of the ESA to use the best available scientific and commercial data available as well as providing creditability and reliability to the model results and thus USFWS management decisions.
- *Maximize user-friendliness.* This objective entailed ensuring the model platform is available to USFWS biologists and the user interface is within the expertise of USFWS biologists. This will engender internal support and thus increase the use of the model in decision making.
- *Maximize model functionality.* This objective entailed ensuring the model: (1) can predict the cumulative probability of extinction and recovery over a range of timeframes and risk levels, (2) can predict such probabilities at multiple spatial scales, (3) allows comparison between baseline and proposed scenarios, (4) includes relevant demographic, spatial, and temporal structure, (5) allows for changes to baseline parameters, (6) accommodates temporal effects (e.g., annual, multi-year, one-time) in parameter values, and (7) adequately incorporates stochasticity and uncertainty.

Alternatives and Consequences for Predictive Model Building Decision

Ideally, in response to the RFP, USFWS would have received several proposals with alternative model constructs and selected the alternative that best met our fundamental objectives. However, USFWS received only 1 proposal. As this proposal addressed nearly all of our objectives and given the urgency, we selected the proposal and did not pursue additional proposals. Thus, there was no need to compare proposals or conduct more decision analysis.

Discussion

Despite not completing a full decision analytical process, we achieved the primary purpose we set out to accomplish. The insights obtained from this decision analysis provided better clarity into the intended uses of the population model, which led to a detailed, objective-oriented RFP and, ultimately, to the development of a useful, scientifically rigorous predictive population model (see Thogmartin et al. 2013 for details of the model). Specifically, by examining decision structures for sample decisions for which the model would be used, we were able to discern more precisely the needs of the model and hence gained a fuller understanding of the objectives for the model. Further, the model provides a tool that comports with the best available science and offers the functionality needed to analyze the effects of actions on the species, both of which were fundamental objectives. The decision structures developed for the case studies provide at-ready decision frameworks (i.e., objectives and alternatives and a model to evaluate the consequences) for use with future project reviews. This was an unexpected time-saving benefit from the decision process.

Although meeting 2 fundamental objectives, the produced model did not satisfactorily achieve the 3rd objective for a user-friendly model, and the development of the model took much longer than anticipated. The prototype did not analyze how well the user-friendly objective would be met. Had we explored different model structures and tool platforms during the prototype phase, we would have certainly better accounted for the user-friendly objective. Similarly, a fuller exploration of candidate tool platforms would have necessarily led to a discussion of current technologies and shortened the time for model development.

Although not perfect, in the end, the decision process gave clarity to how USFWS wanted the model to perform under the various ESA decision contexts and provided for a more precise RFP. We believe the decision process (fig. 18.1) led to a better-calibrated model than a typical RFP (based on general notions of what a population model should accomplish) would have.

ACKNOWLEDGMENTS

The decision process was initiated at an SDM workshop sponsored by the National Conservation Training Center. The workshop team members included Mike Armstrong, Andy King, Robyn Niver, Lori Pruitt, Dale Sparks, and Jennifer Szymanski. Mike Runge and Conor McGowan led the team through the SDM process and helped develop a conceptual framework for the model. Following the SDM workshop, the USFWS team worked with Wayne Thogmartin, Carol Sanders-Reed, and Richard Erikson to develop the model. The team also elicited additional expert input at various stages of the model development. The following individuals provided expert information: Robert Barclay, Eric Britzke, Paul Cryan, Winifred Frick, Scott Johnson, Allen Kurta, and T. O'Shea. The findings and conclusions in this article are those of the authors and do not necessarily represent the views of the US Fish and Wildlife Service.

LITERATURE CITED

Bergeson S. 2017. Multi-scale analysis of roost characteristics and behavior of the endangered Indiana bat (*Myotis sodalis*). PhD thesis. Terre Haute, Indiana: Indiana State University. 188 pp.

Blehert DS, Hicks AC, Behr M, Meteyer CU, Berlowski-Zier BM, Buckles EL, Coleman JTH, Darling SR, Gargas A. Niver R, Okoniewski JC, Rudd RJ, Stone WB. 2009. Bat white-nose syndrome: an emerging fungal pathogen? *Science* 323:227.

Brack V Jr, Stihler CW, Reynolds RJ, Butchkoski CM, Hobson CS. 2002. Effect of climate and elevation on distribution and abundance in the Mideastern United States. Pages 21–28 in Kurta A, Kennedy J, eds. *The Indiana Bat: Biology and Management of an Endangered Species*. Austin, TX: Bat Conservation International.

Greenhall A. 1973. Indiana bat: a cave-dweller in trouble. *National Parks Conservation Magazine* 47:14–17.

Pearson OP, Koford MR, Pearson AK. 1952. Reproduction of the lump-nosed bat (*Corynorhinus rafinesquei*) in California. *Journal of Mammalogy* 33:273–320.

Racey PA. 1982. Ecology of bat reproduction. Pages 57–104 in Kunz TH, ed. *Ecology of Bats*. New York: Plenum Press.

Schowalter DB, Gunson JR. 1979. Reproductive biology of the big brown bat (*Eptesicus fuscus*) in Alberta. *The Canadian Field Naturalist* 93:48–54.

Thogmartin WE, Sanders-Reed C, Szymanski JA, King RA, Pruitt L, McKann PC, Runge MC, Russell RE. 2013 White-nose syndrome is likely to extirpate the endangered Indiana bat over large parts of its range. *Biological Conservation* 160:162–172.

Turner GG, Reeder DM, Coleman JTH. 2011. A five-year assessment of mortality and geographic spread of white-nose syndrome in North American bats and a look to the future. *Bat Research News* 52(2):13–27.

US Endangered Species Act [ESA] of 1973, as amended, Pub. L. No. 93-205, 87 Stat. 884 (Dec. 28, 1973). www.fws.gov/endangered/esa-library/pdf/ESAall.pdf.

US Fish and Wildlife Service [USFWS]. 2007. *Indiana Bat (Myotis sodalis) Draft Recovery Plan: First Revision*. Fort Snelling, MN: US Fish and Wildlife Service Report. http://citeseerx.ist.psu.edu/viewdoc/download?doi=10.1.1.369.1892&rep=rep1&type=pdf.

19

Adam W. Green and
Larissa L. Bailey

Using a Research Experiment to Reduce Key Uncertainty about Managing Vernal Pool Habitats for Obligate Amphibian Species

Climate change and continued urbanization are the main threats to amphibian populations in the mid-Atlantic and eastern regions of the United States. Forested federal lands are becoming isolated and interspersed in a land use matrix that is inhospitable to amphibian species that rely on seasonal (vernal) pools for breeding. National wildlife refuges and national parks are legally obligated to maintain and restore ecosystems, and in the northeastern United States, vernal pool habitats are identified as priority systems for research and monitoring. Refuge and park managers require a flexible research framework for learning about how specific management actions, such as lengthening the hydroperiod of vernal pools, may affect conservation objectives. In this chapter, we highlight how existing monitoring data can help identify research priorities within a structured decision framework. We developed and applied a model of patch occupancy dynamics linked to an objective function and used sensitivity analyses to identify uncertainties critical to achieving managers' objectives. These elements guided cost-effective experimental design aimed at reducing these uncertainties.

Problem Background

Habitat loss or modification and climate change are the main threats to many amphibian populations (e.g., Collins and Storfer 2003). In the mid-Atlantic and eastern regions of the United States, forested federal lands are becoming isolated and interspersed in a matrix that is inhospitable to amphibian species that rely on seasonal (vernal) pools. Populations of these species are likely to decline and may go locally extinct under future climate and land use scenarios (Brooks 2004). Federal land managers are considering active management to conserve amphibian populations on their lands, but considerable uncertainty exists about what management strategies should be implemented to reverse, halt, or prevent amphibian declines. To tackle this uncertainty, we developed an adaptive management framework at a local scale (e.g., an individual refuge or park) that could be adopted or modified to meet the needs and objectives of federal land managers. We chose 2 areas that have relatively large and small numbers of available breeding habitat (vernal pools): Patuxent Research Refuge (PRR) and Rock Creek Park (ROCR) located near Washington, DC, USA (fig. 19.1).

Decision Maker and Their Authority

Under the National Wildlife Refuge Improvement Act of 1997, a primary mission of the National Wildlife Refuge (NWR) system is wildlife conservation. Conservation involves maintaining and restoring

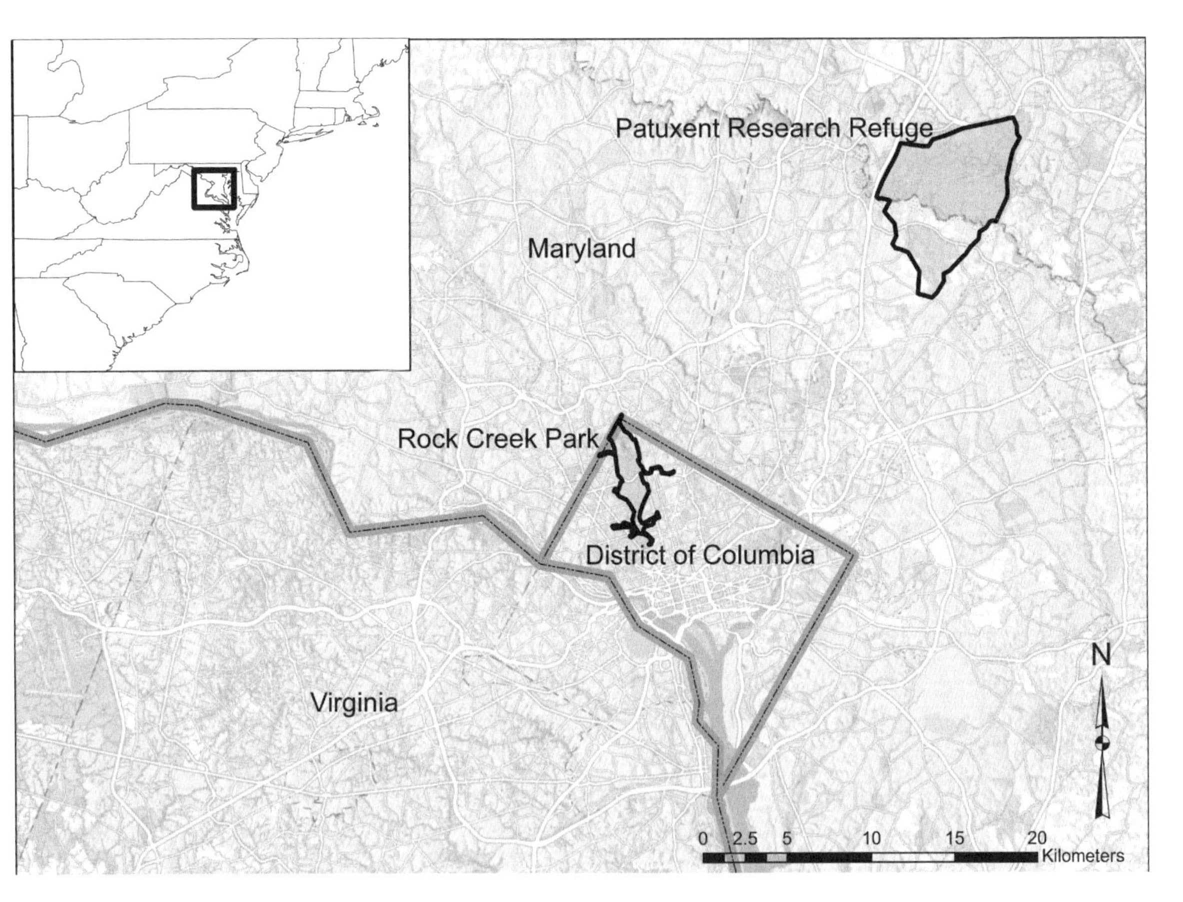

Figure 19.1. Location of study areas within the mid-Atlantic region of the United States.

representative ecosystems. In addition, the act mandates that the US Fish and Wildlife Service develop comprehensive conservation plans for all NWRs, and several Northeastern refuges list forested wetlands, or vernal pools, as a conservation and management priority (e.g., Rachel Carson NWR, Canaan Valley NWR). Likewise, National Park Service (NPS) managers are directed by federal law and NPS policies to determine the status and trends of natural resources under their Inventory and Monitoring Program (National Park Service 2005). Several NPS networks have identified amphibians as a priority taxonomic group (e.g., National Capital Region Network).

Ecological Background

Vernal pools are recognized for their unique assemblages of species, including an array of obligate amphibians (Calhoun and deMaynadier 2008). Climate change and continued urbanization are major threats to these habitats and their associated amphibian populations. Even though human population growth has slowed in the region, land conversion rates continue to accelerate (NAST 2001), and remaining pro tected federal lands are often surrounded by inhospitable urban and suburban development. Climate change models predict increased temperatures and variability of precipitation, which will likely alter pool hydroperiod (number of days per year that a pool is wet) and increase the incidence of amphibian reproductive failure (Taylor et al. 2006).

Wood frogs (*Lithobates sylvatica*) breed almost exclusively in vernal pools near forested upland habitat (Kenney and Burne 2001). Historically widespread, wood frogs decline with increasing urbanization (Rubbo and Kiesecker 2005); they are uncommon in urban detention ponds (Simon 2006) and in impoundments found on federal lands (Julian et al. 2006), primarily because these permanent wetlands contain predators such as fish and other large anuran species. Adult wood frogs migrate to vernal pools in the early spring, mate, and deposit egg masses. The duration of the breeding season is short (1–3 weeks), and each female lays a single egg mass (Lannoo 2005). Eggs develop and hatch within 2 weeks, and metamorphosed animals emerge 4–6 weeks after hatching.

Conscious of potential declines of at-risk amphibians and their habitats, 14 national parks and national wildlife refuges in the northeastern United States initiated programs in the early 2000s to locate vernal pools and monitor wood frog populations (Mattfeldt et al. 2009; Van Meter et al. 2008). Initial analyses suggested that hydroperiod influenced wood frog breeding probability at pools (Mattfeldt et al. 2009). However, some managers expressed concern that the monitoring programs only measured adult breeding and did not distinguish those habitats that may be reproductive sinks (Schlaepfer et al. 2002). For example, at Patuxent Research Refuge (PRR), where metamorph data had also been collected, approximately 25–50% of pools supported wood frog breeding, but only 10–20% produced metamorphs (Green et al. 2013). Importantly, little information existed about what management strategies should be implemented to conserve vernal pool habitats and prevent amphibian declines.

Decision Analysis

First Prototype Workshop

In April 2008, amphibian biologists, hydrologists, and local refuge managers gathered at PRR to develop a prototype adaptive management framework for wood frog populations on PRR that would be flexible enough to be adopted by other parks and refuges. Management objectives were agreed upon quickly but were necessarily vague due to the lack of historical data to determine an appropriate decision threshold, where changes in conditions result in a different optimal management strategy. Managers wanted to ensure the long-term persistence of wood frog populations within the park or refuge, subject to cost constraints. They thought the best way to ensure persistence was to maintain some number of pools (or a percentage of pools) with successful reproduction (metamorphs). However, there was substantial uncertainty regarding the number (or percentage) of pools with annual reproduction needed to sustain the local amphibian metapopulation.

Various alternative actions were discussed, including (1) creation of new vernal pool habitat, (2) manipulation of existing vernal pool habitat, (3) diversion or addition of water to drying pools, and (4) translocation of egg masses from drying pools. Permanent habitat manipulations were favored by managers because they required no annual monitoring and associated actions, but how to create and maintain seasonal vernal pools is not entirely understood. Most attempts to create vernal pools have been unsuccessful (Lichko and Calhoun 2003), resulting in permanent or semipermanent hydroperiods that are subject to invasion by predatory species (Vasconcelos and Calhoun 2006).

Additionally, decision makers lacked demographic information critical to developing models of metapopulation response to the proposed management actions. Estimates of local wood frog colonization and extinction probabilities were almost nonexistent. Understanding the influence of these parameters on metapopulation dynamics would help managers target actions that could influence these processes and focus research experiments to determine the effects of such actions.

We developed a multistate occupancy model to (1) determine the appropriate ecological thresholds for management objectives and (2) identify important processes likely to influence the wood frog metapopulation dynamics (local extinction and colonization probabilities). This information led to an experiment to determine the effects of management actions on wood frog metapopulation processes. Results were incorporated into a utility function that combines ecological and financial objectives to provide management recommendations for local amphibian metapopulations.

Model Building and Research

We developed a second-order Markov occupancy model to estimate the probabilities that a pool supports wood frog breeding and successful reproduction or metamorphosis (Green et al. 2013). The occurrence of breeding frogs at a pool in year t is likely to be influenced by 2 processes: (1) the return of breeding females from the previous year ($t-1$) and (2) new breeders that successfully metamorphosed 2 years prior ($t-2$) and returned to their natal pool. Pool i could support wood frog breeding in year t with probability ψ^1_{it}, given breeding and metamorph occupancy states in the previous years $t-1$ and $t-2$, respectively. If a pool supported breeding in year t, it could support metamorphosis with probability ψ^2_{it} (see Green et al. 2013 for detailed model description).

We fit this model to occupancy data collected at the 2 refuges with drastically different numbers of vernal pools. Multiple surveys were conducted during the 2006–2012 breeding and metamorphosis periods at 53 of the estimated 2,200 vernal pools at PRR (Van Meter et al. 2008) and at all 9 pools at ROCR from 2008 to 2011. We explored the effects of hydroperiod length, spatial arrangement, and our chosen management action (see below) on these parameters. Hydroperiod length included 3 categories: (1) short-hydroperiod pools rarely retained water through early June, (2) medium-hydroperiod pools retained water through early June in some years, and (3) long-hydroperiod pools retained water through early June in most years. Spatial arrangement, explored at PRR only, described whether a pool was near (<300 m) another pool (clustered) or isolated. We found that all factors (previous occupancy states, hydroperiod length, and spatial arrangement) influenced the probability that a pool supported breeding (Green and Bailey 2015). Short-hydroperiod pools that did not support breeding or metamorphosis in previous years had a very low probability of supporting breeding (typically $\psi^1_{it} < 0.15$; tables 19.1 and 19.2). That probability increased at pools that had longer hydroperiods and supported breeding and/or metamorphosis in prior years (tables 19.1 and 19.2). If a pool supported breeding, the probability of successful reproduction was fairly high ($\psi^2_{it} \geq 0.30$) for medium- and long-hydroperiod pools at PRR and long-hydroperiod pools at ROCR (table 19.3).

We initially had no information on the response of wood frogs to potential management actions. Therefore, we used our estimated mean demographic rates to perform a sensitivity analysis to identify which occupancy parameters would be most influential in increasing the proportion of pools producing metamorphs (Green et al. 2011). Sensitivity analyses quantify the change in a parameter of interest (e.g., ψ^2_{it}) when values of lower-level parameters change and can help identify management actions to target these dynamic parameters or processes. Our sensitivity analyses revealed that increasing colonization rates for short-hydroperiod pools that lacked breeding in recent years had the most influence on increasing reproductive success, followed by increasing the conditional probability of metamorphosis (Green et al. 2011).

Accordingly, we experimentally manipulated a subset of existing pools at PRR to extend their hydroperiod length. We intended to lengthen the hydroperiod enough to ensure successful metamorphosis but not enough to create permanent pools, which could result in the colonization of predators, such as green frogs (*Lithobates clamitans*). After metamorphosis in 2009, we deepened 4 poorly producing pools and installed a liner to reduce water loss through seepage and evapotranspiration (Biebighauser 2007; Green et al. 2013). We expected this action would (1) increase the probability of colonization by creating more suitable breeding habitat for dispersing frogs and (2) increase the probability of metamorphosis by allowing more time for tadpoles to develop prior to drying.

Occupancy surveys continued after the installation of liners to test the efficacy of the management action. We fit the occupancy model described above to an appended data set including natural and manipulated pools. We found that breeding occupancy probabilities at manipulated pools were similar to those of long-hydroperiod pools and metamorph occupancy rates were between those of short- and medium-hydroperiod pools (tables 19.1 and 19.3). Because management was only implemented at 4 pools and we only had 1 year of data after management occurred, our estimates of the management effects were imprecise. However, our results suggested that management would increase the proportion of pools producing metamorphs.

Decision Solution

Our experiment and analysis of breeding and metamorph occupancy data helped reduce sources of uncertainty associated with wood frog metapopulation dynamics. Wood frog populations are highly variable (Berven 1990), and several years of data are necessary to understand factors influencing annual varia-

Table 19.1. Means (SD) of posterior predictive distributions of the probability of vernal pools supporting wood frog breeding ($\hat{\psi}^1$) at Patuxent National Research Refuge, Maryland, USA

Hydroperiod	Spatial	r_{t-1}[a]	s_{t-2}[a]	2008	2009	2010
Short	Clustered	0	0	0.058 (0.023)	0.033 (0.017)	0.140 (0.053)
		0	1	0.094 (0.095)	0.059 (0.069)	0.192 (0.152)
		1	0	0.125 (0.065)	0.074 (0.044)	0.270 (0.125)
		1	1	0.168 (0.141)	0.108 (0.111)	0.314 (0.194)
	Isolated	0	0	0.148 (0.064)	0.089 (0.049)	0.307 (0.112)
		0	1	0.200 (0.159)	0.132 (0.127)	0.356 (0.206)
		1	0	0.274 (0.121)	0.176 (0.096)	0.482 (0.161)
		1	1	0.319 (0.194)	0.221 (0.170)	0.511 (0.211)
Medium	Clustered	0	0	0.385 (0.107)	0.258 (0.100)	0.611 (0.122)
		0	1	0.426 (0.207)	0.310 (0.200)	0.627 (0.190)
		1	0	0.569 (0.116)	0.422 (0.122)	0.763 (0.108)
		1	1	0.592 (0.177)	0.457 (0.195)	0.774 (0.136)
	Isolated	0	0	0.615 (0.156)	0.477 (0.172)	0.792 (0.123)
		0	1	0.630 (0.206)	0.505 (0.230)	0.795 (0.150)
		1	0	0.766 (0.118)	0.648 (0.150)	0.886 (0.082)
		1	1	0.776 (0.139)	0.664 (0.182)	0.893 (0.083)
Long	Clustered	0	0	0.976 (0.053)	0.960 (0.082)	0.989 (0.028)
		0	1	0.978 (0.055)	0.963 (0.086)	0.990 (0.028)
		1	0	0.989 (0.023)	0.981 (0.039)	0.995 (0.011)
		1	1	0.991 (0.022)	0.984 (0.038)	0.996 (0.010)
	Isolated	0	0	0.989 (0.028)	0.981 (0.045)	0.995 (0.016)
		0	1	0.990 (0.028)	0.982 (0.046)	0.995 (0.013)
		1	0	0.995 (0.011)	0.992 (0.018)	0.997 (0.006)
		1	1	0.996 (0.009)	0.993 (0.015)	0.998 (0.004)
Managed	Clustered	0	0	–	–	0.959 (0.104)
		0	1	–	–	0.961 (0.099)
		1	0	–	–	0.973 (0.068)
		1	1	–	–	0.976 (0.067)
	Isolated	0	0	–	–	0.978 (0.056)
		0	1	–	–	0.979 (0.047)
		1	0	–	–	0.987 (0.043)
		1	1	–	–	0.988 (0.043)

Source: Reproduced from Green and Bailey 2015 (table 1).

[a] The breeding occupancy status in year $t-1$ is represented by r_{t-1} and the metamorph occupancy status in year $t-2$ by s_{t-2} (0 = absent, 1 = present).

Table 19.2. Means (SD) of posterior predictive distributions of the probability of vernal pools supporting wood frog breeding ($\hat{\psi}^1$) at Rock Creek National Park, Washington, DC, USA

Hydroperiod	r_{t-1}[a]	2009	2010	2011
Short	0	0.061 (0.118)	0.253 (0.246)	0.079 (0.141)
	1	0.098 (0.164)	0.365 (0.336)	0.127 (0.191)
Medium	0	0.478 (0.280)	0.830 (0.191)	0.571 (0.266)
	1	0.555 (0.246)	0.860 (0.182)	0.640 (0.234)
Long	0	0.929 (0.183)	0.991 (0.033)	0.944 (0.166)
	1	0.960 (0.115)	0.993 (0.026)	0.971 (0.090)

Source: Reproduced from Green and Bailey 2015 (table 2).

[a] The breeding occupancy status in year $t-1$ is represented by r_{t-1} (0 = absent, 1 = present).

Table 19.3. Means (SD) of posterior predictive distributions of the probability of vernal pools supporting wood frog metamorphosis ($\hat{\psi}^2$) at Patuxent National Research Refuge, Maryland, USA, and Rock Creek National Park, Washington, DC, USA

Study area	Hydroperiod	2008	2009	2010	2011
Patuxent	Short	0.119 (0.104)	0.188 (0.145)	0.036 (0.045)	–
	Medium	0.677 (0.112)	0.781 (0.110)	0.335 (0.115)	–
	Long	0.743 (0.106)	0.831 (0.091)	0.415 (0.147)	–
	Managed	–	–	0.249 (0.196)	–
Rock Creek	Short	–	0.109 (0.236)	0.306 (0.352)	0.186 (0.277)
	Medium	–	0.152 (0.278)	0.312 (0.242)	0.173 (0.196)
	Long	–	0.936 (0.142)	0.979 (0.086)	0.981 (0.066)

Source: Reproduced from Green and Bailey 2015 (table 3).

tion in occupancy rates. Our initial prototype workshop identified considerable structural uncertainty, which we reduced through research and analyses aimed at understanding important factors driving wood frog breeding and reproductive success. Detection probabilities for wood frog egg masses and tadpoles were high, enhancing our ability to accurately observe the system (high observability). Finally, we used experimentation to quantify the effect of management on wood frog breeding and reproductive success (partial controllability; Williams 1997, 2001).

Using the dynamic parameters estimated above, we revisited the fundamental objective of ensuring long-term persistence of wood frog populations within a park or refuge. It was unclear how many occupied pools supporting breeding and/or metamorphosis were needed to meet this objective. Therefore, we simulated wood frog occupancy dynamics using a population viability analysis (PVA) to evaluate the probability of quasi-extinction (complete reproductive failure in 2 consecutive years) under various circumstances. We considered scenarios using different numbers of pools and investigated their relationship with quasi-extinction probability. We used the estimated parameter distributions described above from each of the 2 study areas (i.e., PRR and ROCR) in combination with 6 levels of pool populations (5, 10, 25, 50 ,75, and 100 pools) to create 12 different simulation scenarios.

For each scenario, we projected metapopulation dynamics over 100 years and 1,000 iterations. Our simulations included environmental uncertainty, and pools were assigned hydroperiod and spatial arrangements consistent with each refuge (see Green and Bailey 2015 for details). For each iteration, we recorded whether the metapopulation reached quasi-extinction. If so, the year of extinction was recorded; if not, we summarized the number of pools producing metamorphs each year.

We found similar results for both the PRR and ROCR wood frog metapopulations: as the number of pools in the system increased, the probability of quasi-extinction decreased, and the mean time to extinction (TTE) increased (fig. 19.2). The probability of quasi-extinction dropped below 5% for systems with ≥50 pools, and metamorph occupancy rates stabilized at 0.16 using estimates from both PRR and ROCR. Conversely, in systems with few pools, quasi-extinction occurred often and early (fig. 19.2). Based on these results, we determined a minimum of 50 pools is necessary to ensure a 0.95 probability of population persistence over 100 years.

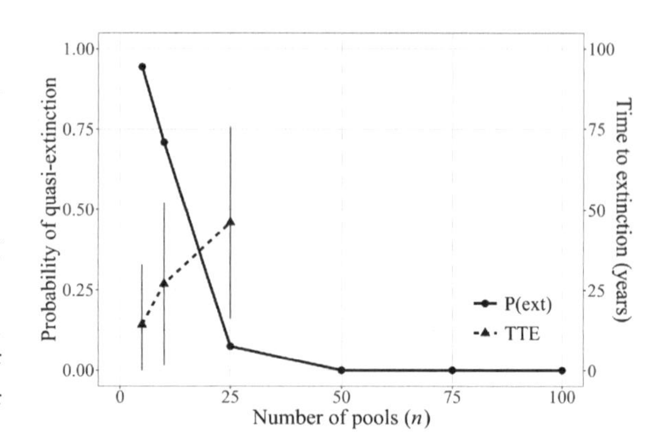

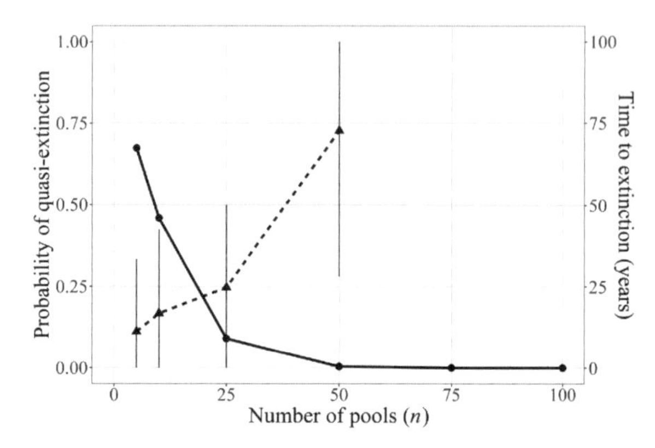

Figure 19.2. The probability of quasi-extinction, *P*(*ext*), and mean time to extinction (TTE) for wood frog metapopulations as a function of the number of vernal pools in the metapopulation, given estimates of breeding and metamorph occupancy at (*top*) Patuxent National Research Refuge and (*bottom*) Rock Creek National Park. Error bars for TTE represent 1 standard deviation. *Reproduced from Green and Bailey 2015 (fig. 1).*

We incorporated these ecological thresholds into a utility function that balanced financial costs and biological benefits of implementing management:

$$V_{a,s} = (1 - \phi_n) \times (1 - 0.05a),$$

where $V_{a,s}$ is the value of implementing action a given a starting state s; ϕ_n is the probability of extinction, given the resulting number of pools in the system, n, after management is implemented; and a is the number of pools managed or created ($a = 1$, 2, or 3 pools per year). Management actions that result in an expected metamorph occupancy rate greater than the threshold determined by our PVA receive a utility value ($U_{a,s}$) equal to $V_{a,s}$. No value is associated with actions resulting in an expected metamorph occupancy rate dropping below the threshold level (i.e., $U_{a,s} = 0$; fig 19.3; see also Green and Bailey 2015).

Using our estimated effect of managing poorly producing, short-hydroperiod pools, we considered 3 management scenarios to explore how decisions might change in response to available options. These scenarios included (1) manipulating up to 3 existing pools per year at PRR, (2) manipulating up to 3 existing pools per year at ROCR, and (3) creating up to 3 new pools per year at ROCR. Because of computational limitations, we limited our time frame for identifying optimal management strategies to 10 years. We considered creating pools at ROCR because it may be necessary to create new habitat to supplement the small number of existing pools at that site (9) and assumed that created pools had the same occupancy probabilities as manipulated pools. The maximum value of 3 managed or created pools per year was chosen through informal consultations with managers of our 2 study areas, and it represents

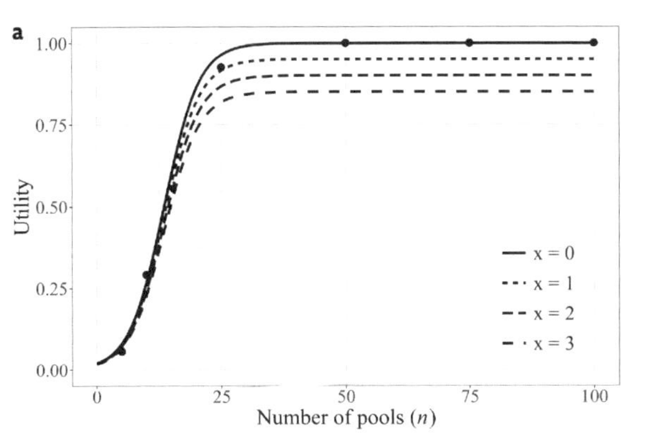

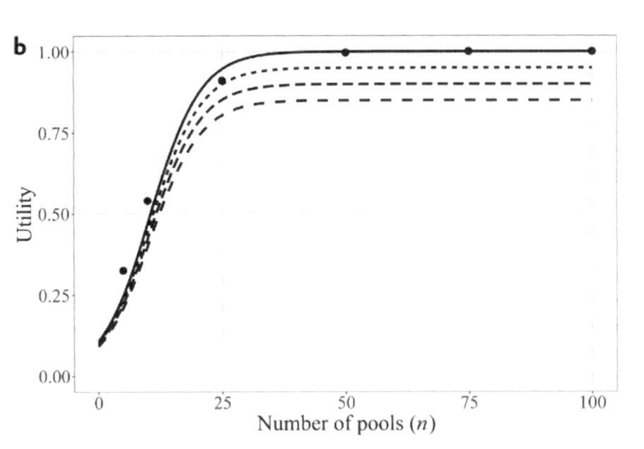

Figure 19.3. The utility value for a metapopulation consisting of a specified number of pools (n), given the number of pools managed (x) and that the metamorph occupancy threshold is met, at (a) Patuxent National Research Refuge and (b) Rock Creek National Park. For metapopulations not meeting the metamorph occupancy threshold, utility is 0. Points represent the probability of persistence observed across 1,000 simulations of metapopulation dynamics, given site-specific occupancy estimates.

the upper limit of effort that would be given to habitat manipulation. Other managers may choose any maximum value to reflect their logistical and habitat constraints.

We used stochastic dynamic programing (SDP; Bellman 1957; Lubow 1995) to identify the optimal management decision for all 3 scenarios. SDP uses the utility function to determine the value of making a particular management decision, given the current state of the system (Runge, chapter 21, this volume). By projecting system dynamics and management decisions, a decision policy can be found that optimizes all objectives over a given time horizon while accounting for stochastic system dynamics, management efficiency, and model uncertainty. Our management actions were permanent, and therefore, the number of pools available to manage was dependent on how many pools have already been managed. Likewise, it was not possible to transition to a state with fewer managed pools than in previous time steps.

Optimal management decisions varied but showed a similar pattern across the 3 scenarios. For PRR managers (scenario 1) and assuming 53 pools in the metapopulations (i.e., the number surveyed), it was optimal to manipulate 3 short-hydroperiod pools each year for the first 4 years and 2 pools in the 5th year, for a total of 14 pools. For ROCR managers, the optimal strategy for scenario 2 (manipulate pools) was to manipulate 2 of the 5 short-hydroperiod pools in the 1st year, and for scenario 3 (create new pools), it was optimal to create 3 pools per year for the first 3 years, 1 pool in year 4, and 3 pools in years 5 and 6. Generally, these optimal strategies ensure that the metamorph occupancy threshold is consistently met ($\psi_2 \sim 0.16$) and the utility is >0. After the threshold is met, there is no value to manipulating pools because it does not increase the probability of persistence. However, if creating new pools is possible, it is optimal to create pools until the costs of management become greater than the increase in the probability of persistence. It is also best to manipulate or create pools early because the permanence of the action assures that the full benefit for maintaining wood frog metapopulations can be achieved.

Discussion

A formal decision analysis approach requires identifying explicit management objectives. By making objectives explicit, stakeholders can more easily enumerate a list of alternative management actions that may influence those objectives. Initially we did not know how to quantify our fundamental objective of long-term persistence of wood frog metapopulations. Our lack of estimates of breeding and metamorphosis probabilities were the impetus for our modeling efforts. Combined with monitoring data, this modeling yielded defendable parameter estimates and an understanding of factors influencing these dynamic processes (e.g., hydroperiod, environmental conditions). We used these estimates within a PVA to quantify ecological thresholds needed to ensure metapopulation persistence, thus quantifying our fundamental objective.

Often decision makers have little information on historical population sizes or ecosystem dynamics relating to management objectives. This uncertainty may lead to reluctance in defining appropriate biological objectives. PVA provides a means to quantify these objectives, even when data are sparse, by projecting the population through time in response to annual changes in environmental conditions or management actions. The iterative nature of adaptive resource management provides a useful framework for updating means objectives by incorporating new information as it is collected and addressing potential future changes in ecological processes or parameter values (Williams et al. 2007; part 6, this volume).

Our sensitivity analysis helped us identify the most influential processes driving wood frog metapopulation dynamics (probabilities of colonization and successful reproduction), allowing us to prioritize the practical actions aimed at increasing these probabilities. We experimentally manipulated existing short-hydroperiod, poorly producing pools by installing liners and quantified the resulting increase in the probability of reproductive success. However, these results were based on a small number of manipulated pools and only 1 year of post-treatment data. Further monitoring is necessary to better understand how wood frogs respond to this action and whether those benefits are maintained over a long time period. Other attempts at creating ephemeral hydroperiods have had limited success (Lichko and Calhoun 2003), resulting in permanent pools that are habitat for predators of wood frog eggs and tadpoles, such as fish and larger anuran species. Continued monitoring in an adaptive management framework would allow for updating the dynamic occupancy parameters associated with management pools and making better management decisions in the future. Additional experimental studies could test the efficacy of other management actions.

The 2 study areas highlighted in this chapter are part of a larger effort to better manage vernal pool species (Van Meter et al. 2008). Our utility function, while realistic, simply represents a template that includes biological and logistical objectives whose values will vary among management units. For example, the penalty accounting for the financial costs of management may change as the relative importance of costs varies across units. Managers can also identify the point at which the gain in persistence is negligible (e.g., $\phi > 0.85$, $\phi > 0.99$), such that the increase in persistence resulting from management does not represent an increase in the utility. Alternative utility functions should be developed to represent appropriate stakeholders' values for each objective in a particular scenario.

The optimal management strategies identified using our approach were specific to our study sites and the management scenarios considered, but the approach can be easily modified to reflect objectives identified by stakeholders at other national parks and wildlife refuges. The modeling framework can incorporate site-specific occupancy estimates to better reflect the regional variation in wood frog metapopulation dynamics. Future efforts of management to maintain wood frogs and other vernal pool–obligate amphibian species should incorporate refuge- or park-specific elicitation of objectives and their relative values to determine appropriate utility functions.

ACKNOWLEDGMENTS

The authors would like to thank Jim Nichols, Evan Grant, Sandi Mattfeldt, and Holly Obrecht for their contributions to the development and implementation of this project. Evan Grant, Adrienne Brand, Eric Dalallio, Andrew Bennett, and all field technicians were invaluable in carrying out the field work. US Fish and Wildlife Service Patuxent Research Refuge staff provided logistical support, and Youth Conservation Corps assisted in management experiment implementation. Mevin Hooten assisted with some analyses, and Barry Noon, Chris Funk, and Michael Runge provided valuable feedback on previous drafts. US Geological Survey ARMI provided funding for this project.

LITERATURE CITED

Bellman R. 1957. *Dynamic Programming*. Princeton, NJ: Princeton University Press.

Berven KA. 1990. Factors affecting population fluctuations in larval and adult stages of the wood frog (*Rana sylvatica*). *Ecology* 71:1599–1608.

Biebighauser TR. 2007. *Wetland Drainage, Restoration, and Repair*. Lexington, KY: University of Kentucky Press.

Brooks RT. 2004. Weather-related effects on woodland vernal pool hydrology and hydroperiod. *Wetlands* 24:104–114.

Calhoun AJK, deMaynadier PG, eds. 2008. *Science and Conservation of Vernal Pools in Northeastern North America*. New York: CRC Press.

Collins JP, Storfer A. 2003. Global amphibian declines: sorting the hypotheses. *Diversity and Distributions* 9:89–98.

Green AW, Bailey LL. 2015. Using Bayesian population viability analysis to define relevant conservation objectives. *PlOS ONE* 10.12:e0144786.

Green AW, Bailey LL, Nichols JD. 2011. Exploring sensitivity of a multistate occupancy model to inform management decisions. *Journal of Applied Ecology* 48:1007–1016.

Green AW, Hooten MB, Grant EHC, Bailey LL. 2013. Evaluating breeding and metamorph occupancy and vernal pool management effects for wood frogs using a hierarchical model. *Journal of Applied Ecology* 50:1116–1123.

Julian JT, Snyder CD, Young JA. 2006. The use of artificial impoundments by two amphibian species in the Delaware Water Gap National Recreation Area. *Northeastern Naturalist* 13:459–468

Kenney LP, Burne MR. 2001. *A Field Guide to the Animals of Vernal Pools*. Westborough, MA: Massachusetts Division of Fisheries and Wildlife.

Lannoo MJ, ed. 2005. *Amphibian Declines: The Conservation Status of United States Species*. Berkley, CA: University of California Press.

Lichko LE, Calhoun AJK. 2003. An evaluation of vernal pool creation projects in New England: project documentation from 1991–2000. *Environmental Management* 32:141–151.

Lubow BC. 1995. SDP: generalized software for solving stochastic dynamic optimization problems. *Wildlife Society Bulletin* 1995:738–742.

Mattfeldt SD, Bailey LL, Grant EHC. 2009. Monitoring multiple species: estimating state variables and exploring the efficacy of a monitoring program. *Biological Conservation* 142: 720–737.

National Assessment Synthesis Team [NAST]. 2001. *Climate Change Impacts on the United States: The Potential Consequences of Climate Variability and Change*. Report for the US Global Change Research Program. Cambridge, UK: Cambridge University Press.

National Park Service. 2005. *Long-Term Monitoring Plan for Natural Resources in the National Capital Region Network*. Washington, DC: Inventory and Monitoring Program, Center for Urban Ecology.

Rubbo MJ, Kiesecker JM. 2005. Amphibian breeding distribution in an urbanized landscape. *Conservation Biology* 19:504–511.

Schlaepfer MA, Runge MC, Sherman PW. 2002. Ecological and evolutionary traps. *Trends in Ecology and Evolution* 17:474–480.

Simon JA. 2006. Relationships among land cover and amphibian use of constructed wetlands. MS thesis. Towson, MD: Towson University.

Taylor BE, Scott DE, Gibbons JW. 2006. Catastrophic reproductive failure, terrestrial survival, and persistence of the marbled salamander. *Conservation Biology* 20:792–801.

Van Meter R, Bailey LL, Grant EHC. 2008. Methods for estimating the amount of vernal pool habitat in the northeastern United States. *Wetlands* 28:585–593.

Vasconcelos D, Calhoun AJK. 2006. Monitoring created seasonal pools for functional success: a six-year case study of amphibian responses, Sears Island, Maine, USA. *Wetlands* 4:992–1003.

Williams BK. 1997. Approaches to the management of waterfowl under uncertainty. *Wildlife Society Bulletin* 25:714–720.

———. 2001. Uncertainty, learning, and the optimal management of wildlife. *Environmental and Ecological Statistics* 8:269–288.

Williams BK, Szaro RC, Shapiro CD. 2007. *Adaptive Management: The US Department of the Interior Technical Guide*. Washington, DC: US Department of the Interior.

20 Prioritizing Uncertainties to Improve Management of a Reintroduction Program

Sarah J. Converse

The success of wildlife reintroduction efforts rests on the demographic performance of released animals. Whooping Cranes in the eastern migratory population—reintroduced beginning in 2001—demonstrate adequate survival but poor reproduction. Managers and scientists have used an iterative process of learning and management to respond to this management challenge, but by 2015, uncertainty about the causes of reproductive failure remained substantial. An expert judgment–driven process was used to develop and refine competing hypotheses for reproductive failure and to evaluate the impact of various management actions on components of reproduction (nesting success and fledging success) in light of the various hypotheses. I used that information to calculate value of information, the expected improvement in management performance associated with an increase in knowledge, which suggests research and monitoring priorities for the future.

Problem Background

Reintroduction of Whooping Cranes (*Grus americana*) has been a major focus of recovery efforts for the endangered species since the mid-1970s (Canadian Wildlife Service and US Fish and Wildlife Service 2005; Converse et al. 2018a; French et al. 2018). The eastern migratory population (EMP) was established through releases of captive-reared birds beginning in 2001. The population migrates between Wisconsin and the southeastern United States. The first nesting attempts occurred in 2005, but widespread reproductive failure has been observed in most years since (Converse et al. 2013c; Runge et al. 2011; Urbanek et al. 2010b).

Reproductive failure in the EMP has spawned nearly a decade of research and management efforts to determine and respond to the causes of failure (see Converse et al. 2018b for a review). There are substantial challenges in managing reproductive failure in a population that exhibits delayed reproduction, carries out its life history in a complex ecological setting, and has been reintroduced from captive-reared stock. By early 2015, while much had been learned about reproductive failure in the population, many questions remained. Therefore, a meeting was organized in March 2015 to initiate a process to evaluate the state of knowledge, reassess the challenges limiting growth of the EMP, and develop recommendations for managers. Given the severe uncertainty, substantial focus was placed on identifying priorities for learning to improve decision making over time.

Decision Maker and Their Authority

The EMP is managed by a partnership of public agencies and private organizations known as the Whooping Crane Eastern Partnership (WCEP; www.bringbackthecranes.org). The WCEP research and science team organized the March 2015 meeting. While WCEP itself has no legal authority per se, individual partners have both mandates and resources to undertake various management actions collaboratively or individually. The US Fish and Wildlife Service is one member of WCEP and holds ultimate management authority for the Whooping Crane. The EMP is a designated experimental, nonessential population (Endangered Species Act Section 10[j]). In addition, the US Fish and Wildlife Service owns Necedah National Wildlife Refuge (Necedah NWR), the original release site where most nesting had taken place through 2015. Other partners include the Wisconsin Department of Natural Resources, which manages wildlife in Wisconsin; the US Geological Survey's Patuxent Wildlife Research Center, which has conducted Whooping Crane research and raised Whooping Cranes for release; the International Crane Foundation, which has conducted research, led monitoring efforts for WCEP, and raised Whooping Cranes for release; and Operation Migration, which led ultralight flights beginning in 2001 to teach young Whooping Cranes the migration route.

Ecological Background

Whooping Cranes are the only endemic crane in North America. They are also the rarest crane in the world and are listed as endangered in the United States, Canada (Canadian Wildlife Service and US Fish and Wildlife Service 2005), and on the International Union for Conservation of Nature Red List (iucnredlist.org). Wetland obligates, Whooping Cranes were driven nearly to extinction through habitat loss and overhunting (French et al. 2018). By the early 1940s, fewer than 30 individuals remained, and the population that is the source of all extant Whooping Cranes numbered about 15 individuals. In addition to conservation actions focused on management of this remnant population, the Aransas Wood Buffalo Population, reintroduction has been a major component of the recovery efforts for the species, with releases to 1 or more of 4 different reintroduced populations carried out nearly every year since the mid-1970s (Converse et al. 2018a). As of 2015, 2 reintroduced populations were receiving releases, the EMP and the Louisiana Non-Migratory Population (French et al. 2018).

Releases to establish the EMP were initiated in 2001, with ultralight aircraft leading migrations of costume-reared birds, during their first autumn, from Necedah NWR in central Wisconsin to the Gulf coast of Florida (Urbanek et al. 2005). Costume rearing is a form of rearing in which costumed humans care for birds, often using puppets designed to resemble adult cranes. This approach disguises the human form, substantially reducing the risk of birds imprinting on humans. Releases to the EMP continued with this method through 2015. In 2005, direct autumn releases were initiated, with costume-reared birds released in Wisconsin in their first autumn to follow older birds south on migration (Servanty et al. 2014; Urbanek et al. 2010a). In 2013, WCEP initiated autumn release of birds reared by captive pairs (known as parent rearing). WCEP partners released parent-reared birds near other Whooping Cranes to encourage association with the older birds.

Reproductive failure has been the major challenge to the EMP reintroduction effort (Converse et al. 2018b; Runge et al. 2011; Servanty et al. 2014). In 2005, the first nest was initiated, and in 2006, the first chick successfully fledged. However, all other nests in 2006, and all nests in 2007 and 2008, failed to hatch. Nesting failure was highly synchronous across the population, with most nests failing within 1 or 2 days of each other.

Impetus for Decision-Analytic Process

In early 2009, the first workshop was undertaken to elicit and evaluate competing hypotheses and develop a response plan for the problem of reproductive failure (Runge et al. 2011). That process resulted in a value of information analysis. Value of information is a set of decision-analytic tools designed to evaluate the expected increase in management performance with reduction of uncertainty (Canessa et al. 2015; Runge et al. 2011; Williams et al. 2011). Value of information can be used to understand how much uncertainty impedes decision making and thus to assess the value of learning. The analysis conducted in 2009 suggested the most fruitful avenue for investigation was the impact of blood-feeding black flies (*Simulium spp.*) on crane nest desertion. The black fly hypothesis was one of several hypotheses posed to explain the observed reproductive failure (Converse et al. 2013c; King et al. 2013; Urbanek et al. 2010b).

From the summer of 2009 through 2013, WCEP partners and collaborators carried out intensive research to evaluate the black fly hypothesis (Barzen et al. 2018; Converse et al. 2013c, 2018b). A bacterial larvicide, *Bacillus thuringiensis israelensis* (*Bti*), was applied experimentally to reduce black fly populations, and nesting success in application years was high enough to produce chicks in numbers that allowed another problem to become evident: fledging failure. The large majority of chicks died in the first month or two after hatching.

By March 2015, WCEP had taken multiple research and management actions to address the problem of reproductive failure, based on different hypotheses. Most notably, the release site for young captive-reared birds had been relocated to eastern Wisconsin with the goal of avoiding poor habitat at the original release site, including black flies (Converse et al. 2018a). In addition, WCEP had implemented parent rearing, allowing learning about hypothesized relationships between rearing method, early learning, and reproductive success. However, important uncertainties remained.

While a strong link had been established between black flies and nesting failure, questions remained about whether black flies were the ultimate cause of nesting failure or merely a proximate cause, which would suggest some larger behavioral challenges with the released birds. There was also a growing awareness that nesting failure was not the only problem: even when nests successfully produced chicks, those chicks rarely fledged.

The optimal path forward for Whooping Crane management in the EMP was highly dependent on the cause of reproductive failure. Different hypothesized causes for reproductive failure suggested substantially different approaches to solving the problem. For example, if reproductive failure was due entirely to black flies, moving the releases to a different area without large numbers of black flies would be a reasonable approach. If instead something about the rearing of birds produced breeders with poor nest attendance behavior, a new release site would not likely solve the problem.

Decision Structure

When deciding under uncertainty, it is possible to take action in the face of uncertainty, to pause before acting and use that opportunity to reduce uncertainty, or to manage and learn simultaneously through adaptive management. Value of information analyses evaluate the expected improvement in performance on objectives to be gained from reducing uncertainty prior to acting versus acting in the face of the full uncertainty. Also, value of information analyses can be used to determine which sources of uncertainty to address to most improve performance on objectives. With so many hypotheses for reproductive failure, it was difficult to know which to address first or even whether to address uncertainty before acting. Therefore, the decision-analytic process initiated in the March 2015 workshop was designed to determine the value of conducting research before deciding on a course of action for management of reproductive failure in the EMP and to prioritize research questions.

Decision Analysis

The general decision-analytic structure and procedures described here are also described in Converse et al. (2018b). Here I delve into how uncertainty in expert judgments influences inference about the most valuable research and management pathways.

Objectives

While Whooping Crane reintroduction management is certainly a multiple-objective problem (Converse et al. 2013b, 2018a), the primary fundamental objective of WCEP is the establishment of a migratory Whooping Crane population in eastern North America with a low probability of extinction in the absence of further releases. In the 2015 workshop, the focus was on reproductive failure as the major impediment to that fundamental objective. Though reproductive success is a means, rather than fundamental, objective, because of its importance it was treated as fundamental in the analyses. The reproductive success objective was broken into 2 separate components for the purposes of identifying hypotheses and management actions: nesting success and fledging success. Nesting success is the probability that a nest with an egg produces at least 1 hatchling, and fledging success is the probability that a hatchling survives to fledge, which is assumed to occur on 1 October in the autumn following nesting.

Hypotheses

During the 2015 workshop, participants were assembled, from inside and outside WCEP, with expertise in endangered species reintroduction, wetland and waterbird management, captive breeding, and Whooping Crane ecology. After a day of presentations to establish baseline knowledge on the available data, research outcomes, and management context to date, participants were asked to develop a list of hypotheses to address the problem of reproductive failure. A brainstorming exercise was conducted, and all ideas were recorded for further consideration. Participants divided into groups, and each group focused on a set of the brainstormed hypotheses to be refined and described in detail, including whether the hypotheses were relevant to nesting failure, fledging failure, or both. After this process, experts provided an initial weighting of hypotheses by distributing 100 points among the hypotheses (separately for the nesting failure hypotheses and the fledging failure hypotheses). Only those hypotheses with at least 5% weight were retained. Lists were created of the retained hypotheses for nesting failure and for fledging failure. Many of the hypotheses in the nesting failure list had analogues in the fledging failure list, and vice versa. The lists of hypotheses for nesting and fledging failure are shown in table 20.1 and table 20.2, respectively.

Alternative Actions

For each of the 2 lists of hypotheses, groups at the workshop identified management actions that could be taken conditional on the various hypotheses being true. The full list of actions, given here, includes notations about which of the hypotheses the expert groups were attempting to address in developing a given action from among the nesting failure hypotheses (N) and the fledging failure hypotheses (F) provided in table 20.1 and table 20.2, respectively.

1. Do nothing different—current impediments will resolve with time. (N3, F4)
2. Implement post-release training in adults to instill predator defense behavior. (F1)
3. Implement pre-release training in young to instill predator defense behavior. (F1)
4. Optimize impoundment drawdown timing to provide chick-rearing habitat (i.e., appropriate water depths) away from the area of woody vegetation that provides predator cover at wetland margins. (F1)
5. Implement *Bti* treatment. (N1)

Table 20.1. Hypotheses about the causes of nesting failure in the eastern migratory population of Whooping Cranes

Number	Hypothesis
N1	Black flies are an acute stress for Whooping Cranes that cause nest abandonment when/where black fly abundances are high.
N2	Low food availability and/or quality on the breeding area, combined with black fly stress, causes nest abandonment.
N3	Lack of post-release experience with nesting, due to age or opportunity, reduces nesting success.
N4	Genetic structure of the current captive population affects nesting success after release. Mechanisms could include founder effects, inbreeding depression, genetic drift, and/or captive selection.
N5	Current costume-rearing methods do not impart information necessary for post-release nesting success, via inadequate learning experiences regarding nest defense, incubation constancy, resource selection, or behavioral plasticity.
N6	The social structure (i.e., groups) of costume-reared cohorts is detrimental to the development of normal behaviors due to muting of natural behavioral variation.

Note: Hypotheses developed at a March 2015 workshop.

Table 20.2. Hypotheses about the causes of fledging failure in the eastern migratory population of Whooping Cranes

Number	Hypothesis
F1	Predator effects are high, resulting in poor fledging success for Whooping Cranes.
F2	Low food availability and/or quality on the breeding area, combined with black fly stress, reduces chick survival.
F3	Low food availability (limited abundance or distribution) and/or poor nutrition profile of food during the chick-rearing period limits reproductive success. Mechanisms could include poor soil productivity, impoundment management, or invasive species limiting the quality and/or availability of food resources.
F4	Lack of post-release experience with chick rearing, due to age or opportunity, reduces chick survival.
F5	Genetic structure of the current captive population affects chick survival after release. Mechanisms could include founder effects, inbreeding depression, genetic drift, and/or captive selection.
F6	Current costume-rearing methods do not impart information necessary for post-release chick survival, via inadequate learning experiences regarding nest defense, incubation constancy, resource selection, or behavioral plasticity.
F7	The social structure (i.e., groups) of costume-reared cohorts is detrimental to the development of normal behaviors due to muting of natural behavioral variation.

Note: Hypotheses developed at a March 2015 workshop.

6. Implement forced renesting (remove first nests to encourage renesting outside of peak black fly period). (N1)
7. Conduct habitat management in the Yellow River (the primary source of black flies on Necedah NWR) to reduce black fly populations at the source: manage impoundment releases or improve water and stream quality to increase habitat quality for black fly larval predators like stone flies (order Plecoptera). (N1)
8. Conduct wetland management to improve food resources: alter hydrology, etc. (N2, F2, F3)
9. Conduct wet meadow restoration to improve food resources, i.e., forms of engineering such as ditch plugging. (N2, F2, F3)
10. Provide dummy egg, then return nest's eggs at the end of incubation. (N3, F4)
11. Bring eggs from the Aransas Wood Buffalo Population into the captive population as breeders or release directly. (N4, F5)

12. Rear and release parent-reared chicks in mid-September. (N5, F6)
13. Put fertile eggs from captivity in all nests that are further along in incubation. (N5, F6)
14. Chick adoption: put very young chicks from captive facilities with failed breeding pairs (chicks released at 1–2 weeks old). (N5, F6)
15. Costume rear 1–2 birds at a time (by a keeper), do not socialize birds in groups (as previously done for ultralight and direct autumn release birds), and release in these numbers. (N6, F7)

Predictions

In August 2015, a subset of 8 experts that attended the March meeting participated in an in-depth elicitation process, using a modified Delphi process (Burgman et al. 2011; Martin et al. 2011). Over a period of time that included 2 conference call meetings interspersed with days allotted for individual work, the experts provided their judgment of the effect of each of the actions on nesting success and fledgling success. Experts were asked to make predictions *conditional* on each of the hypotheses for nesting failure or fledging failure being true. That is, experts were asked, "What would the nesting success rate be if one took action *X*, assuming that hypothesis *Y* were the true cause of nesting failure?" Here I focus on the results of their elicitation considering predicted responses at 10 years after implementation of an action (see also Converse et al. 2018b for further details). Experts were asked to focus on the problem of reproductive failure at Necedah NWR (the original release site, where most nesting occurs). In 2015, while limited nesting had occurred outside Necedah NWR, sample sizes were too small to determine whether reproductive failure was a challenge elsewhere. The experts also provided their relative belief in the hypotheses, again by distributing 100 points among the hypotheses for nesting failure and another 100 points among the hypotheses for fledging failure.

Value of Information Analysis

Following Converse et al. (2018b), I calculated expected value of perfect information (EVPI; the improvement in expected management outcomes if uncertainty could be fully resolved) for both the nesting success and fledging success objectives. I also calculated expected value of partial information, EVPXI, the improvement in expected management outcomes if a subset of the uncertainty could be resolved (e.g., either confirming or refuting individual hypotheses). See Runge et al. (2011) for a description of the methods for calculating EVPI and EVPXI and more detail about the concept of value of information. To explore differences in judgments across experts, I calculated EVPI and EVPXI for each individual expert, using only their mean predictions and their hypothesis weights, and I calculated mean EVPI and EVPXI using the mean responses across experts and the mean hypothesis weights across experts (with experts weighted equally).

Results of Analysis

Forced renesting was the optimal action for improving nesting success under uncertainty, based on the mean of expert judgments (alternative 6; table 20.3). However, that action was optimal for only 2 of the experts when considered individually. Treatment with *Bti* (alternative 5) was optimal for 3 experts, pre-release predator training (alternative 3) was optimal for 1, parent rearing was optimal for 1 (alternative 12), and placing captive produced eggs in nests was optimal for 1 (alternative 13). The EVPI varied substantially, between an expected gain of 0.02 to 0.20 in nesting success, across experts.

The EVPXI is useful for ranking sources of uncertainty in terms of the expected management performance to be gained. Based on the mean responses, the hypothesis associated with the highest EVPXI for improving nesting success was the genetic structure hypothesis (nesting failure hypothesis 4; table 20.4). That hypothesis was followed closely by the black fly

Table 20.3. Expected value of perfect information (EVPI), for improving Whooping Crane nesting success 10 years after management implementation

Level of analysis	Optimal action under uncertainty[a]	Expected value of optimal action under uncertainty	Expected value with uncertainty resolved	EVPI	Proportional EVPI
Mean	Forced renesting (6)	0.282	0.368	0.086	0.307
Expert 1	Pre-release training (3)	0.300	0.335	0.035	0.117
Expert 2	*Bti* (5)	0.483	0.608	0.125	0.259
Expert 3	Captive eggs in nests (13)	0.460	0.480	0.020	0.043
Expert 4	*Bti* (5)	0.220	0.420	0.200	0.909
Expert 5	Forced renesting (6)	0.272	0.352	0.080	0.294
Expert 6	Forced renesting (6)	0.204	0.335	0.131	0.644
Expert 7	Parent rearing (12)	0.242	0.350	0.108	0.446
Expert 8	*Bti* (5)	0.373	0.472	0.099	0.265

Note: based on judgments elicited from experts ($n = 8$). Information includes the optimal action to take given uncertainty and the expected value of that action; the expected value with uncertainty resolved; the EVPI, equal to the difference between the expected value with uncertainty resolved and the expected value under uncertainty; and the proportional EVPI, equal to the EVPI divided by the expected value under uncertainty. Results include the mean response over experts and expert-specific responses.

[a] Optimal action to take under uncertainty is keyed (parenthetically) to the list of alternative actions in the main text.

Table 20.4. Expected value of partial information (EVPXI), for improving Whooping Crane nesting success 10 years after implementation of management

	Hypotheses[a]					
Level of analysis	Black flies	Low food & black flies	Lack of experience	Genetic structure	Costume rearing	Group rearing
Mean	0.315 (2)	0.282 (6)	0.291 (5)	0.318 (1)	0.309 (3)	0.296 (4)
Expert 1	0.315 (1)	0.310 (2)	0.300 (3)	0.310 (2)	0.300 (3)	0.300 (3)
Expert 2	0.560 (2)	0.483 (4)	0.483 (4)	0.508 (3)	0.583 (1)	0.483 (4)
Expert 3	0.475 (1)	0.465 (2)	0.460 (3)	0.460 (3)	0.460 (3)	0.460 (3)
Expert 4	0.300 (2)	0.220 (3)	0.220 (3)	0.320 (1)	0.320 (1)	0.220 (3)
Expert 5	0.295 (2)	0.273 (5)	0.293 (3)	0.303 (1)	0.288 (4)	0.288 (4)
Expert 6	0.232 (3)	0.204 (6)	0.230 (4)	0.241 (2)	0.266 (1)	0.209 (5)
Expert 7	0.344 (1)	0.242 (4)	0.242 (4)	0.248 (3)	0.242 (4)	0.294 (2)
Expert 8	0.469 (1)	0.373 (5)	0.442 (2)	0.378 (4)	0.390 (3)	0.390 (3)

Note: Based on judgments elicited from experts ($n = 8$). The EVPXI is the expected improvement in management performance if individual sources of uncertainty were resolved (i.e., each hypothesis either confirmed or refuted). Results include the mean response over experts and expert-specific responses. Included parenthetically is the rank of the EVPXI of the hypothesis for each expert.

[a] Details on hypotheses can be found in the list of nesting failure hypotheses, table 20.1.

hypothesis (nesting failure hypothesis 1). Responses varied across experts, with judgments of different experts suggesting that different hypotheses should be prioritized for evaluation. Based on their individual input, the EVPXI was highest for the genetic structure hypothesis for 1 expert, the black fly hypothesis for 4 experts, and the costume-rearing hypothesis (nesting failure hypothesis 5) for 2 experts. For the remaining expert, the genetic structure hypothesis and the costume-rearing hypothesis were tied at the highest rank.

For fledging success, the optimal action under uncertainty based on the mean across experts was parent rearing (alternative 12; table 20.5). Again, though, that action was optimal for only 2 of the experts, while pre-release training (alternative 3) was optimal for 3, egg swapping (alternative 10) was optimal for 2, and post-release training (alternative 2) was optimal for 1. EVPI again varied substantially from an expected gain of 0.006 to 0.138 in fledging success, across experts.

To improve fledging success, the mean of expert judgments indicated it would be most valuable to address the genetic structure hypothesis (fledging failure hypothesis 5), followed by the predator hypothesis (fledging failure hypothesis 1; table 20.6). Perhaps not surprisingly, given fewer years of research focused on the problem, expert judgments suggested greater uncertainty about the priority hypotheses for testing. While for 4 experts the genetic structure hypothesis ranked first in EVPXI, and for 1 expert the low food availability hypothesis (fledging failure hypothesis 3) ranked first, 3 experts had ties in their top ranking. These included a tie between genetic structure and costume rearing (fledging failure hypothesis 6), a tie between predators and

Table 20.5. Expected value of perfect information (EVPI) for improving Whooping Crane fledging success 10 years after management implementation

Level of analysis	Optimal action under uncertainty[a]	Expected value of optimal action under uncertainty	Expected value with uncertainty resolved	EVPI	Proportional EVPI
Mean	Parent rearing (12)	0.319	0.351	0.033	0.102
Expert 1	Post-release training (2)	0.310	0.335	0.025	0.081
Expert 2	Pre-release training (3)	0.406	0.412	0.006	0.015
Expert 3	Egg swapping (10)	0.450	0.490	0.040	0.089
Expert 4	Egg swapping (10)	0.257	0.395	0.138	0.537
Expert 5	Parent rearing (12)	0.270	0.287	0.017	0.065
Expert 6	Pre-release training (3)	0.304	0.351	0.047	0.155
Expert 7	Parent rearing (12)	0.418	0.458	0.040	0.096
Expert 8	Pre-release training (3)	0.328	0.335	0.007	0.021

Note: Based on judgments elicited from experts ($n = 8$). Information includes the optimal action to take given uncertainty and the expected value of that action; the expected value with uncertainty resolved; the EVPI, equal to the difference between the expected value with uncertainty resolved and the expected value under uncertainty; and the proportional EVPI, equal to the EVPI divided by the expected value under uncertainty. Results include the mean response over experts and expert-specific responses.

[a] Optimal action to take under uncertainty is keyed (parenthetically) to the list of alternative actions in the main text.

Table 20.6. Expected value of partial information (EVPXI), for improving Whooping Crane fledging success 10 years after implementation of management

	Hypotheses[a]						
Level of analysis	Predators	Low food & black flies	Low food	Lack of experience	Genetic structure	Costume rearing	Group rearing
Mean	0.328 (2)	0.320 (4)	0.323 (3)	0.319 (5)	0.333 (1)	0.319 (5)	0.323 (3)
Expert 1	0.310 (3)	0.310 (3)	0.310 (3)	0.310 (3)	0.320 (1)	0.320 (1)	0.315 (2)
Expert 2	0.406 (2)	0.406 (2)	0.412 (1)	0.406 (2)	0.406 (2)	0.406 (2)	0.406 (2)
Expert 3	0.450 (3)	0.450 (3)	0.460 (2)	0.460 (2)	0.465 (1)	0.460 (2)	0.460 (2)
Expert 4	0.301 (1)	0.257 (4)	0.263 (3)	0.301 (1)	0.279 (2)	0.279 (2)	0.301 (1)
Expert 5	0.270 (3)	0.278 (2)	0.270 (3)	0.270 (3)	0.280 (1)	0.270 (3)	0.270 (3)
Expert 6	0.323 (1)	0.306 (5)	0.316 (3)	0.304 (6)	0.317 (2)	0.323 (1)	0.307 (4)
Expert 7	0.418 (3)	0.428 (2)	0.418 (3)	0.418 (3)	0.448 (1)	0.418 (3)	0.418 (3)
Expert 8	0.328 (4)	0.330 (2)	0.329 (3)	0.330 (2)	0.333 (1)	0.328 (4)	0.328 (4)

Note: Based on judgments elicited from experts ($n = 8$). The EVPXI is the expected improvement in management performance if individual sources of uncertainty were resolved (i.e., each hypothesis either confirmed or refuted). Results include the mean response over experts and expert-specific responses. Numbers in parentheses show the rank of the EVPI of the hypothesis for each expert.

[a] Details on hypotheses can be found in the list of fledging failure hypotheses, table 20.2.

costume rearing, and a 3-way tie between predators, lack of experience (fledging failure hypothesis 4), and group rearing (fledging failure hypothesis 7).

Decision Implementation

In 2016, WCEP terminated ultralight-led releases for the EMP reintroduction (www.bringbackthecranes.org/whooping-cranes/changes-are-hatching-in-the-whooping-crane-eastern-partnership/). In the years since, WCEP has placed greater emphasis on parent rearing, which expert judgment suggested was the optimal action to take in the face of uncertainty to improve fledging success (based on the mean response across experts). In addition, forced renesting has been used on a broad scale for the last several years; this was the optimal action identified for addressing nesting failure under uncertainty (again, based on the mean response across experts). Limited evaluation of the genetic structure (specifically, captive selection, see Converse et al. 2018b) hypothesis has been conducted, based on observational data (Barzen et al. 2018), and has indicated some weak evidence in favor of the hypothesis. A stronger test could be undertaken by experimentally releasing individuals from the Aransas Wood Buffalo Population (e.g., hatched from collected eggs), rather than the captive population, into the EMP. As of this writing, such an experiment—which would need to be spearheaded by the International Whooping Crane Recovery Team (made up of US and Canadian members)—has not been undertaken.

Discussion

Many problems—and virtually all reintroductions—suffer from severe uncertainty (Converse et al. 2013a). Value of information emphasizes that not all uncertainty is equally important in improving management performance. An earlier value of information analysis (Runge et al. 2011) focused on the problem of reproductive failure in this population just as the problem was emerging. That analysis indicated that evaluating the black fly hypothesis was most valuable (i.e., had the highest EVPXI). Subsequent research has indicated strong support for an effect of black flies on nesting success (Barzen et al. 2018; Converse et al. 2013c, 2018b; King et al. 2013). The reason for continuing to consider alternative hypotheses for nesting failure is the issue of proximate vs. ultimate causes. While black flies certainly appear to be the proximate cause of much of the nesting failure in this population, it is not yet clear whether the ultimate cause lies deeper. For example, birds with inadequate early learning experiences—perhaps imparted through costume rearing—might be less attentive of nests and thus abandon nests relatively quickly when subjected to the stress of blood-feeding black flies.

Overall, expert judgment suggested that addressing hypotheses about characteristics of the birds themselves, versus characteristics of the reintroduction environment, would be most valuable. Specifically, the hypothesis with the greatest EVPXI, based on the mean response across experts and for both the nesting success and fledging success responses, was the genetic structure hypothesis. Captive selection—selection for traits in captivity that are nonadaptive after release—is the mechanism that has been most frequently cited as a potential genetic issue in this population (Barzen et al. 2018; King et al. 2013).

Variation in judgments and resulting inference among experts in this analysis was substantial. While expert judgments can be biased due to various cognitive traps (Burgman 2004; Burgman et al. 2011; McBride et al. 2012), the differences demonstrated here may indicate true differences in the interpretation of available information. Including experts with a diversity of judgments should result in more robust conclusions. Continuing to engage in an iterative process of evaluating new information and identifying priority sources of uncertainty will be important as managers and scientists learn more about reproductive ecology in this reintroduced population. The between-expert uncertainty also suggests that a multipronged approach to research may be warranted.

In other words, examining multiple hypotheses simultaneously may be the best approach to identifying, and addressing, the causes of reproductive failure in this population.

ACKNOWLEDGMENTS

The author thanks B. N. Strobel and J. A. Barzen for their help in organizing the March 2015 meeting and postmeeting process and the International Crane Foundation for hosting the meeting, and for collaborating on an earlier publication from this process. The following individuals participated in the 2015 workshop: *A. Lacy, *A. Pearse, *B. Brooks, B. Hartup, B. Tarr, C. Bowden, C. Sadowski, D. Gawlik, *D. Lopez, G. Archibald, G. Olsen, J. Duff, J. French, J. Howard, J. Langenberg, M. Mace, *M. McPhee, M. Wellington, N. Lloyd, P. Fasbender, P. Miller, P. Nyhus, *S. Hereford, S. King, *S. Matteson, S. Warner, and W. Harrell (*indicates a workshop participant that also participated in the postworkshop elicitation). Reviews by D. A. Armstrong, C. T. Moore, and D. R. Smith improved the manuscript.

LITERATURE CITED

Barzen JA, Converse SJ, Adler PH, Lacy A, Gray E, Gossens A. 2018. Examination of multiple working hypotheses to address reproductive failure in reintroduced Whooping Cranes. *The Condor* 120:632–649.

Burgman M. 2004. Expert frailties in conservation risk assessment and listing decisions. Pages 20–29 in Hutchings P, Lunney D, Dickman C, eds. *Threatened Species Legislation: Is It Just an Act?* Mosman, Australia: Royal Zoological Society of New South Wales.

Burgman MA, McBride MF, Ashton R, Speirs-Bridge A, Flander L, Wintle B, Fidler F, Rumpff L, Twardy C. 2011. Expert status and performance. *PLOS ONE* 6:e22998.

Canadian Wildlife Service, US Fish and Wildlife Service. 2005. *International Recovery Plan for the Whooping Crane.* Ottawa, Canada: Recovery of Nationally Endangered Wildlife (RENEW), and US Fish and Wildlife Service, Albuquerque, New Mexico.

Canessa S, Guillera-Arroita G, Lahoz-Monfort JJ, Southwell DM, Armstrong DP, Chadès I, Lacy RC, Converse SJ. 2015. When do we need more data? A primer on calculating the value of information for applied ecologists. *Methods in Ecology and Evolution* 6:1219–1228.

Converse SJ, Moore CT, Armstrong DP. 2013a. Demographics of reintroduced populations: estimation, modeling, and decision analysis. *Journal of Wildlife Management* 77:1081–1093.

Converse SJ, Moore CT, Folk MJ, Runge MC. 2013b. A matter of tradeoffs: reintroduction as a multiple objective decision. *Journal of Wildlife Management* 77:1145–1156.

Converse SJ, Royle JA, Adler PH, Urbanek RP, Barzen JA. 2013c. A hierarchical nest survival model integrating incomplete temporally varying covariates. *Ecology and Evolution* 3:4439–4447.

Converse SJ, Servanty S, Moore CT, Runge MC. 2018a. Population dynamics of reintroduced Whooping Cranes: research and management application. Pages 139–160 in French JB, Jr., Converse SJ, Austin JE, eds. *Whooping Cranes: Biology and Conservation.* Biodiversity of the World: Conservation from Genes to Landscapes. San Diego, CA: Academic Press.

Converse SJ, Strobel BN, Barzen JA. 2018b. Reproductive failure in the eastern migratory population: the interaction of research and management. Pages 161–178 in French JB, Jr., Converse SJ, Austin JE, eds. *Whooping Cranes: Biology and Conservation.* Biodiversity of the World: Conservation from Genes to Landscapes. San Diego, CA: Academic Press.

French JB, Jr., Converse SJ, Austin JE. 2018. Whooping Cranes past and present. Pages 3–16 in French JB, Jr., Converse SJ, Austin JE, eds. *Whooping Cranes: Biology and Conservation.* Biodiversity of the World: Conservation from Genes to Landscapes. San Diego, CA: Academic Press.

King RS, Trutwin JJ, Hunter TS, Varner DM. 2013. Effects of environmental stressors on nest success of introduced birds. *Journal of Wildlife Management* 77:842–854.

Martin TG, Burgman MA, Fidler F, Kuhnert PM, Low-Choy S, McBride MF, Mengersen K. 2011. Eliciting expert knowledge in conservation science. *Conservation Biology* 26:29–38.

McBride MF, Fidler F, Burgman MA. 2012. Evaluating the accuracy and calibration of expert predictions under uncertainty: predicting the outcomes of ecological research. *Diversity and Distributions* 18:782–794.

Runge MC, Converse SJ, Lyons JE. 2011. Which uncertainty? Using expert elicitation and expected value of information to design an adaptive program. *Biological Conservation* 144:1214–1223.

Servanty S, Converse SJ, Bailey LL. 2014. Demography of a reintroduced population: moving toward management models for an endangered species, the Whooping Crane. *Ecological Applications* 24:927–937.

Urbanek RP, Fondow LEA, Satyshur CD, Lacy AE, Zimorski SE, Wellington M. 2005. First cohort of migratory

Whooping Cranes reintroduced to eastern North America: the first year after release. *Proceedings of the North American Crane Workshop* 9:213–223.

Urbanek RP, Fondow LEA, Zimorski SE. 2010a. Survival, reproduction, and movements of migratory whooping cranes during the first several years of reintroduction. *Proceedings of the North American Crane Workshop* 11:124–132.

Urbanek RP, Zimorski SE, Fasoli AM, Szyszkoski EK. 2010b. Nest desertion in a reintroduced population of migratory Whooping Cranes. *Proceedings of the North American Crane Workshop* 11:133–141.

Williams BK, Eaton MJ, Breininger DR. 2011. Adaptive resource management and the value of information. *Ecological Modelling* 222:3429–3436.

PART VI ADDRESSING LINKED AND DYNAMIC DECISIONS

21 Introduction to Linked and Dynamic Decisions

Michael C. Runge

Often, a decision maker is faced with a series of linked decisions, rather than an isolated one-off decision. In natural resource management, it is common to make a similar type of decision on a regular basis (e.g., annually). Such linked decisions have at least 1 of 2 important properties: they may be dynamic, that is, the actions taken early affect both the immediate outcomes and the effects of actions taken later; and they may be adaptive, that is, early actions might generate learning that can be applied to later actions. This situation has given rise to the development of methods for adaptive management. This chapter provides an overview of the decision analytical methods available to support framing and solving dynamic decisions and briefly reviews the case studies that follow.

Introduction

One of the most iconic examples of wildlife management is the annual setting of hunting regulations for ducks in the United States (Johnson et al. 1997). Using well-designed surveys to understand the current status of the duck population, managers choose hunting regulations that seek to provide a large, sustainable harvest. Each annual decision cannot be treated in isolation; the managers must anticipate how this year's decision will affect next year's population, which affects the following year's decision, and so on. Thus, a key feature of this decision context is that it is a set of decisions that are linked over time.

A land trust that purchases land or accepts conservation easements for the long-term protection of a particular ecosystem or species faces a similar type of decision problem. Each parcel of land that comes up for sale or each landowner offering an easement represents an opportunity, and the land trust must consider those opportunities in the larger context. How does the protection of one parcel of land affect the connectivity of the other pieces, how does it affect the cash reserves or workload of the organization, and how does it open up opportunities to protect other land in the future? Conversely, how does the choice not to protect a parcel of land reduce the conservation value of the network of parcels, free up resources to protect other parcels in the future, or close off future opportunities? In this case, the individual decisions are linked over both space and time.

These types of decisions, often called linked or dynamic decisions, have at least 1 of 2 unique features. First, they may be dynamic, in that a decision made at one point in time affects the status of the system and thus decisions that will be made in the future. Second, they may provide the opportunity for learning. Monitoring the results of an early decision may resolve some of the uncertainty, thereby allowing the decision maker to make better choices in the future: this is the hallmark of *adaptive management* (Walters 1986). In other words, linked decisions can be linked through the *system dynamics* (how the managed system changes through time, including as a result of taking action) or through the *information dynamics* (how our knowledge of the system and its responses to action change through time); often they are linked in both ways.

Three new scientific tasks are unique to linked decisions. First, the predictions now need to anticipate how the *system state* at time t (the description of the status of the population, habitat, or ecosystem being managed) is modified by the action at time t to produce a new state of the system at some later point in time, as well as any immediate rewards that accrue as a result of taking action at time t. That is, a dynamic predictive model is required. Second, a new concept arises—that of the *information state*, which describes what we know about the system dynamics. Using concepts from Bayesian decision theory, the new task includes forecasting how the information state may change as a result of any action taken. Third, the search for an optimal solution requires highly technical optimization methods, because of the need to explicitly accommodate time within the problem.

No new value-based tasks are unique to linked decision problems, although the iterative nature of the decision provides not only an opportunity to update the information state but also an opportunity to update the policy and governance elements of the decision framing (fig. 21.1; Pahl-Wostl 2009). For example, after several cycles of decisions, the decision makers and stakeholders may realize the original statement of objectives did not fully capture what they care about, so they might update the set of objectives (so-called *double-loop learning*). Even more profoundly, the agencies involved might realize that the current institutions are not best poised to achieve the objectives and may consider changes to the governance structure (*triple-loop learning*).

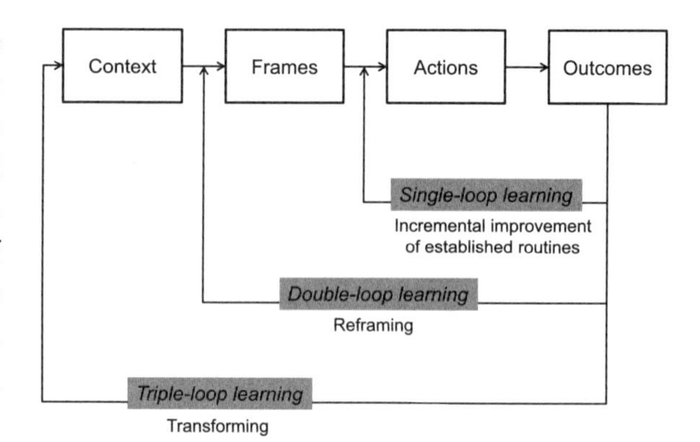

Figure 21.1. Diagram of single-, double-, and triple-loop learning. *Pahl-Wostl 2009.*

Survey of Tools for Addressing Dynamic Decision Problems

The tools for addressing dynamic decision problems fall into 2 categories: structuring tools and technical tools. In our rush to use the fancy technical tools, we sometimes overlook the importance of structuring, but as in all decision problems, the structure matters. Both categories of tools are discussed below.

Structuring a Dynamic Decision Problem

Articulating the structure of a dynamic decision is challenging: The primary elements are the same as in other classes of decisions (Runge and Bean, chapter 1, this volume, fig. 1.1), but there are several new elements, and all the elements require consideration of the dynamic context of the problem. As in other decision classes, early attention to thoughtful structuring is one of the most valuable steps in decision analysis.

The objectives in a dynamic decision need to consider either the outcomes that accumulate over the whole series of actions or the outcome that is realized after the last action. In the case of harvest management, for example, the objective may be to maximize the cumulative harvest over an indefinite time horizon. In the case of a land trust's decisions, the objective may concern the functioning of the target ecosystem 100 years from now, perhaps as measured by the area of intact, contiguous habitat. That is, the objectives do not focus on the immediate returns (this year's harvest) but on the cumulative or long-term rewards. The sequence of these rewards, however, might matter; typically, we value things that are closer in time, so we might discount future rewards relative to current rewards, but for conservation objectives, we may care about long-term outcomes. The development of a *temporal discount function* is an important element of the decision maker's values (Fisher and Krutilla 1975; Henderson and Sutherland 1996). There may also be multiple objectives for a linked decision, so the tools of multi-criteria decision analysis might need to be integrated. An interesting consideration, which has received scant attention, is whether we might use different discount rates for different objectives.

There are several ways to think about the alternatives in a dynamic decision. At the level of a single decision point, the alternatives are the individual actions available at that time. The set of actions may not be the same at all points in time: some actions may require previous actions to have taken place (e.g., animals cannot be taken into captivity until a captive facility is built); some actions may be state dependent (e.g., the number of animals that can be translocated is limited by the number of animals alive); and some actions may be time dependent (e.g., seasonal). Another way to think about the alternatives is as temporal portfolios—that is, the alternative sequences of actions taken over all points in time, a framing that emphasizes the linkages across the individual decision points. A third way to think about the alternatives is as state-dependent policies that specify what action is to be taken as a function of the state of the system at any point in time (these policies may or may not also be time-dependent). For most problems pertaining to real systems, the number of possible state-dependent policies is very large, and the decision maker's task is to select the policy that is expected to best achieve the objectives.

The evaluation of consequences in a dynamic problem, then, can be seen as the forecasting of the accrued long-term outcomes as a function of the full temporal sequence of actions taken. As with articulation of objectives and alternatives, this task is often broken down, in this case by focusing on predicting (1) the change in the state of the system from time t to time $t+1$ as a function of the action taken at time t, and (2) the immediate reward, if any, earned from taking a particular action at time t. The *state variables* need to be identified prior to developing such a predictive model. The state variables are the measures of the status of the system that are relevant for forecasting and making decisions. Dynamic problems that can be expressed in this way as a sequence of one-time-step events are known as *Markov decision problems* (MDPs, Marescot et al. 2013).

Uncertainty is typically a concern in dynamic problems, for 2 reasons. First, aleatory uncertainty (irreducible uncertainty, see Runge and Converse, chapter 13, this volume) may compound over time, so that forecasts become more uncertain farther in the future. For this reason, system models typically incorporate stochastic elements. State-dependent policies can often cope with stochastic variation because they specify what actions to take if the system drifts away from what is expected. Second, epistemic uncertainty (uncertainty arising from the limits of knowledge, see Runge and Converse, chapter 13, this volume) may matter to the decision, and it may be possible to reduce it over time. That is, a dynamic problem may have an embedded information problem (see Smith, chapter 17, this volume). Thus, clear articulation of the critical uncertainty, perhaps as alternative system models, shows the way toward better decision making over time.

One of the new elements in a dynamic decision problem is a *monitoring system*. Monitoring provides 2 important pieces of information that are relevant to the decision maker. First, monitoring allows measurement of the current system state, which is used by the decision maker to choose the best current action; this is particularly important when there is large stochastic variation. Second, monitoring the effects of the actions provides a check on the predictive model that underlies the decision framework; this feedback may allow uncertainty to be reduced, improving predictions and hence the choice of action at future points in time. The design of the monitoring system depends on the structure of the problem: on the objectives, on the state variables, and on the critical uncertainties.

The second new element in a dynamic decision problem is an *information model*. Just as the system model predicts changes in the system state as a function of the actions, the information model predicts changes in the information state as a function of the actions and the current system state. When there is critical epistemic uncertainty, some actions may reduce uncertainty more than others (e.g., implementing actions about which little is known), and it might be valuable to predict those differences, to evaluate whether it is worth taking action for the sake of generating learning. *Bayes' theorem*, either in discrete or continuous form, is usually used to update the information state based on new information. Bayes' theorem provides a quantitative way to express posterior knowledge as a function of prior knowledge and new data. A predictive information model works like a *Bayesian preposterior analysis* (Berger 1985), anticipating how the information state could change depending on the outcome observed.

Solving a Dynamic Decision Problem

At its core, a dynamic decision problem is a *linked decision tree*, in which there are a series of decision nodes connected through time, possibly with uncertain processes (chance nodes) between them. Figure 21.2 shows how quickly the complexity of the system can proliferate: with only 2 actions available at each time point, and 3 states of the system, there are already 18 possible pathways by time $t=2$ (and 648 by $t=4$). For small trees, brute force calculations are possible, in which the consequences of all pathways are calculated, but even moderate numbers of actions and states quickly make such calculations prohibitive. Instead, linked decision trees are usually solved by the *rollback method*, working from right to left, where expected values (or expected utilities, see Runge and Converse, chapter 13, this volume) are applied at the chance nodes, and maximizations (or minimizations, depending on the objective) are applied at decision nodes. This calculation method reveals an interesting insight: in linked decisions, you cannot figure out what to do today unless you've already figured out what you're going to do in the future.

The extension of the rollback method for moderate to large Markov decision problems is called *stochastic dynamic programming* (SDP, Puterman 1994). This set of algorithms is used to find state- and time-dependent policies that optimize the objective function (a single mathematical expression of the objectives). Some of the algorithms can be used to find *stationary policies*—state-dependent, but not time-dependent, policies. The solutions from implementation of SDP are exact but suffer from the "curse of dimensionality" (Bellman 1957), in that the number of calculations required rises exponentially with the size of the state and action space. For large problems, approximation methods, like *reinforcement learning*, are used instead (Fonnesbeck 2005).

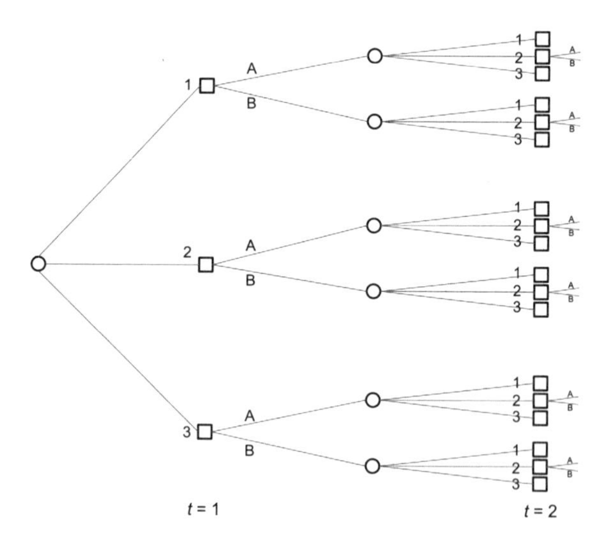

Figure 21.2. Linked decision tree. The system can be in 1 of 3 states (1, 2, or 3), and 2 actions (A, B) are available at each time. Decision nodes are shown with squares; chance (stochastic) nodes are shown with circles.

When epistemic uncertainty is an important component of a dynamic decision, the solution needs to anticipate not only how any action affects the system state but also how it could affect the information state; this is known as the "dual-control" problem, a central question in adaptive management. One partial solution, known as *passive adaptive management*, does not fully solve the dual-control problem: the optimization does not anticipate a change in the information state, but the implementation does incorporate it. The full solution, known as *active adaptive management* in the field of natural resource management, includes the information state in the SDP algorithm (Williams 1996). Recent work has shown the solution of active adaptive problems to be a special case of *partially observable Markov decision processes* (POMDPs; Chadès et al. 2012; Williams 2011).

Case Studies

The next 3 chapters present case studies featuring dynamic decisions and showcase decision analytical tools available to help decision makers structure and analyze problems that have linkages over time and space.

In chapter 22, Victoria Hunt and colleagues discuss their work with the US Fish and Wildlife Service (USFWS) to restore wetlands in the Prairie Pothole region of central North America. More than half of the wetlands in this region were converted into agricultural land between 1950 and 1990, in some cases by filling them with upland soil. Restoration of any individual wetland basin is a one-off decision made in the face of uncertainty, but learning and adaptation can occur through spatial replication. The authors used multiple predictive models to capture uncertainty about the effectiveness of sediment removal as a restoration technique and updated the relative belief in these models based on post-restoration monitoring data, using Bayes' theorem. Each year, new wetland basins are considered for restoration, so the updated knowledge about the effectiveness from early restorations is applied to later restorations. Eight years into the program, important insights are emerging about which types of wetland can be most successfully restored through excavation methods. Because so many of the existing examples of adaptive management rely on temporal replication on a single management unit, this case study provides important insight about applying adaptive management through spatial replication.

In chapter 23, Clinton Moore and colleagues discuss the Native Prairie Adaptive Management program (NPAM), an extensive effort led by USFWS in the north-central United States to restore and manage native prairie habitat. Even in remnant patches of prairie, there has been habitat degradation as introduced cool-season grasses have invaded. Restoration consists of defoliation treatments, like burning and grazing, but there is considerable uncertainty about the optimal choice and frequency of these interventions, as well as the value of taking no action at times. In this application of adaptive management, individual patches of remnant prairie serve as the management units, with temporal replication within units and spatial replication across units. Expert elicitation was used to build an initial predictive model and to specify the critical uncertainties. Active adaptive dynamic programming is used to assign treatments to units, and Bayes' theorem is used annually to update the belief weights on the alternative models, integrating monitoring data into subsequent predictions. In addition to improving management as a result of learning, this program has increased communication and cooperation among an extensive group of collaborators.

In chapter 24, Conor McGowan and colleagues describe a long-term collaboration among state and federal agencies to manage horseshoe crab (*Limulus polyphemus*) harvest in Delaware Bay. Harvest regulations for horseshoe crabs are set annually and also affect shorebirds that rely on horseshoe crab eggs for energetic requirements during migration. Uncertainty about the effect of horseshoe crab egg abundance on red knot (*Calidris canutus rufa*) survival impedes the fishery management decision. A formal adaptive management process, using passive adaptive SDP, is used to set regulations annually while allowing for learning to accrue. The implementation of this program is now mature enough that a double-loop process has been initiated to review the framing of the decision.

Open Questions and Challenging Issues in Adaptive Management

Since Carl Walters's seminal book on adaptive management (Walters 1986), there has been an explosion of literature on the topic, within which two contrasting themes arise: one theme extols the value of adaptive management and urges its widespread application; the other theme lists all the ways that adaptive management has failed. The middle ground lies in a sober recognition that adaptive management is appropriate and valuable in some circumstances, but not all decision problems are suited to it. The guide to adaptive management published by the Department of the Interior (Williams et al. 2007) provides a rubric for the conditions that warrant an adaptive management approach: (1) when there is a mandate to take action in the face of uncertainty; (2) when there is institutional capacity and commitment to undertake and sustain an adaptive program; (3) when a real management choice is to be made; (4) when there is an opportunity to apply learning; (5) when clear and measurable management objectives can be identified; (6) when the value of information for decision making is high; (7) when uncertainty can be expressed as a set of testable models; and (8) when a monitoring system can be established to reduce uncertainty. More careful examination of these conditions is warranted as decision makers and analysts consider whether a particular decision should be cast in an adaptive management framework.

As conditions 4 and 6 indicate, adaptive management is both a dynamic problem and an information problem (Smith, chapter 17, this volume). The question that should motivate adaptive management is whether the critical uncertainty needs to be resolved (Runge et al. 2011), but we have found relatively few programs that purport to implement adaptive management have done a formal value-of-information analysis (that is, they have not evaluated condition 6). The motivation, design, and implementation of adaptive management could be improved by attention to this connection among decision analysis methods.

Other connections between dynamic decisions and other classes of decisions also deserve more exploration. In particular, there are many examples of multi-criteria decision analysis in natural resource management (see Converse, chapter 5, this volume) and many examples of adaptive management, but there are very few examples of adaptive management that specifically grapple with trade-offs among multiple objectives. Likewise, we have found insufficient examination of risk tolerances in the context of adaptive management and expect that considerations of multiple objectives and risk are as present in dynamic decisions as they are in one-off decisions.

The discussion of adaptive management in this chapter and the case studies to follow focuses on a decision-theoretic understanding of adaptive management. Another large line of discussion in the adaptive management literature focuses on governance issues—the institutional relationships that affect the management of a natural resource—and how they can be considered in an adaptive process (e.g., Gunderson and Light 2006). To some extent, these 2 schools of thought regarding adaptive management remain quite separated (McFadden et al. 2011). There is great potential to improve our management of natural resources by integrating decision analysis and sociopolitical approaches.

LITERATURE CITED

Bellman R. 1957. *Dynamic Programming*. Princeton, NJ: Princeton University Press.

Berger JO. 1985. *Statistical Decision Theory and Bayesian Analysis*. New York: Springer.

Chadès I, Carwardine J, Martin TG, Nicol S, Sabbadin R, Buffet O. 2012. MOMDPs: A solution for modelling adaptive management problems. *Proceedings of the AAAI Conference on Aritifical Intelligence* 26:267–273.

Fisher AC, Krutilla JV. 1975. Resource conservation, environmental preservation, and the rate of discount. *The Quarterly Journal of Economics* 89:358–370.

Fonnesbeck CJ. 2005. Solving dynamic wildlife resource optimization problems using reinforcement learning. *Natural Resource Modeling* 18:1–40.

Gunderson L, Light SS. 2006. Adaptive management and adaptive governance in the Everglades ecosystem. *Policy Sciences* 39:323–334.

Henderson N, Sutherland WJ. 1996. Two truths about discounting and their environmental consequences. *Trends in Ecology & Evolution* 11:527–528.

Johnson FA, Moore CT, Kendall WL, Dubovsky JA, Caithamer DF, Kelley JR, Jr., Williams BK. 1997. Uncertainty and the management of mallard harvests. *Journal of Wildlife Management* 61:202–216.

Marescot L, Chapron G, Chadès I, Fackler PL, Duchamp C, Marboutin E, Gimenez O. 2013. Complex decisions made simple: A primer on stochastic dynamic programming. *Methods in Ecology and Evolution* 4:872–884.

McFadden JE, Hiller TL, Tyre AJ. 2011. Evaluating the efficacy of adaptive management approaches: Is there a formula for success? *Journal of Environmental Management* 92:1354–1359.

Pahl-Wostl C. 2009. A conceptual framework for analysing adaptive capacity and multi-level learning processes in resource governance regimes. *Global Environmental Change* 19:354–365.

Puterman ML. 1994. *Markov Decision Processes: Discrete Stochastic Dynamic Programming*. Hoboken, NJ: John Wiley and Sons.

Runge MC, Converse SJ, Lyons JE. 2011. Which uncertainty? Using expert elicitation and expected value of information to design an adaptive program. *Biological Conservation* 144:1214–1223.

Walters CJ. 1986. *Adaptive Management of Renewable Resources*. New York: Macmillan.

Williams BK. 1996. Adaptive optimization of renewable natural resources: solution algorithms and a computer program. *Ecological Modelling* 93:101–111.

———. 2011. Resolving structural uncertainty in natural resources management using POMDP approaches. *Ecological Modelling* 222:1092–1102.

Williams BK, Szaro RC, Shapiro CD. 2007. *Adaptive Management: The US Department of the Interior Technical Guide*. Washington, DC: Adaptive Management Working Group, US Department of the Interior.

22 Restoration of Wetlands in the Prairie Pothole Region

Victoria M. Hunt,
Melinda G. Knutson,
and Eric V. Lonsdorf

In the Prairie Pothole region of central North America, thousands of small wetlands provide critical habitat for migratory waterfowl. Unfortunately, more than half of Prairie Pothole wetlands were converted to agricultural lands between 1950 and 1990, resulting in significant habitat loss and degradation. Many wetland basins were deliberately drained and filled with upland soils, and other basins accumulated eroded sediment from adjacent agricultural lands. In recent decades, the US Fish and Wildlife Service (USFWS) has worked with other government agencies and private landowners to restore affected wetlands and learn about the effectiveness of sediment removal as a restoration strategy. This chapter describes a large-scale, adaptive management effort in which decisions about whether to remove sediment are iterated over space. Through this work, USFWS managers seek to address the question of whether sediment excavation, in addition to the usual practice of restoring basin hydrology, improves wetland restoration outcomes. The hypothesis tested was that removing sediment from wetland basins exposes the native seed bank, promotes revegetation by native wetland plants, and enhances wetland function. Preliminary results for the few wetlands that were restored at least 8 years ago indicate that wetland type is a strong external driver of restoration outcomes. Regardless of whether or not sediment excavation was conducted, seasonal wetlands have relatively improved outcomes in terms of habitat quality compared to temporary wetlands.

Problem Background

Decision Maker and Their Authority

In recent decades, US Fish and Wildlife Service (USFWS) personnel have worked to restore thousands of Prairie Pothole wetlands. The USFWS restores wetlands on Service-owned lands (e.g., Waterfowl Production Areas). In Minnesota alone, about 250 wetlands are restored annually by the USFWS program (Sheldon Myerchin, personal communication). In many cases, restorations are conducted cooperatively with other government agencies and with private landowners via the USFWS Partners for Fish and Wildlife Program. The Partners for Fish and Wildlife Program is crucial because the majority of wetlands in the Prairie Pothole region are in private ownership. The program has engaged more than 45,000 landowners and resulted in restoration of more than a million acres of wetland habitat in the last 25 years (US Fish and Wildlife Service 2018).

Other federal and state agencies, such as the USDA Natural Resources Conservation Service and the Minnesota Department of Natural Resources, also engage in wetland restoration on both private and public lands. Wetland restoration techniques routinely used by federal and state land managers (henceforth referred to as land managers) in the Prairie Pothole region include plugging or filling of drainage ditches, breaking up tile lines used for drainage, and constructing berms. The principal purpose of such efforts is to disrupt drainage mechanisms that were constructed to facilitate agriculture. Following the restoration of hydrology at a given wetland basin, restorations are often simply revegetated via natural recolonization (Galatowitsch and van der Valk 1996). Alternatively, sometimes land managers perform plantings or seedings to encourage the re-establishment of native vegetation.

Ecological Background

Prairie Pothole wetlands are small, depressional wetlands located throughout central North America. These wetlands, formed by glaciers more than 10,000 years ago, provide critical habitat for wildlife including migratory waterfowl (US Environmental Protection Agency 2018). The Great Plains and Prairie Pothole region are considered the most important threatened waterfowl habitat on the continent. Millions of waterfowl breed in the Prairie Pothole region each year (Bellrose 1976). Over 50% of North American migratory waterfowl, including many important game species, rely on this habitat (US Environmental Protection Agency 2018). Waterfowl species that breed in this region include pintails *(Anas acuta)*, mallards *(Anas platyrhynchos)*, gadwalls *(Anas strepera)*, blue-winged teals *(Anas discors)*, shovelers *(Anas clypeata)*, canvasbacks *(Aythya valisineria)*, and redheads *(Aythya americana)* (Bellrose 1976). Other migratory species that rely on this wetland habitat include many marsh birds and shorebirds (Niemuth et al. 2006), such as marbled godwits *(Limosa fedoa)* (Ryan 1982) and willets (*Tringa semipalmata*) (Ryan and Renken 1987). Unfortunately, only a small fraction of the important wetland habitats in the Prairie Pothole region remains pristine today, and about half have disappeared entirely (US Environmental Protection Agency 2018). In some areas, as much as 90% has been lost due to conversion to agriculture (Doherty et al. 2013).

Wetlands that have a history of agricultural use, as is common in the Prairie Pothole region, present unique restoration challenges. One major challenge is posed by accumulated sediment. Sediment accumulates from tillage and wind erosion, and soil is also deliberately leveled when wetlands are converted to agriculture. Sediment accumulation is generally detrimental to overall wetland function (Gleason et al. 2003). Sediment buries the native plant seed bank and creates soil conditions conducive to invasive species, including reed canarygrass (*Phalaris arundinacea*) and cattails *(Typha spp.)* (Smith et al. 2016). Sedimentation influences soil density and organic matter at the soil surface, increases nitrogen and phosphorous inputs, and reduces water depth and duration of flooding, all of which contribute to invasion of non-native plant species (Green and Galatowitsch 2002; Smith et al. 2016). Prairie Pothole wetlands containing accumulated sediment have reduced capacities to provide flood control by buffering against surges of rainwater, flood waters, and snow melt (US Environmental Protection Agency 2018). Via multiple mechanisms, sedimentation is associated with reduction in waterfowl and wildlife habitat quality and with reduced native species diversity of wetland plants (Smith et al. 2016). The most obvious indicator of a degraded wetland basin is overgrowth of cattails or reed canarygrass, which robs the basin of open water that is needed by most waterfowl and other wetland species, such as amphibians.

The effectiveness of sediment removal via excavation, in terms of restoring habitat quality and wetland function, is largely unknown. This uncertainty currently complicates restoration efforts in the Prairie Pothole region. Excavated wetlands show vegetation trends that are more similar to wetlands that do not have a history of agricultural use, compared to wetlands where sediment has not been excavated (Smith et al. 2016). However, if sediment removal does not improve restoration outcomes, there are reasons to avoid it: removal of sediment adds to the time and cost of restoration, and it requires technical expertise and special equipment.

Prior to excavation, the depth of excess sediment must be mapped throughout the basin. Even within a single basin, there may be considerable variation in sediment depth, and sediment can be patchy. Mapping is performed on the basis of color, texture, and nutrient and pH tests from soil cores. Some basins accumulate a meter of sediment or more. Once the sediment depth has been mapped and a plan (prescription) for removal has been formulated, sediment is excavated using heavy equipment, e.g., bulldozers. Although the detrimental effects of accumulated sediment on wetland function are well documented (Gleason et al. 2011), it is currently unknown if the process of removing sediment can reverse those detrimental effects, thus yielding superior restoration outcomes.

Prairie Pothole wetland restoration is further complicated by the possible confounding effects of structural differences among wetlands. Prairie Pothole wetlands can be temporary, seasonal, or semipermanent, following definitions in Stewart and Kantrud (1971). It is unknown if the effectiveness of sediment removal depends on the hydrological classification of the wetland undergoing restoration, but there is reason to suspect that it does. For example, relative to large permanent wetlands, seasonal wetlands, which tend to be smaller and shallower, are more likely to dry quickly and accumulate detrimental sediment which results in low restoration recovery rates (Bartzen et al. 2010; Smith et al. 2016). Hydrological classification is also important when considering quality of waterfowl habitat. For example, semipermanent wetlands supported the greatest number of bird use days relative to wetland area (O'Neal et al. 2008).

Decision Structure

Wetland basins were screened on multiple criteria before they could be enrolled in this project (fig. 22.1). For example, basins with a previous history of agricultural cropping were included, and basins with flow-through rivers or streams that would carry in sediment were excluded. Due to the relatively long periods of time required to assess restoration outcomes (8 years per basin), landowners entered into a long-term agreement ensuring their multiyear participation in the project.

The agency land manager (henceforth land manager) who decides whether to excavate sediment at the start of a given wetland restoration considers multiple objectives. This decision is iterated over space by different land managers, who share information in a network for the purposes of policy development and increasing speed of learning (Moore et al. 2013). If the restorations are conducted on private land via easement agreements, the private landowner also participates in restoration decisions. In the network, land managers learn from the outcomes of each other's restorations by participating in an adaptive management framework (Walters 1986). Learning accrues to multiple decision makers over space and time, ultimately resulting in more informed decision making and improved wetland restoration outcomes (Moore et al. 2013). The project began in 2008 and continues to the present day. Additional wetland basins are enrolled in the project each year.

Decision Analysis

Objectives

We elicited objectives from state, federal, and NGO wetland managers involved in wetland restoration during a workshop. The fundamental objective of Prairie Pothole wetland restoration is to provide healthy, functioning wetland habitat across the landscape for waterfowl and migratory birds. This fun-

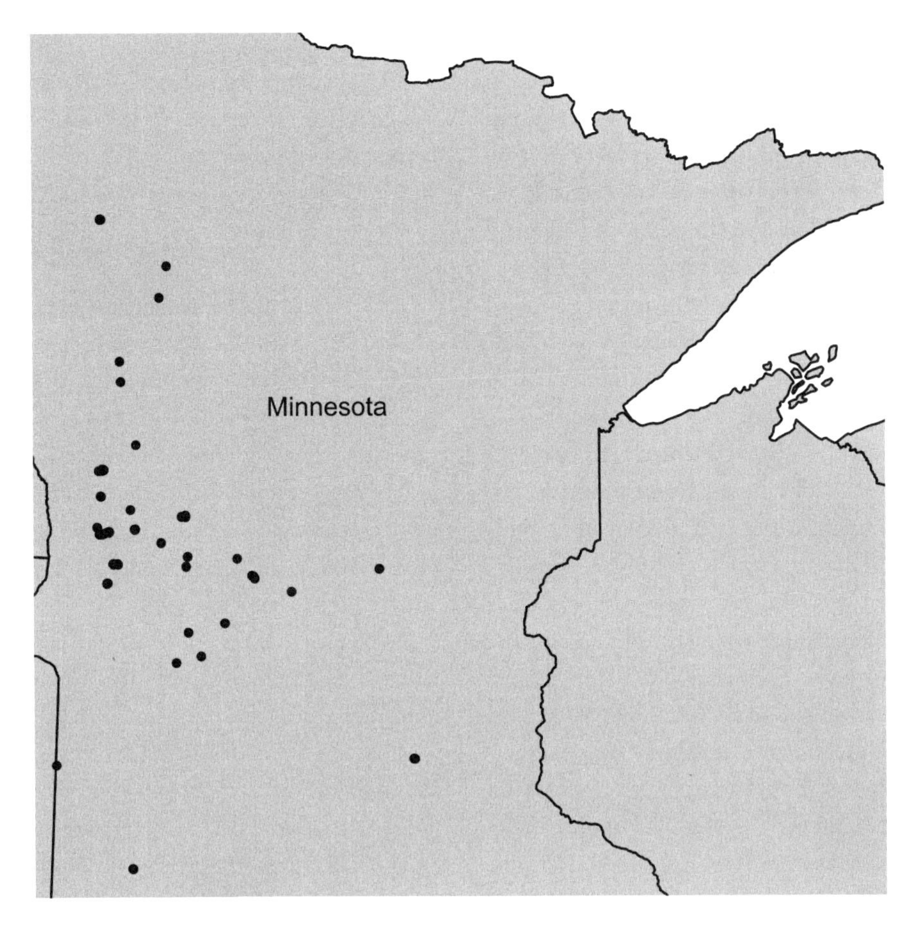

Figure 22.1. Map of wetland basins participating in this project.

damental objective can be broken down into 2 means objectives: (1) restore appropriate hydrology for the wetland type (i.e., temporary, seasonal, or semipermanent), and (2) restore native plant species and vegetative structure to create high-quality bird habitat. Basins are restored to our best assessment of the wetland type that was present prior to agricultural tillage, as determined via historical records, maps, and photographs. Although cost is a consideration, the additional cost of excavation at a site is difficult to estimate; heavy equipment contractors are usually involved with or without sediment excavation because bulldozers are needed to restore hydrology. Land managers were concerned with understanding the effect of sediment excavation on wetland outcomes, regardless of cost. Therefore, cost was not explicitly included in the decision structure.

Alternatives

Basins that met the selection criteria were assigned 1 of 2 possible treatment options by the land manager at the start of restoration. If the manager assigned sediment excavation, this management action accompanied a standard suite of other restoration actions including plugging drainage ditches and tile lines and constructing berms to restore degraded and drained wetland basins. If the land manager assigned no excavation, all of the aforementioned actions were carried out, with the exception of sediment excavation. Therefore, there is a status quo procedure for conducting wetland restorations; the question this research seeks to address is whether the addition of sediment excavation improves outcomes for waterfowl habitat quality. Participating mangers were advised that for each general area where 1 or more basins were excavated, at least 1 wetland undergoing restoration should be left unexcavated, to achieve a more balanced study design and to optimize information gain. However, for this project to be feasible within the operational, financial, and time constraints managers face, the managers were not required to randomly assign treatments to sites, nor were there any requirements concerning the geographic distribution of sites enrolled in the project. Participation in the project was entirely voluntary on the part of the managers; those who participated expected to learn how to improve their wetland restorations. Even for those participating in the project, most restoration sites under their supervision were not enrolled in the project due to time constraints.

Details collected about basins prior to restoration included basin land use histories, catchment land-use histories, and soil survey information. Photographs, aerial maps, and detailed site descriptions were also compiled. Selected wetland basins were monitored once before restoration (preassessment) and 1, 2, 3, 4, 6, and 8 years following restoration (postassessments), with the 8-year postassessment being most important.

Consequence Analysis

Project leadership stipulated from the outset that the monitoring assessment protocol must be rapid. The participants in this project faced constraints on staff time and availability. Often a single biologist was responsible for conducting all of the monitoring assessments associated with multiple wetlands, distributed over many counties. Indeed, most wetland restorations receive no standardized monitoring at all due to these constraints.

To ensure that the protocol was rapid and easy to use, assessments were made using visual observation only, no additional equipment was required, and all variables were categorical. At each monitoring event, the observer recorded the values of 5 monitoring metrics. These metrics directly related to the project's means objectives (fig. 22.2).

The measured attributes were (1) an estimate of how much standing water was present in the basin,

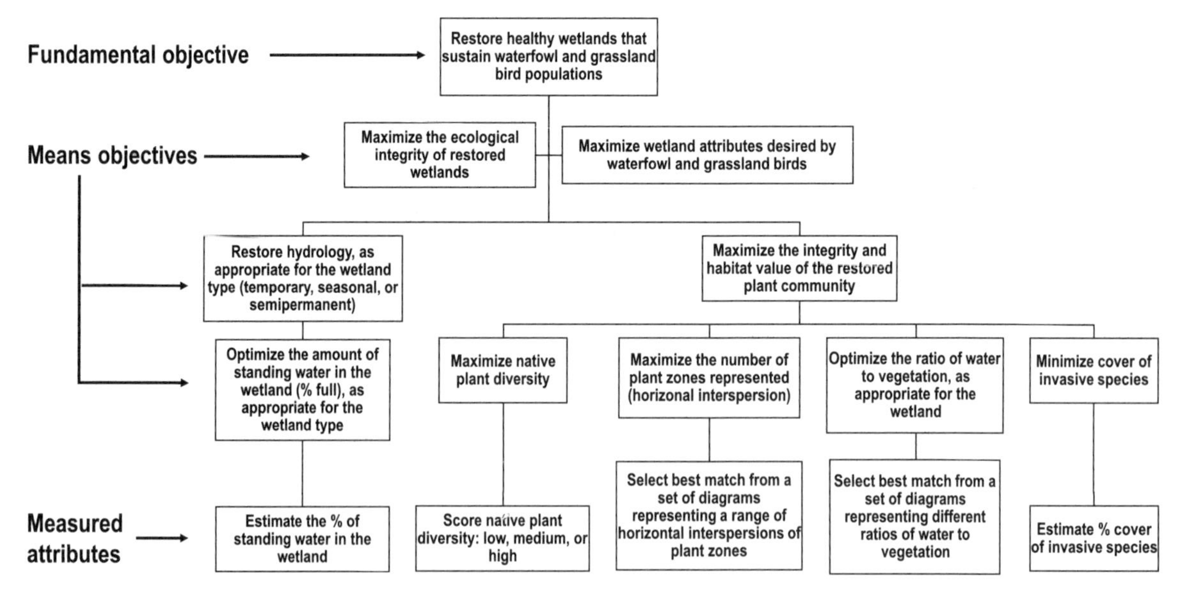

Figure 22.2. Objectives hierarchy and associated measured attributes.

relative to its total capacity, (2) horizontal interspersion (a metric pertaining to plant community structure; Mack 2001), (3) native plant diversity, (4) percentage cover of invasive plant species (e.g., reed canarygrass), and (5) cover type (a categorical variable that describes amount of open water, bare soil, and vegetation, which is important for birds). The values of monitoring metrics determined the overall success of the restoration and quantified changes over time.

Although the fundamental objective is to create high-quality habitat for waterfowl and migratory birds, no direct measurements were made of bird population-level responses. We wanted to evaluate outcomes that directly link to restoration decisions. Wetland vegetation and hydrology were deemed useful proxies for the fundamental objective; breeding habitat associations for wetland birds are well understood, especially for waterfowl (Naugle et al. 2001). Migratory bird population dynamics are influenced by many factors beyond the scope of this project, including habitat quality along the entire flyway, the weather, year-to-year variations in population sizes of migratory flocks, etc. In addition, logistical considerations precluded monitoring bird populations at these sites; the optimal timing of bird surveys (spring) and vegetation surveys (summer) did not coincide, which would require multiple monitoring visits to a site within a year.

Monitoring data were used to reduce uncertainty regarding the effectiveness of sediment excavation. To this end, hypotheses about the effects of sediment excavation are represented via competing models. The competing models are composed of transition matrices, which specify the probability of beginning in one system state and ending in another system state under a specific hypothesis, i.e., under the hypothesis that sediment excavation is the optimal action. Project leadership agreed that restorations with and without sediment excavation could both potentially yield optimal outcomes, and the reverse might also be true. Therefore, competing hypotheses were considered in pairs: restorations that included sediment removal could lead to improvement or degradation, and restorations without sediment removal could also lead to either improvement or degradation. To account for differences on the basis of hydrology, there were different versions of transition matrices for the 3 types of wetlands (temporary, semipermanent, and seasonal). All possible states also had an associated utility ("happiness score") indicating proximity of wetlands to meeting project objectives (fig. 22.2).

Transition matrices were built on the basis of expert opinion at the outset of the project (table 22.1). This process began with a structured decision-making workshop (Reynolds et al. 2016), in which partici-

Table 22.1. Example transition matrices for percent full, an indicator of how much water is in the wetland relative to total capacity

A. Sediment excavation will lead to improvement

	End state				
Start state	0%	1–25%	26–50%	51–75%	76–100%
0%	0.3	0.5	0.1	0.05	0.05
1–25%	0.3	0.4	0.2	0.05	0.05
26–50%	0.2	0.3	0.4	0.05	0.05
51–75%	0.1	0.1	0.1	0.5	0.2
76–100%	0.1	0.1	0.2	0.2	0.4

B. Sediment excavation will lead to degradation

	End state				
Start State	0%	1–25%	26–50%	51–75%	76–100%
0%	0.6	0.2	0.1	0.05	0.05
1–25%	0.4	0.4	0.1	0.05	0.05
26–50%	0.1	0.4	0.4	0.05	0.05
51–75%	0.1	0.1	0.4	0.3	0.1
76–100%	0.1	0.1	0.1	0.4	0.3

Note: The transition matrices contain the probabilities associated with transitioning from every possible starting state to every possible ending state. Matrices represent the expected probabilities for wetland basins 8 years after excavating sediment, for seasonal wetlands, under 2 competing hypotheses A and B. Shading corresponds to the probability, with darker shading representing relatively greater probabilities.

pants discussed project goals and logistics. Managers worked with a modeling consultant (Lonsdorf) and workshop facilitator (Knutson) to develop a set of transition matrices for expected probabilities of transitions between different states, based on their experiences with past restorations. We used an iterative approach via conference calls and meetings to derive a set of transition matrices that best reflected the managers' collective expectations. This was among the most challenging aspects of project setup because the managers were unaccustomed to thinking about restoration outcomes in terms of transition matrices, and the 8-year time scale required anticipating outcomes nearly a decade into the future.

Decision Solution

Model weights for all competing models began at 0.5, representing an uninformed prior. Model weights were subsequently updated using monitoring data, via application of Bayes' theorem:

$$P(A|B) = \frac{P(B|A) \times P(A)}{P(B)}$$

where $P(A)$ and $P(B)$ are the probabilities of observing occurrence A and B, respectively, $P(A|B)$ is the conditional probability of observing A given that B is true, and $P(B|A)$ is the probability of observing B given that A is true. For example, starting with model weights of 0.5 (uninformed priors), if a seasonal wetland were to transition from 1–25% full to 26–50% full 8 years after excavation, the transition probabilities associated with that occurrence were 0.2 in the model reflecting the hypothesis that sediment excavation is optimal, and 0.1 in the model reflecting that sediment excavation is nonoptimal. Using these new data, model weights are updated according to Bayes' theorem as follows:

$$\frac{0.2 \times 0.5}{0.2 \times 0.5 + 0.1 \times 0.5} = 0.67$$

Therefore, in this example, data were collected that more closely aligned with the hypothesis that sediment excavation was the optimal action. As a consequence, the model weight for the model representing this hypothesis increased from 0.5 to 0.67.

Monitoring metrics are used to assess the system-state transitions before and after management. These observed transitions are then used to update the model weights of competing models using Bayesian updating, as described in the preceding section. This updating could occur after any new information was collected (e.g., after a basin reached the 8-year mark after restoration and was monitored), although in practice updating occurred annually before the project's annual stakeholder meeting in the summer.

Decision Implementation

This project is a long-term commitment for managers; 8 years is expected to be required before definitive outcomes will be observed. For example, a wetland restoration can have open hydrology and primarily native species for a few years but later be dominated by invasive species, which in turn reduce or eliminate open water. Basins were first enrolled in this research project beginning in 2009. Therefore, for the first time, in 2017, we are beginning to see results of basins that have reached the 8-year mark since restoration. There are currently 16 basins that have reached the 8-year mark: 6 seasonal, 3 semipermanent, and 7 temporary basins. The effort is ongoing, with continued active recruitment and new sites being added each year.

Amassing adequate sample sizes to draw meaningful and generalizable conclusions is an administrative challenge. Long-term outcomes are simply not available yet for all but a few basins. There are currently 103 enrolled basins, and monitoring data have been recorded for 547 pre- and post-restoration assessments. However, we present results from only the 16 basins enrolled since 2009 because results before the 8-year mark can be misleading; due to

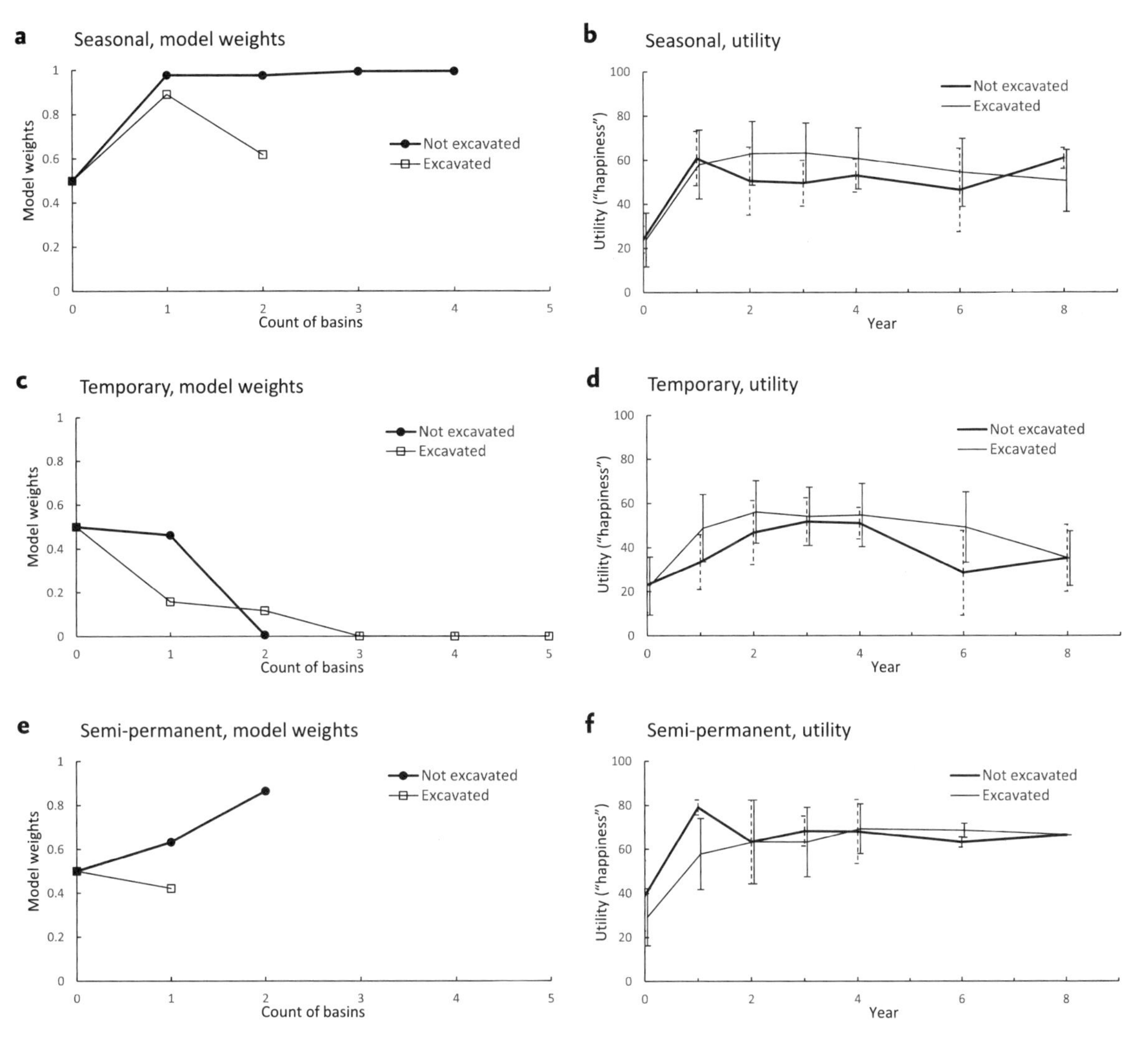

Figure 22.3. Change in model weights over the course of the project and change in utilities over time. Panels (*top left*) and (*top right*) show model weights and utilities for seasonal wetlands, respectively; panels (*middle left*) and (*middle right*) show model weights and utilities for temporary wetlands; and panels (*bottom left*) and (*bottom right*) show model weights and utilities for semipermanent wetlands.

increasing invasive species (cattail) dominance over time, all restorations tend to look good in the first few years.

The competing models at this preliminary stage do not yet indicate that overall better outcomes are achieved for excavated basins relative to unexcavated basins at the 8-year mark (fig. 22.3). However, they do indicate that there are differences based on the hydrological classification of the wetland undergoing restoration. For seasonal wetlands restored in 2009 (2 excavated and 4 unexcavated wetlands) excavation has resulted in improvements of habitat quality, and therefore model weights for the model representing this prediction have increased above 0.5. However, for temporary wetlands, the opposite is observed. For temporary wetlands, restorations have not resulted in predicted levels of habitat improvement, for either excavated (2) or unexcavated (5) basins. Very few monitored semipermanent wetlands have reached the 8-year mark (2 excavated and 1 unexcavated), which underscores the need for a larger sample size and a longer time horizon.

Discussion
Challenges and Recommendations

This project's primary challenges were operational and logistical. In this discussion, we divide the challenges into 4 main interrelated categories, with associated recommendations for each: (1) social and institutional challenges, (2) lack of detailed monitoring data, (3) long temporal horizon, and (4) decision structure that partitioned data.

SOCIAL AND INSTITUTIONAL CHALLENGES

Participating managers faced constraints on staff time, and the natural resource management agency land managers that worked on this project had many other concurrent responsibilities. This particular challenge applies frequently to long-term adaptive management efforts (Westgate et al. 2013). Managing and restoring are core functions of land management agencies, but monitoring is often considered a luxury (La Peyre et al. 2001). Indeed, external (grant) funding for ecosystem restoration is often restricted to direct restoration costs; funding for monitoring, especially long-term monitoring, is frequently excluded (Lovett et al. 2007).

Participation in this project remains entirely voluntary; upper level managers have not required stations to participate because to do so would constrain local flexibility in responding to changes in staffing, operating budgets, and responding to competing agency priorities. Therefore, despite ongoing recruiting, we were not able to enroll enough sites to ensure large sample sizes for all wetland types, especially at the beginning of the project. We recommend that funding sources allow for or even require long-term monitoring as part of the funding package. Agencies engaged in adaptive management should seek to enroll enough sites at the beginning of a project to yield useful results in the shortest possible time.

LACK OF DETAILED MONITORING DATA

To make data collection rapid, we only used categorical data from visual assessments. It is a significant challenge to maintain rigorous, high-quality monitoring over long time frames in a management agency (Gitzen et al. 2012). The conservation agencies engaged in wetland restorations find even the simple requirements of the project (preassessment, random assignment of treatments, writing and overseeing a site-specific excavation prescription, and postrestoration monitoring) to be significant obstacles to participation. Even for participating agencies, only a small proportion of their wetland restorations are enrolled in the project due to the time commitment of monitoring. One possible approach to dealing with this challenge would be to have multiple protocols associated with different levels of technical skill and time. This would allow participants with the time and expertise to collect more detailed information, but other participants with more restrictive time constraints could still participate. We also recommend simply accepting and acknowledging the trade-off associated with rapid, relatively easy data collection; the rapid assessment protocol will make it impossible to detect subtle system-state changes. This project has attracted the attention of University of Minnesota researchers, and at least 1 graduate student is engaged in more focused research, using the existing information for sites enrolled in the project as a foundation. Such companion projects that delve into specific research questions are another way to expand learning opportunities.

LONG TIME HORIZON

Another challenge encountered in this project was that the project had a long time step. Managers expected that it would take 8 years after restoration before outcomes could be fully assessed. This is a very long time for management agencies to sustain a project without feedback. The long time frame virtually guarantees that staff will leave and enter the project. Staff turnover presents unique challenges for long-term monitoring because project continuity is threatened and institutional memories may be lost when project leaders leave or transfer. This project has had 3 different field-based project coordinators and 3 regional chiefs of refuges (regional program management). Each transition requires orientation of new leaders and review of the project's history, purpose, and progress. Strategies for dealing with staff turnover include cultivating a "champion" for the project at each transition (Walters 2007) and carefully documenting the project's history, protocols, data management and data analysis procedures so that information and momentum are maintained. A recent challenge for this long-term project is that standard practice for wetland restorations has evolved in the 8 years since initiation. Now in 2017, most small wetland restorations employ excavation because costs are relatively low, so there are fewer opportunities to enroll unexcavated seasonal wetland basins.

Our recommendation for dealing with the challenge of a long temporal horizon is to check on progress at intermediate time points and make corrections to the project trajectory as needed. For example, we assessed progress by performing preliminary Bayesian updating after 4 years and assessed utility per year to determine the trajectory of improvement over time for each basin. We hold annual coordination meetings for the cooperators to share the latest results and provide a forum for discussing problems, questions, and concerns. These annual meetings maintain interest and engagement among participants separated by distance and individual agency affiliations.

DATA PARTITIONING

Our data set was subdivided into 3 hydrological classifications, and for each there were competing models for excavated and unexcavated basins, which meant that relatively few data were available to update model weights for any of the competing models. Additionally, there are few excavations on the largest and deepest (semipermanent) wetlands due to the high cost of removing and finding depositories for large amounts of sediment. Our recommendation would be to periodically reconsider if the decision structure can be streamlined to accommodate more efficient learning. For example, an alternative competing model would be to use the utility score as the only metric and view the curve of utility over time as the response, rather than an 8-year time horizon. Thus one could gain insight into the trajectory of the wetlands each year. Also, if we had anticipated obstacles in obtaining adequate sample sizes for semipermanent wetlands, we could have removed this wetland type from the project and focused the available resources on the other wetland types.

Conclusions and Next Steps

At the present time, cost and logistics are driving restoration practices in the absence of current, relevant information on outcomes of restoration practices. In a management agency, these factors also influence the assignment of treatments to basins, complicating interpretation of outcomes. So, managers face a dilemma; adaptive management takes time and collective patience to stay the course, but unless those investments are made, we cannot improve management in the future. Such learning also requires more attention to details, such as site selection and randomization of treatments, and more collaborative engagement, administrative support, and monitoring than have traditionally been practiced in a management setting. Without well-documented, quantitative information that links management practices to outcomes, ineffective or harmful management practices may persist when more effective alternatives are available or could be developed. This is the cost of not engaging in adaptive management.

Our project provides the simplest framework for adaptive management (Kendall and Moore 2012) that we are aware of: there are only 2 decision alternatives and 1 fundamental objective, and the monitoring protocols are straightforward and rapid. The decision analysis process led to development of a common protocol for assessment of Prairie Pothole wetlands. Most managers conduct some visual inspections of their wetland restorations; following the protocol allowed managers to assess how well their visual inspections aligned with some relatively simple, but standardized, metrics of wetland waterfowl habitat quality. The protocol was comparable across sites and was rapid and feasible to carry out for managers operating under time constraints. Most importantly, the monitoring led to learning from the outcomes of prior restorations.

We tested the hypothesis that sediment excavation yielded superior restoration outcomes in terms of waterfowl habitat and wetland function. Our data so far do not indicate that excavation is strongly superior to simply restoring hydrology, especially for seasonal wetlands, the category for which we have the most data. Preliminary evidence suggests that wetland type may be a strong external driver on the restoration outcomes: seasonal wetlands show improvement in habitat quality 8 years after restoration, whether or not they have been excavated, whereas the opposite is observed with temporary wetlands. The differences between excavated and unexcavated basins, if any, may require more data to detect or be too subtle to be detected by our monitoring methods. The utilities over time appear to show that "happiness" with temporary wetlands is lower overall compared to that of seasonal wetlands, which is important to consider when evaluating these preliminary results. Our recommendation is to continue the project for the full 8 years for the >100 basins currently enrolled, and continue to enroll sites to increase sample sizes, since increasing the rigor and specificity of the monitoring is not a feasible option.

This project serves as a rare example of successful long-term adaptive management, implemented by management agencies working together. The project is generating valuable insights for land managers to support conservation delivery. Despite the challenges, we have demonstrated that it is possible for a management agency to establish and sustain adaptive management as a tool for reducing uncertainty. In our case, we focused on a very common restoration practice employed by multiple conservation partners and affecting hundreds of wetland basins annually.

ACKNOWLEDGMENTS

The authors thank Lori Stevenson, Shawn Papon, John Riens, and Sheldon Myerchin for their leadership of the project. Todd Sutherland, Wayne Thogmartin, Michael Conroy, Patricia Heglund, Sara Jacobi, Christina Hargiss, and Lori Nordstrom also made significant contributions. The authors thank the many people who enrolled sites, performed the restorations and site assessments, and collected and entered monitoring data. The findings and conclusions in this article are those of the authors and do not necessarily represent the views of the US Fish and Wildlife Service.

LITERATURE CITED

Bartzen BA, Dufour KW, Clark RG, Caswell FD. 2010. Trends in agricultural impact and recovery of wetlands in prairie Canada. *Ecological Applications* 20:525–538.

Bellrose FC. 1976. *Ducks, Geese, and Swans of North America*. Harrisburg, PA: Stackpole Books.

Doherty KE, Ryba AJ, Stemler CL, Niemuth ND, Meeks WA. 2013. Conservation planning in an era of change: state of the U.S. Prairie Pothole Region. *Wildlife Society Bulletin* 37:546–563.

Galatowitsch SM, van der Valk AG. 1996. The vegetation of restored and natural prairie wetlands. *Ecological Applications* 6:102–112.

Gitzen RA, Millspaugh JJ, Cooper AB, Licht DS, eds. 2012. *Design and Analysis of Long-Term Ecological Monitoring Studies*. Cambridge, UK: Cambridge University Press.

Gleason RA, Euliss NH, Hubbard DE, Duffy WG. 2003. Effects of sediment load on emergence of aquatic invertebrates and plants from wetland soil egg and seed banks. *Wetlands* 23:26–34.

Gleason RA, Euliss NH, Tangen BA, Laubhan MK, Browne BA. 2011. USDA conservation program and practice effects on wetland ecosystem services in the Prairie Pothole region. *Ecological Applications* 21:S65–S81.

Green EK, Galatowitsch SM. 2002. Effects of *Phalaris arundinacea* and nitrate-N addition on the establishment of wetland plant communities. *Journal of Applied Ecology* 39:134–144.

Kendall WL, Moore CT. 2012. Maximizing the utility of monitoring to the adaptive management of natural resources. Pages 74–98 in Gitzen RA, Millspaugh JJ, Cooper AB, and Licht DS, eds. *Design and Analysis of Long-Term Ecological Monitoring Studies*. Cambridge, UK: Cambridge University Press.

La Peyre MK, Reams MA, Mendelssohn IA. 2001. Linking actions to outcomes in wetland management: An overview of US state wetland management. *Wetlands* 21:66–74.

Lovett GM, Burns DA, Driscoll CT, Jenkins JC, Mitchell MJ, Rustad L, Shanley JB, Likens GE, Haeuber R. 2007. Who needs environmental monitoring? *Frontiers in Ecology and the Environment* 5:253–260.

Mack JJ. 2001. *Ohio Rapid Assessment Method for Wetlands v. 5.0*. Columbus, OH: Ohio Environmental Protection Agency, Division of Surface Water. www.epa.state.oh.us/portals/35/401/oram50um_s.pdf.

Moore CT, Lonsdorf EV, Knutson MG, Laskowski HP, Lor SK. 2011. Adaptive management in the US National Wildlife Refuge System: Science-management partnerships for conservation delivery. *Journal of Environmental Management* 92:1395–1402.

Moore CT, Shaffer T, Gannon J. 2013. Spatial education: improving conservation delivery through space-structured decision making. *Journal of Fish and Wildlife Management* 4:199–210.

Naugle DE, Johnson RR, Estey ME, Higgins KF. 2001. A landscape approach to conserving wetland bird habitat in the prairie pothole region of eastern South Dakota. *Wetlands* 21:1–17.

Niemuth ND, Estey ME, Reynolds RE, Loesch CR, Meeks WA. 2006. Use of wetlands by spring-migrant shorebirds in agricultural landscapes of North Dakota's Drift Prairie. *Wetlands* 26:30–39.

O'Neal BJ, Heske EJ, Stafford JD. 2008. Waterbird response to wetlands restored through the conservation reserve enhancement program. *Journal of Wildlife Management* 72:654–664.

Reynolds JH, Knutson MG, Newman KB, Silverman ED, Thompson WL. 2016. A road map for designing and implementing a biological monitoring program. *Environmental Monitoring and Assessment* 188:1–25.

Ryan MR. 1982. Marbled godwit habitat selection in the northern prairie region. Paper 8384. Ames, IA: Iowa State University.

Ryan MR, Renken RB. 1987. Habitat use by breeding Willets in the northern Great Plains. *The Wilson Bulletin* 99(2):175–189.

Smith C, DeKeyser ES, Dixon C, Kobiela B, Little A. 2016. Effects of sediment removal on prairie pothole wetland plant communities in North Dakota. *Natural Areas Journal* 36:48–58.

Stewart RE, Kantrud HA. 1971. Classification of natural ponds and lakes in the glaciated prairie region. Federal Government Series No. 92. US Fish and Wildlife Service, Bureau of Sport Fisheries and Wildlife.

US Environmental Protection Agency. 2018. Prairie Potholes. www.epa.gov/wetlands/prairie-potholes.

US Fish and Wildlife Service. 2018. Partners for Fish and Wildlife Program. www.fws.gov/partners.

Walters CJ. 1986. *Adaptive Management of Renewable Resources*. Caldwell, NJ: Blackburn Press.

———. 2007. Is adaptive management helping to solve fisheries problems? *AMBIO: A Journal of the Human Environment* 36:304–307.

Westgate MJ, Likens GE, Lindenmayer DB. 2013. Adaptive management of biological systems: a review. *Biological Conservation* 158:128–139.

23 An Adaptive Approach to Vegetation Management in Native Prairies of the Northern Great Plains

CLINTON T. MOORE,
JILL J. GANNON,
TERRY L. SHAFFER, AND
CAMI S. DIXON

The extent of native prairie throughout the north-central United States has sharply declined since European settlement, and much that remains has been invaded by introduced cool-season grasses, reducing floristic diversity and quality. On lands under its ownership, the US Fish and Wildlife Service is working to restore native prairie integrity by reducing occurrence of introduced species, principally smooth brome and Kentucky bluegrass, under the Native Prairie Adaptive Management (NPAM) program. Restoration involves defoliation at various times and in various forms, including burning and grazing. Ecological system dynamics and response to defoliation treatments are poorly understood. Each year, managers face a difficult decision about whether to defoliate a prairie, and if so, which defoliation treatment to employ given the current type and degree of invasion and recent history of defoliation. As such, managers of native prairies desire to learn about effectiveness and efficiency of these approaches through the decision-making and implementation process. The NPAM program provides real-time decision support to managers with the objective of increasing the cover of native grasses and forbs. The program operates under a framework that systematically reduces uncertainty about biological response to management, leading to improved future management decisions. This chapter showcases an example of adaptive management that is jointly implemented across a geographically dispersed network of participants.

Problem Background

The prairies of the US Northern Great Plains evolved in response to climate, recurring fire, and herbivory by free-ranging native ungulates (Samson et al. 2004; fig. 23.1). Soil characteristics and climate partition the ecosystem into distinct regions of mixed-grass (westerly region of drier climate) and tallgrass prairie (easterly region of moister climate). European settlement brought about the suppression of natural defoliation events (i.e., fire and grazing by free-roaming herbivores), the conversion of native grasslands to annual crop fields, and the establishment of introduced cool-season grasses that compete with native plants. Consequently, native prairie across the range has declined both in extent and quality (Samson et al. 2004).

The National Wildlife Refuge System (NWRS) of the US Fish and Wildlife Service (Service) is considered a conservation "reservoir" of native prairie across the Northern Great Plains. Following the devastating Dust Bowl era and consequent loss of habitat for migratory waterfowl, the Service acquired

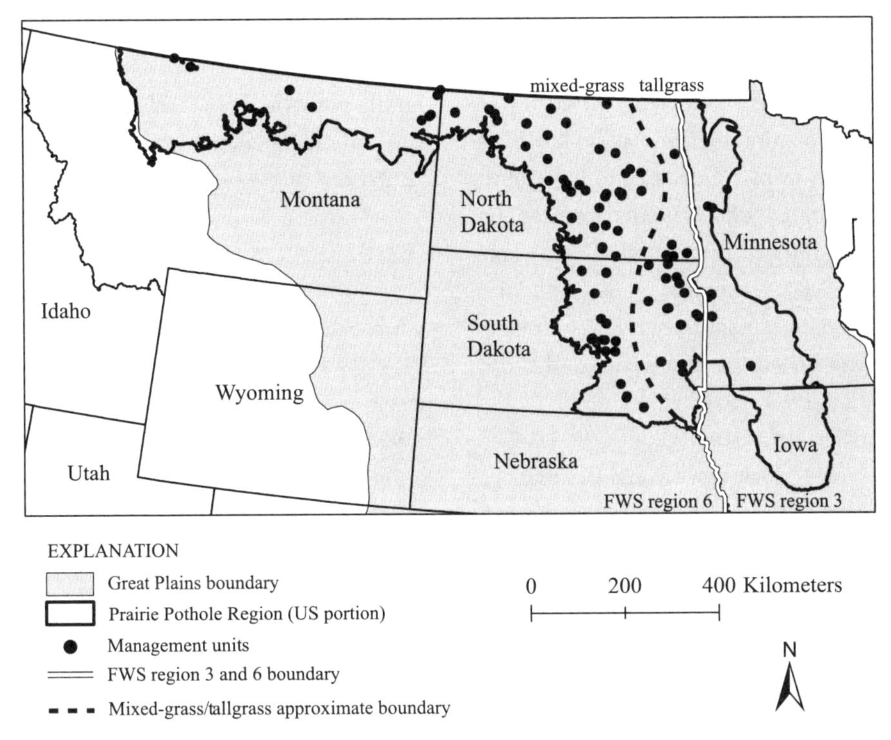

Figure 23.1. Extent of the mixed-grass and tallgrass prairies in the Great Plains of the northern United States. The Service management units enrolled in the NPAM program are contained within the US portion of the Prairie Pothole Region, an ecoregion of soils and topography formed by glacial activity. The dashed line is the approximate demarcation of the mixed-grass (westerly) and tallgrass prairie (easterly) systems. Refuges from 2 Service regions (3 and 6) participate in the NPAM program.

lands to establish refuges for waterfowl in the form of national wildlife refuges and waterfowl production areas (refuges). The procurements included intact native prairie, wetlands, and former agricultural lands. To provide abundant nesting cover for ducks, the Service broadly adopted policies of long-term rest for the prairies under its stewardship (Dixon et al. 2019). An increase of introduced cool-season grasses, principally smooth brome (*Bromus inermis*) and Kentucky bluegrass (*Poa pratensis*), had been reported on Service-owned lands in the region. An initial study completed on refuge lands in North Dakota demonstrated the detriments of managing native prairie as relatively static, late successional systems, notably by documenting higher smooth brome invasion on refuge lands compared with private lands (Murphy and Grant 2005). Outcomes from this study spurred conversations among biologists and managers across the Dakotas, providing the impetus for a broader-scale survey. In 2006, Service managers and biologists across northeastern Montana, North Dakota, South Dakota, and western Minnesota, along with grassland scientists, convened a workshop and agreed that a coordinated, multipartner solution was needed to guide the selection of actions on the ground and to systematically learn from experiences across the range (Grant et al. 2009). One outcome of this workshop was completion of a comprehensive survey of Service lands in the Dakotas, firmly establishing that most native prairie units were significantly impaired by introduced cool-season grasses. These data indicate that, combined, smooth brome and Kentucky bluegrass make up more than 50% of vegetation in native prairies in North Dakota, South Dakota, and parts of Montana (Grant et al. 2020). Although Service land managers generally acknowledged that the reintroduction of defoliation events may be key to reducing vulnerability of native prairies to invasion by these grasses, restoration efforts were disjointed, undocumented, and often unsuccessful. The Dakota-wide survey provided the data to document the magnitude of the introduced cool-season grass problem, and it clarified the need for immediate and broad-scale collaboration.

In 2008, the Service partnered with the US Geological Survey (USGS) to design an adaptive management framework for the selection of actions on individual management units of native prairie on Service lands. The manager at each participating refuge is the decision maker in this application. While this person possesses ultimate authority to choose courses of action on lands under their purview, all managers participating in the effort agreed to a common operational framework, including a shared statement of objectives. Consensus on objectives was facilitated by policies and programmatic guidance under which the Service operates.

The Native Prairie Adaptive Management (NPAM) framework provides recurrent, annual decision guidance targeted to individual management units, based on the current condition assessed on the unit (Gannon et al. 2013). In this context, a "unit" is the smallest parcel of land that receives a single management action over its entire extent (128 units enrolled in 2017; median unit size 27 ha; range 3–241 ha). Units enrolled in NPAM must contain native sod (intact, remnant grassland with no cultivation history) and have smooth brome and Kentucky bluegrass as the dominant invasive species of concern. Ideally, they possess the infrastructure and resources allowing them to receive any one of the full menu of management actions considered (e.g., cattle fencing for grazing, fire breaks for burning). "Current conditions" include measurements of the current vegetation cover, as determined by annual monitoring, and the defoliation history over the past 7 years (Gannon et al. 2013; Grant et al. 2004). Participants commit to monitor, carry out management actions, record details of the actions, and enter the annual vegetation and management action details in a centralized database.

Decision Analysis

A first step in structuring the decision process for NPAM was the formation of a science team composed of the USGS scientists and a small group of NWRS biologists. The biologists represented the larger community of managers and biologists (participants) across the region. Their participation in the process was vital, as they brought into the discussions the viewpoints of their colleagues, perspectives on the abilities and constraints shared by the NWRS community, and insight about the biology of the system and its expected response to actions. Within the science team, one Service biologist served as NPAM coordinator, assuring continuity of operation after the USGS development role diminished. As the program developed and neared implementation, a second Service biologist served as database coordinator.

The science team convened annual workshops with participants to elicit ideas, seek consensus on design and methods, and report progress. At the workshops, strategic decisions were made, such as the decision to approach management on a single-year cycle of monitoring and decision-making, rather than at some longer time scale. The participants also recognized the importance of developing parallel decision structures for the mixed-grass and tallgrass systems, pointing out that different types of management actions are practiced in the 2 systems, and that the systems might respond differently to actions. Outside of the annual workshops, the team met on its own as needed to discuss and reach consensus on more granular concerns, such as hypothesis development and expert elicitation for the parameterization of framework components.

Objective

The first NPAM workshop in 2008 yielded consensus on the management objective. After evaluating several proposals, including those that considered vegetative structure or focal vertebrate and invertebrate organisms as outcomes of management, the participants focused on a working objective of increasing cover of native prairie grasses and forbs at least cost. The rationale for the objective was recognition that the increase in native cover was intrinsically valuable itself and was thought to be a precondition for achieving other biological goals. In subsequent work, the science team formalized the objective in terms of quantitative utilities evaluated at each annual time step, with the ultimate goal of finding actions that maximized the utilities accumulated over time (Gannon et al. 2013). The utility model was derived through an elicitation exercise with Service members of the science team, in which they were challenged to place value on achieving specific native cover outcomes dependent on starting cover condition and the action used to obtain the outcome. Because the action taken was explicitly recognized, management cost was formally accommodated in the utility model. For example, a highly valued outcome would be one in which high native vegetation cover was observed 1 year after applying a low-cost action to a degraded prairie. Conversely, transitioning from a high-quality prairie to one of low quality 1 year after applying an expensive action would be highly undesirable. Intermediate scenarios received intermediate values. During this exercise, the science team understood that the point was only to assess the value of each specific scenario, not to evaluate the likelihood of occurrence of the scenario: assessment of likelihood would occur in the consequence analysis step.

Decision Alternatives

The first workshop also achieved consensus on a simple set of discrete decision alternatives in the form of management actions. The alternatives focused on whether to carry out defoliation actions on a unit, and if so, in what form (e.g., grazing, fire, grazing and fire, or mechanical means). Because decisions were to be made on a 1-year cycle, alternatives were constrained to those that could be carried out between September 1 and August 31 in the following year. Participants desired a suite of alternatives with which they had practical experience, enabling them to implement any of the potential actions with a predictable lead time for planning after receiving the recommendation. Finally, much discussion at the workshop revolved around proper timing and intensity of defoliation actions. The group recognized that formal consideration of timing and intensity in the decision process would introduce innumerable alternatives that would be difficult to replicate, resulting in little or no meaningful learning about effects of actions. Consequently, the participants chose a general form of defoliation action as the management recommendation. The decision about timing and intensity was left to the discretion of each decision maker, but within parameters established by the group (e.g., an action involving cattle grazing could nevertheless be classified as "no action" if a minimum grazing utilization threshold were not achieved), and with the understanding that specific details of the action would be documented and permanently recorded. Under this approach, participants understood that coarse control of the action would yield variation in response and therefore slower learning about efficacy of an action. However, this was deemed a necessary trade-off for keeping the decision structure simple and the overall problem tractable.

To accommodate differences in conventional management practices between mixed-grass and tallgrass prairies and to facilitate learning around hypotheses unique to those systems, distinct sets of decision alternatives were established for each grassland type (Gannon et al. 2013). For mixed-grass units, the permissible decision alternatives were no action (*rest*), grazing in some form (*graze*), fire in some form (*burn*), or fire combined with grazing in the same management year (*burn/graze*). In the tallgrass system, the permissible alternatives were no action (*rest*), grazing within a specific phenological window for growth of smooth brome (*graze/w*), burning within the smooth brome phenological window (*burn/w*), or either (a) grazing or burning outside of the phenological window or (b) mechanical defoliation at any time (*defoliate*). The notion of treating with respect to a phenological window was proposed by Willson and Stubbendieck (2000), who experimentally determined the development stages when smooth brome was most vulnerable to fire in tallgrass prairies. The window varies by site and by year; thus, individual managers monitor phenological stage and make site-specific determinations of the window timing each year.

Consequence Analysis

State-and-transition models form the prediction infrastructure for NPAM (Gannon et al. 2013). Based on results of the monitoring program, each management unit is classified into 1 of 16 discrete vegetation states. The states result from combinations of 4 categories of percentage cover by native grasses and forbs (0–30%, 30–45%, 45–60%, and 60–100%) and 4 categories of the dominant undesired plant: smooth brome, Kentucky bluegrass, brome-bluegrass co-dominant, and other nondesirable vegetation (remainder). Categorization is revisited every year; thus, units transition from one state to another in response to management actions and to uncontrolled events, such as weather. The state-and-transition model is a matrix of probabilities that expresses the likelihood of transitioning from one state to another in 1 year.

Every decision alternative induces a distinct matrix of probabilities, allowing actions to differentially affect transitions in vegetation composition. Furthermore, participants believed that management history may affect how a unit responds to management actions; therefore, units were also classified each year into 3 discrete levels (low, medium, high) based on a constructed index that jointly characterizes the recency and frequency of defoliation actions (Gannon et al. 2013). A set of transition probability matrices for each combination of decision alternative and management history level constitutes a single model in the NPAM framework.

A fundamental problem faced by the science team was the matter of providing 3,072 (16 × 16 vegetation transitions × 4 actions × 3 defoliation history levels) model parameters when no data were available for any of the transitions. We structured an elicitation exercise in which the responding expert was presented with a limited number of specific scenarios about current composition (percentage cover by native grasses and forbs and type of dominant undesired species), management history, and a proposed management action (Gannon et al. 2013). For each scenario, the respondent was asked to describe in simple but quantifiable terms the distribution of expected composition outcome, as informed by their experience and expert knowledge. By limiting the exercise to key transitions within the entire space of scenarios, we gained sufficient information to interpolate values into other parts of the model structure through analysis of the responses in a linear model framework (Gannon et al. 2013). Service members of the science team nominated experts from within the group to participate in the exercise. We conducted the exercise separately for mixed-grass and tallgrass systems.

To address longstanding uncertainty about how prairies respond to defoliation actions, the science team cast key uncertainties about management in the form of distinct hypotheses along some of the lines proposed by Grant et al. (2009). For both grassland types, biologists were uncertain whether relative efficacy of actions was dependent on the type of dominant invading grass, whether a recent history of defoliation played any role in the outcome of a current action, and whether an invasion threshold existed that, when exceeded, reduced efficacy of all forms of defoliation action. Around these uncertainties, we built 4 competing models of increasing complexity by directionally manipulating the baseline responses from the elicitation exercise and interpolating parameters as before (Gannon et al. 2013; table 23.1). Biologists working in the tallgrass system had additional uncertainties about substitutability of treatments within and outside of the smooth brome phenological window; thus, for tallgrass prairie, we formulated 2 additional models as specific hypotheses (Gannon et al. 2013; table 23.1).

Learning about competing hypotheses is reflected in relative probabilities attached to each model, which we take as measures of relative credibility of each model (Gannon et al. 2013). Credibility weights change through time as management actions are taken, data on composition response are gathered, and the models' predictions are confronted by the evidence reflected in the response. In 2009, the first year of implementation of the framework, equivocal belief in the models was assumed by assigning equal credibility weight to each model.

An accounting for partial controllability was built into the NPAM framework. From the start, participants expressed concern that recommended actions could not always be implemented, whether for reasons of access logistics, lack of financial or staffing resources, or for other reasons. During the framework development phase and based on an elicitation exercise distributed to all participants, we built a model for partial controllability that provided a probability of executing each alternative action given recommendation of a particular action (Gannon et al. 2013). In addition, during the operational phase, managers were asked to record the reason why a recommendation was not followed so that the

Table 23.1. Predictions about native vegetation response to management action

Model	Predictions
Mixed-grass	
1	All forms of defoliation are equally effective and superior to rest (no action) regardless of vegetation and defoliation history state
2	All forms of defoliation are superior to rest, but response of the vegetation state to management actions depends on type of dominant invader
3	Response of the vegetation state is dependent on type of dominant invader as well as past defoliation history; defoliation actions are more effective and rest is less detrimental on a site that has been frequently defoliated compared with a site that has not
4	Response of the vegetation state is dependent on type of dominant invader and defoliation history, but efficacy of defoliation actions progressively declines with increased level of invasion
Tallgrass	
1	All forms of defoliation are equally effective and superior to rest regardless of vegetation and defoliation history state
2	All forms of defoliation are superior to rest, but response of the vegetation state to management actions depends on type of dominant invader
3	Response of the vegetation state is dependent on type of dominant invader as well as past defoliation history; defoliation actions are more effective and rest is less detrimental on a site that has been frequently defoliated compared with a site that has not
4	Response of the vegetation state is dependent on type of dominant invader and defoliation history, but efficacy of defoliation actions progressively declines with increased level of invasion
5	Variation on model 3: Grazing within the phenological window is as effective on smooth brome–dominated sites as is burning within phenological window
6	Variation on model 3: Any form of defoliation outside of the smooth brome phenological window has the same efficacy as rest

Source: Excerpted from Gannon et al. 2013.

Note: Expressed as competing models under the Native Prairie Adaptive Management framework for mixed-grass and tallgrass systems.

model could be periodically updated with experiential evidence.

Decision Solution

The utility model, the set of decision alternatives, the partial controllability model, and the competing models of consequences were integrated in the decision solution step. For each grassland type, we used active adaptive stochastic dynamic programming to compute action policies that were indexed to current biological state (vegetation composition and management history) and current knowledge state (distribution of credibility weight among competing models; fig. 23.2). To facilitate specific desired behaviors of the models that trigger with the passage of time (e.g., a trigger to force recurrent defoliation action for model 1; a trigger to reduce likelihood of burn implementation following defoliation in the previous year), we also kept track of a state variable that indicated number of years since last application of a defoliation action (3 levels: 1, 2–4, or ≥5 years). We computed the policies for the objective of maximizing the sum of expected utility over a practically

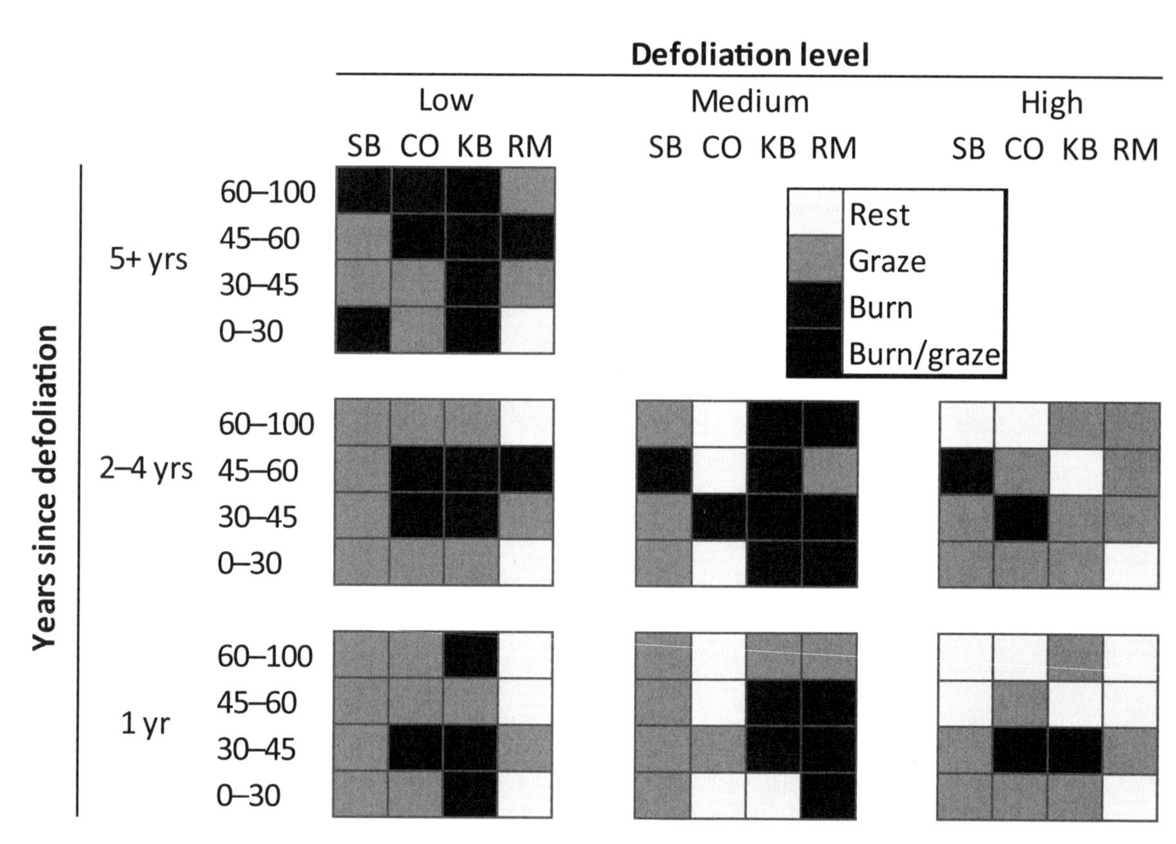

Figure 23.2. A portion of the full decision policy computed by active adaptive stochastic dynamic programming for the mixed-grass system. The policy is first divided among 3 discrete levels of defoliation (low, medium, and high), which describe the management history of a unit by capturing the recency and frequency of defoliation actions, subdividing by 3 discrete times since the last defoliation action occurred on a unit (1 year ago, 2–4 years ago, 5 or more years ago), then subdividing among combinations of 4 levels of native grass-forb cover (0–30%, 30–45%, 45–60%, 60–100%), and lastly subdividing by 4 categories of invasion dominance (smooth brome, co-dominant brome-bluegrass, Kentucky bluegrass, remainder). Cell darkness indicates optimal action for the objective of maximizing long-term expected sum of utility. The policy portion shown here reflects optimal actions for 0.375 credibility weight placed on model 1, 0.125 weight on model 2, and 0.25 weight on each of models 3 and 4; other distributions of credibility weight link to other portions of the full policy.

infinite (1,000-year) time frame, applying a discounting rate to favor gains in native prairie utility sooner than later (Gannon et al. 2013). The optimization accounted for 3 important considerations in decision-making: (1) dynamic behavior of the biological system, in that actions taken at the current time step propagate consequences to be accounted for in later time steps; (2) dynamic behavior of learning, in that actions taken at the current time step may be more or less informative for resolving uncertainty about system behavior than other potential actions; and (3) partial controllability, in that actions are rarely executed in the way they were recommended, leading to greater variation in the response.

Considered individually, accepting each model as "truth" in turn, the generated decision policy comports with management intuition. For example, under the hypothesis positing that any form of defoliation is adequate to promote amount of native cover regardless of the type and amount of invading plant species (model 1), the decision policy indicates grazing (the cheapest defoliation option) on a recurring (5-year) basis. Alternatively, under the model that proposes that native cover maintenance is sensitive to the type of invading plant (model 2), the recommended action is tied to the type and percentage cover of invading species: graze, burn, and burn/graze tend to be recommended actions when the dominating undesired species is smooth brome, Kentucky bluegrass, and both concurrently, respectively. Under the model in which defoliation history is also influential (model 3), the management policy further differs by how recently and how frequently management has been applied: under more recent defoliation actions, recommended actions tend to be "lighter" and cheaper (i.e., actions involving resting or grazing) than when defoliation was conducted farther in the past. In contrast, under model uncertainty, that is, when 2 or more models receive some degree of belief (fig. 23.2), the corresponding decision policy reveals patterns of action that may not be easily understood. In this case, the policy is achieving a "balance" among all the models according to their belief weights while simultaneously indicating actions designed to discredit hypotheses (i.e., reduce uncertainty) as quickly as possible. While this complex computational operation renders patterns in the policy that are not readily intuitive, the operation is ultimately driven by the same mathematical principles that produce intuitive policies for specific models.

Given current status of the management unit and current knowledge state of the competing models, the participant receives an optimal recommendation, drawn from the decision alternatives. The participant then makes a decision and carries out an action, recommended or otherwise. Following each action, the management unit is monitored, and the biological condition is reassessed. The data are used under a Bayesian framework to update the knowledge state based on a comparison of realized vegetation outcome to each model's prediction of the same (Gannon et al. 2013; Moore et al. 2013).

Framework Infrastructure

The NPAM program has important infrastructural elements to enable decision support to be efficiently distributed to many participants dispersed across the region. Participants receive decision guidance, and they contribute toward ecological understanding through a coordinated and well-communicated multistep process. The elements also contribute to ensuring the program has sufficient longevity to accomplish its purpose of increasing native plant cover.

The program is administered by a Service program coordinator who annually updates model credibility weights and distributes management recommendations, oversees the enrollment of management units, troubleshoots issues with participants, provides training sessions, and communicates both with participants and higher-level Service program administrators. The program coordinator is assisted by a database coordinator who maintains the supporting databases and works with participants to provide training and to resolve data issues.

A protocol notebook serves as a comprehensive user's guide to the program, helping to assure that the program survives personnel turnover at the coordinator and participant levels. Communication about the program is emphasized, both internally through annual meetings/webinars with participants and externally through presentations to administrators and other audiences. The program is administered through a web portal, where supporting documents, forms, and reports are maintained. Annual training in monitoring techniques and use of the data entry portal is provided to participants.

Data entry, data management, and data analysis functions were recently integrated in an application that automates many steps of the process (Hunt et al. 2016). The application includes a web-based interface that facilitates the collection and quality checking of vegetation monitoring data and data on implemented actions. A database platform operated by the program coordinator provides data validation, runs algorithms to update model belief weights, searches policy tables for new management recommendations, and provides tabular program reports and statistics.

In 2012, the Service assumed full operational control of the program as NPAM transitioned from its startup phase into its implementation phase. At the same time, the science team that assembled the program transitioned into an advisory team whose primary mission is to provide operational oversight and to assure program longevity through the consideration of improvements or the response to emerging problems or challenges.

Decision Implementation

The NPAM program was implemented in a pilot phase in 2009; participants followed the established monitoring protocols and entered observations into a prototype database. No management recommendations were issued in the pilot phase, but participants made selections from the set of management alternatives. Initial models were completed in 2010, allowing decision guidance for the first time that year. It was also the first opportunity to test the operational schedule of NPAM. The protocol calls for entering monitoring data by the third week of August, followed by a 1-week time window to assemble data, validate values and resolve inconsistencies, run algorithms to update knowledge state, and distribute new management recommendations by September 1. Participant feedback to initial models led to revision; the final version of the model sets was developed and in use by 2011.

The program is currently implemented on more than 120 management units on 20 NWRS refuges in 4 states and 2 Service regions (fig. 23.1). Enrollment of non-Service units began with the involvement of Audubon Dakota in 2016 and the North Dakota Game and Fish Department in 2017.

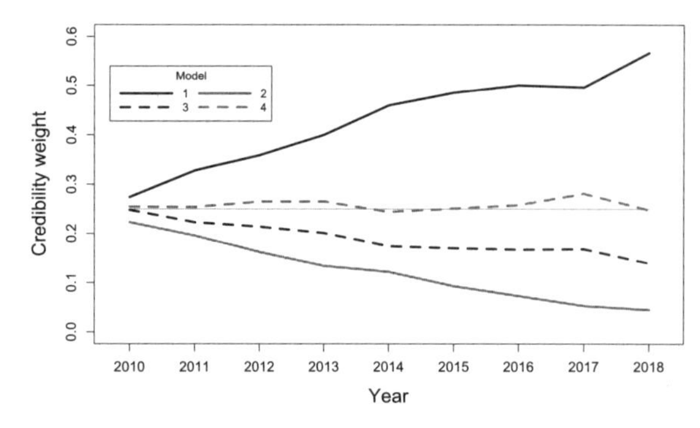

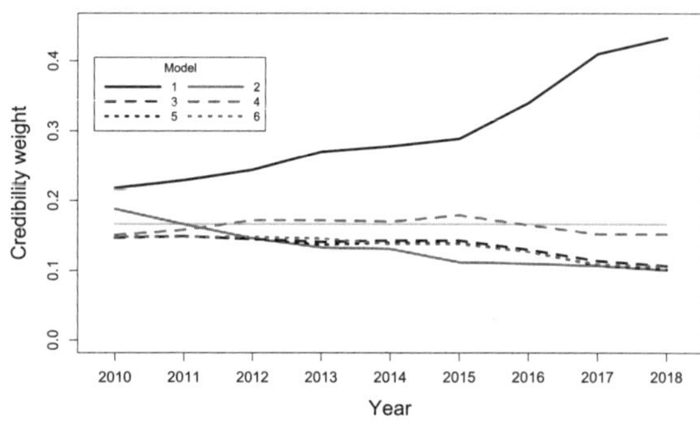

Figure 23.3. Time series of relative credibility weights for models used in mixed-grass systems (*upper*) and tallgrass systems (*lower*) for the Native Prairie Adaptive Management program. Horizontal reference lines (gray) reflect the weight all models would receive if equal confidence were placed on each.

As of 2018, 9 cycles of monitoring, action, and learning have been completed for mixed-grass units, and 7 have been completed for tallgrass units (fewer due to incorporation of the smooth brome phenological window in models and decision alternatives in 2011). Accordingly, relative weights of credibility have accrued for some models and have diminished for others (fig. 23.3). In both systems, credibility weight has increased for the simplest model (model 1), remained unchanged for the most complex model (model 4), and decreased for all other models of intermediate complexity.

After adjusting for mitigating circumstances, such as prior-year precipitation, average proportion cover of native grasses and forbs has increased significantly (5.1% average annual increase in odds ratio for percentage cover; 95% credible interval 3.3–7.1%) in both mixed-grass and tallgrass prairies and across the program as a whole since the inception of NPAM (fig. 23.4). The rate of increase is slow, but it is encouraging to see signs of reversal of a decades-long trend. Thus, early indications are that the program may be meeting its ecological objective.

Discussion

Our experience with the Native Prairie Adaptive Management program has provided lessons that may be useful in other decision settings. NPAM is an example of decision support in which exchange of information occurs across organizational scales. Managers on the ground receive management recommendations, implement actions, and record and enter data at the scale of the management unit. Data from all units are aggregated and translated into knowledge at the program scale. Updated knowledge is then passed down to the unit level in the form of improved decision guidance. The NPAM program may serve as a template wherever an organization of individual practitioners share a common decision problem, common objectives, and a common set of management tools but are hindered in their decision making due to

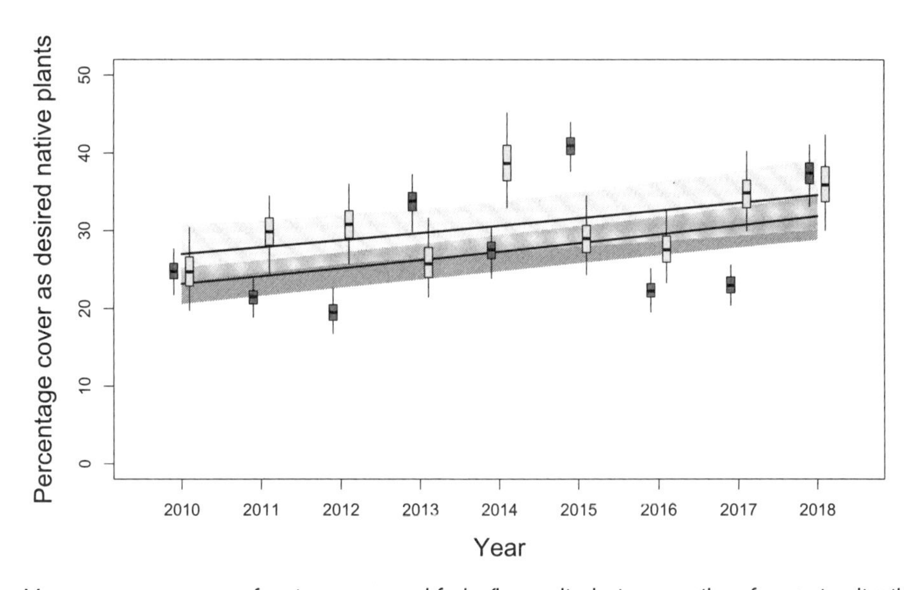

Figure 23.4. Mean percentage cover of native grasses and forbs (boxes displaying quartiles of posterior distribution; whiskers displaying 95% credible intervals) on mixed-grass (lower trend; darker boxes) and tallgrass (upper trend; lighter boxes) management units enrolled in the Native Prairie Adaptive Management program, 2009–2018. Means reflect annual residual effects after adjustment for factors including site effects, treatments, prior-year precipitation, and prior-year native percentage cover.

uncertainty regarding the behavior of the managed system.

The use of replicated sites in decision making offers the potential of learning about large-scale systems in an accelerated manner (Moore et al. 2013). Despite the programmatic and analytical challenges that it introduces, site replication is an asset of this program, as no single site under a single management authority is representative of all prairie management.

The NPAM program represents a type of decision problem in which participants are spatially dispersed but all contribute to collective learning about the larger system. We have learned that standardization and centralization of information and the processes that promote learning must be balanced with maintaining local autonomy in implementing management recommendations. Both are necessary to make a learning-based effort operationally sustainable into the indefinite future. Given the spatially dispersed nature of the effort, the need for programmatic consistency to facilitate learning and decision support, the importance of communication, and the temporally defined structure of the decision-making framework, it is difficult to overstate the importance of steadfast support of the lead organization and dedicated operational staff.

The dispersed nature of the operation, the need for timely decision support on a recurrent basis, and the lack of specialized technical capacity within the management agency pointed to the need for tools that automated and centralized many of the processes. In NPAM, this need was met by a suite of web-based and locally installed applications.

All components of NPAM, including the predictive models, were parameterized entirely on the basis of expert knowledge. While many reasons may exist to postpone developing a formal structure for decision making about an important issue, in our experience, absence of empirical information is not one of them. The effort put into standardizing and automating the collection and entry of data now makes possible the replacement of expert-elicited quantities with ones that are empirically based, a focus of current work.

A decision process with as many moving parts as NPAM will rarely be without problems upon implementation. We experienced a setback following the first decision cycle in 2009–2010 when the science team identified deficiencies in the decision structure. We completed an overhaul of the utility model, state definitions, prediction models, and decision alternatives prior to the next decision cycle. In the language of adaptive management, we completed a first exercise in double-loop learning. In the near future, we will likely complete another double-loop effort as we replace expert-derived model parameters with those based on empirical information.

The NPAM program uses an active adaptive optimization approach in which the gain of information for learning about system behavior is embedded in the recurrence equation of system dynamics (Walters and Hilborn 1978). An advantage of this approach is that the decision policy recognizes those conditions in which more experimental actions may return useful information leading to improved future decision making, as well as those conditions where experimentation would be unacceptably risky (Moore and Conroy 2006). To our knowledge, this approach has not been used in another operational natural resource management system.

Besides evidence that the program is bringing desired results in terms of native grass composition, we believe that the program is also providing returns in social capital. Participants in the effort have perceived a purpose in their monitoring activity and can see the direct linkage between data collection and data use. Participants have adopted an "in it together" perspective as opposed to the "go it alone" attitude that marks so many traditional approaches. Also, NPAM provides a forum for communicating both across (e.g., fire program) and within (e.g., participants and leadership) Service programs, and it fosters external partnerships for supplemental research and collaboration.

ACKNOWLEDGMENTS

The NPAM program is supported by the efforts of dozens of Service employees who are dedicated to the hard work of collecting data and conducting management efforts. The authors thank science team members Kim Bousquet, Pauline Drobney, Vanessa Fields, Bridgette Flanders-Wanner, Todd Grant, and Sara Vacek, who were coarchitects of the decision framework. Jennifer Zorn serves as the database coordinator of NPAM and provides critical day-to-day technical support. Victoria Hunt, Sarah Jacobi, and Kevin McAbee designed and built the data structures, web interface, and analytical code that mostly automate the organization, validation, and processing of data. Amy Symstad and Mitchell Eaton provided insightful and helpful reviews of this chapter. The NPAM program was developed through financial support from the USGS Refuge Cooperative Research Program and from the Service. The findings and conclusions in this article are those of the authors and do not necessarily represent the views of the US Fish and Wildlife Service.

LITERATURE CITED

Dixon C, Vacek S, Grant T. 2019. Evolving management paradigms on US Fish and Wildlife Service lands in the Prairie Pothole Region. *Rangelands* 41:36–43.

Gannon JJ, Shaffer TL, Moore CT. 2013. Native Prairie Adaptive Management: a multi region adaptive approach to invasive plant management on Fish and Wildlife Service owned native prairies. US Geological Survey Open File Report 2013-1279. https://doi.org/10.3133/ofr20131279.

Grant TA, Flanders-Wanner B, Shaffer TL, Murphy RK, Knutsen GA. 2009. An emerging crisis across northern prairie refuges: prevalence of invasive plants and a plan for adaptive management. *Ecological Restoration* 27:58–65.

Grant TA, Madden EM, Murphy RK, Smith KA, Nenneman MP. 2004. Monitoring native prairie vegetation: the belt transect method. *Ecological Restoration* 22:106–112.

Grant TA, Shaffer TL, Flanders B. 2020. Patterns of smooth brome, Kentucky bluegrass, and shrub invasion in the northern Great Plains vary with temperature and precipitation. *Natural Areas Journal* 40(1).

Hunt VM, Jacobi SK, Gannon JJ, Zorn J, Moore CT, Lonsdorf EV. 2016. A decision support tool for adaptive management of native prairie ecosystems. *Interfaces* 46:334–344.

Moore CT, Conroy MJ. 2006. Optimal regeneration planning for old-growth forest: addressing scientific uncertainty in endangered species recovery through adaptive management. *Forest Science* 52:155–172.

Moore CT, Shaffer TL, Gannon JJ. 2013. Spatial education: improving conservation delivery through space-structured decision making. *Journal of Fish and Wildlife Management* 4:199–210.

Murphy RK, Grant TA. 2005. Land management history and floristics in mixed-grass prairie, North Dakota, USA. *Natural Areas Journal* 25:351–358.

Samson FB, Knopf FL, Ostlie WR. 2004. Great Plains ecosystems: past, present, and future. *Wildlife Society Bulletin* 32:6–15.

Walters CJ, Hilborn R. 1978. Ecological optimization and adaptive management. *Annual Review of Ecology and Systematics* 9:157–188.

Willson GD, Stubbendieck J. 2000. A provisional model for smooth brome management in degraded tallgrass prairie. *Ecological Restoration* 18:34–38.

24 Decision Implementation and the Double-Loop Process in Adaptive Management of Horseshoe Crab Harvest in Delaware Bay

Conor P. McGowan,
James E. Lyons, and
David R. Smith

Horseshoe crab harvest in the Delaware Bay on the mid-Atlantic coast of the United States was unregulated and increased substantially through the 1990s. Subsequently, shorebirds that rely on spawned horseshoe crab (HSC) eggs during spring migration, especially red knots, exhibited steep population declines, and conservationists attributed the declines to decreased HSC egg availability. Due to strong controversy and ecological uncertainty, the decision maker, the Atlantic States Marine Fisheries Commission (ASMFC), turned to structured decision making and adaptive management to help make annually recurrent harvest decisions. We formed an adaptive management working group to work with the ASMFC to develop objective statements, utility functions, alternative action sets, and system models for predicting management consequences. The decision analysis was successfully implemented, and the ASMFC used harvest recommendations from the decision analysis management plan for 4 successive annual decisions. In 2016, we implemented a double-loop process to evaluate revisions to the original components of the decision analysis. As part of the double loop we assessed the predictive model set, considered revisions to the utility functions, and considered incorporating into the decision process horseshoe crab mortality caused by a biomedical industry. Decision analysis and the double-loop process helped to make great strides in a formerly intractable problem by focusing on science, uncertainties, and stakeholder values to increase transparency of the decision process.

Problem Background

The Atlantic horseshoe crab (*Limulus polyphemus*) is a biologically, evolutionarily, and ecologically fascinating species with a long history of use by humans (Smith et al. 2017). In recent years the horseshoe crab population in the mid-Atlantic states (Delaware, New Jersey, Maryland, and Virginia, USA) has been at the center of a controversial and difficult management problem (McGowan et al. 2015b; Niles et al. 2009). Horseshoe crabs (hereafter HSC) are harvested as bait for use in eel (*Anquilla rostrata*) and whelk (*Busycon sp.*) fisheries, and HSC blood is an important component of Limulus amebocyte lysate (LAL), a vital biomedical product used for testing for bacterial contamination in medical equipment (Smith et al. 2017). LAL production requires catching and extracting blood from HSC, which may cause a 5 to 30% mortality rate for HSC subjected to bleeding (Leschen and Correia 2010). HSC also provide a vital food resource for migrating shorebirds that rely heavily on HSC eggs spawned by the billions on the sandy beaches of Delaware Bay each spring.

HSC spawning—dependent on water temperature, weather, and lunar cycles—typically coincides with northward migration of hundreds of thousands of shorebirds moving from their wintering grounds to their arctic breeding grounds (Niles et al. 2009).

HSC bait harvest in the region escalated in the 1990s and peaked in 1998 (Millard et al. 2015). Shorebird populations that stop over in Delaware Bay during spring migration declined starting in the mid-1990s and continuing to the early 2000s. The decline was especially apparent for red knots (*Calidris canutus rufa*) and was attributed to decreases in HSC eggs and a failure of birds to acquire the necessary energy resources during migration stopover to complete their migration to the arctic (Baker et al. 2004; Niles et al. 2009). Monitoring data indicated that red knot populations fell from over 100,000 birds in the 1980s to a mere ~15,000 in 2003 (Niles et al. 2008). In 1998, the Atlantic States Marine Fisheries Commission (ASMFC) began regulating horseshoe crab harvest to prevent overexploitation and to help support migrating shorebird populations (Millard et al. 2015). The ASMFC is a quasigovernmental entity that is authorized, along with 14 member states, to set interstate fisheries regulations and quotas via votes at annual meetings. The ASMFC uses a science-focused decision process; the management board is informed by technical committees, which provide scientific assessments and data analyses, and by advisory panels for each managed species, which describe socioeconomic concerns and other information. While the ASMFC management board is the ultimate decision maker with respect to horseshoe crab harvest allowances, there are numerous stakeholders, including state fisheries management agencies, a local community of watermen, the bait packing industry, state wildlife agencies, conservation NGOs, and the US Fish and Wildlife Service. These stakeholders have vested interests and high concern for how the harvest is managed, especially with respect to shorebird conservation. From 1998 onward the ASMFC managed the HSC harvest with an explicit objective for harvest sustainability and an implicit objective for red knot conservation, but by 2008 the ASMFC management approach was mired by stakeholder advocacy, impeding science-based and inclusive multiple-objective decision making. At that time, the management board agreed to adopt a structured decision-making (SDM) approach to increase transparency and develop a decision process driven by objectives in which science could play an imperative role (Gregory and Keeney 2002).

The SDM process focused on annually recurrent harvest quota decisions and the challenges presented by multiple objectives and substantial ecological uncertainty. Given the annual harvest decisions and the persistent ecological uncertainty, we built an adaptive management program with the anticipation of engaging in a double-loop learning process as described by Williams et al. (2007), allowing for periodic updating of the decision structure (fig. 24.1). The double-loop process has 2 phases, the *setup phase*, when the basic components of the decision analysis are deliberated and designed, and the *iterative phase*, when recurrent decisions are made and monitoring evaluates how the system responds to management actions. The double-loop process allows decision makers to periodically evaluate and potentially revise the decision analysis components. In the setup phase, we focused on HSC bait harvest allowances and did not consider biomedical incidental take due to confidentiality agreements between the biomedical companies and the ASMFC. Also, we did not consider specific red knot conservation actions in the initial setup phase because ASMFC has no authority to manage shorebirds or their habitat.

Decision Analysis

Objectives

ASMFC committee structure includes technical committees and the advisory panels, which represent stakeholder groups (McGowan et al. 2015a, 2015b). State and federal agencies with fisheries and non-game wildlife expertise were members on the horseshoe crab or shorebird technical committees.

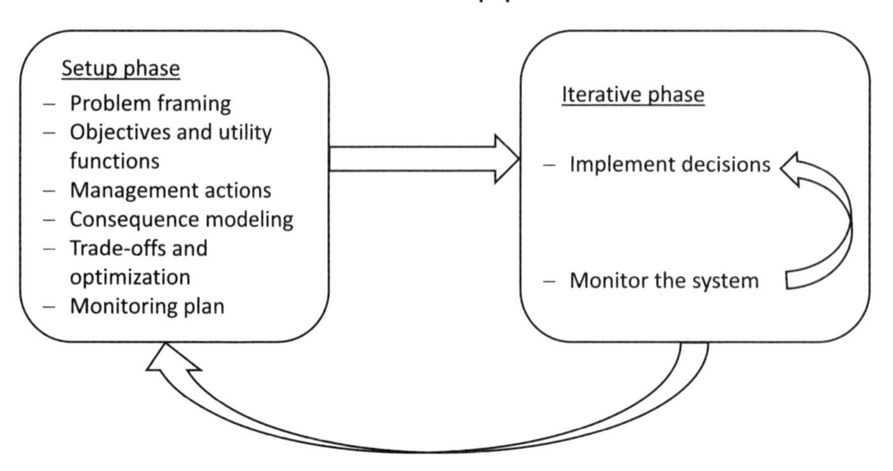

Figure 24.1. A depiction of the double-loop process for adaptive management. *Based on the figure in Williams et al. 2007.*

Academic experts and representatives of conservation NGOs and fishermen's groups with technical expertise were also appointed to 1 or more of these committees. The ASMFC convened a decision analysis and modeling team known as the adaptive management working group (AMWG), and with the ASMFC committees, this team held a series of public meetings for general public outreach and input. The committees and work group governed through consensus to resolve the decision analysis components, such as objective functions or predictive models. Consensus was considered achieved when no additional discussion was offered on any topic at hand (McGowan et al. 2015b).

To facilitate the identification of specific and measurable objectives, we started with a qualitative statement that represented our fundamental objective: "Manage harvest of horseshoe crabs in the Delaware Bay to maximize harvest but also to maintain ecosystem integrity and provide adequate stopover habitat for migrating shorebirds." Given this overarching goal, constrained optimization was used to maximize horseshoe crab harvest constrained by horseshoe crab abundance relative to carrying capacity, population sex ratio as a proxy for fecundity, and a desired level of red knot abundance. Stakeholders helped identify specific abundance targets that served as thresholds in our optimization algorithms; below the identified thresholds, harvest had no utility (Martin et al. 2009; McGowan et al. 2015b).

Alternatives

Five alternative harvest packages specified sex-specific harvest quotas. The quotas ranged from a full moratorium on both sexes to 500,000 males and 240,000 females. The alternatives were considered potentially sustainable and politically feasible.

Consequence Analysis

We developed a projection model specifically linking red knot demographics to female HSC spawning abundance (for details see McGowan et al. 2011a, 2011b, 2015b; Smith et al. 2013; Sweka et al. 2007). To incorporate ecological uncertainty and reflect stakeholder hypotheses about species interactions, we used 3 competing parameterizations of the projection models: (1) HSC abundance does not affect red knots, (2) HSC abundance affects knot productivity, and (3) HSC abundance affects both red knot productivity and survival (McGowan et al. 2015b). Prior belief weights for the 3 competing models were elicited from the ASMFC technical committees as initial model weights for Bayesian model weight updating. The 3 models accrue or lose weight over time, reflecting how well a model predicts the population responses to management actions. Based on a simulation study, Smith et al. (2013) concluded that model weight convergence will likely be protracted.

Optimization and Decision Solution

We identified the harvest package that optimized a reward function given the stage- and age-specific abundances for each species through use of a passive optimization algorithm implemented in the ASDP optimization software (Lubow 2001; McGowan et al. 2015b). The reward function was

$$R = \sum_{t=1}^{\infty} u_t^F 2H_t^F + u_t^M H_t^M$$

where R is the reward, u is the utility of female (F) or male (M) harvest, H is the harvest amount for females (F) or males (M), and t is time. Harvest of females is economically twice the value of males on the bait market because of their larger size (Smith et al. 2009). Utility of male and female harvest was a binary function determined by a set of logical evaluations. Utility of female harvest (u_t^F) was constrained by a red knot abundance threshold (initially set at 45,000 red knots) or a female HSC abundance threshold relative to carrying capacity. The "or" nature of the female harvest utility function reflected some of the ecological uncertainty in the system. Female HSC harvest is constrained by red knot population abundance, but if red knots don't respond to harvest restrictions even though HSC do respond, then eventually HSC female harvest will be valued when female abundance reaches some very high threshold (set at 80% of estimated carrying capacity, 11.2 million females). If the abundance of either species exceeds the stated threshold, utility of female harvest (u_t^F) equals 1, otherwise it is 0. Utility of male HSC harvest is constrained by observed HSC population sex ratio, such that when the male-to-female sex ratio exceeds a point where additional males do not contribute to population growth, male harvest utility (u_t^M) equals 1, otherwise it is 0. Under conditions where the populations are below the utility thresholds, the utility is 0, and the reward from harvesting in those conditions is 0.

Monitoring Plans

Monitoring is integral to an adaptive management plan (Lyons et al. 2008). Horseshoe crab male, female, and pre-breeder abundances were monitored from an offshore trawl survey (Hata and Hallerman 2009). Though there are some efforts to tag crabs and release them, too small a proportion of the population is tagged (Merritt 2015), and recapture efforts are insufficient to estimate abundance in a mark-recapture model (Williams et al. 2002). The offshore trawl survey assumes constant catchability and representative sampling of the Delaware Bay population, and the adaptive management plan encouraged research and assessment of the trawl survey methodology.

Red knot abundance was monitored using aerial survey and ground counts of foraging flocks. These counting methods risk undercounting total population because birds are constantly arriving and departing the migration stopover site, and the observable population changes daily. The adaptive management plan recommended that the monitoring protocols utilize the mark-recapture and resighting data to estimate abundance and implement a superpopulation model to estimate abundance of the flow-through population at Delaware Bay (Lyons et al. 2016). When the monitoring method was changed, it required a reassessment of the utility threshold, since the original threshold, 45,000 knots, had been set using data from aerial surveys in the early 1990s. We used both monitoring methods for 3 years, estimated a ratio of aerial counts to mark-recapture estimates, and then applied that ratio to the original threshold to get the new abundance threshold, 89,000 knots. The mark-recapture-resight data collected by Delaware and New Jersey required observers to visit each potential site at least once every 3 days during the stopover period to record marked individuals and count marked to unmarked ratios. In some years one state or the other sacrificed resighting effort, which requires a significant investment, in favor of other monitoring and research projects on the bird populations, and spatial and temporal coverage suffered. The AMWG continues to communicate with the field crews in each state to ensure the monitoring efforts proceed as planned.

Decision Implementation

The AMWG reported to the technical committees of the ASMFC on the objective statements, models, utility functions, and other components of the decision analysis as the setup progressed (fig. 24.1). After the draft adaptive management plan was finalized, the ASMFC convened a panel to review the plan (Jones et al. 2009). ASMFC requires peer review of all species stock assessments. The review panel suggested revisions and further analyses to consider but suggested those potential changes be evaluated and implemented as part of a double-loop process (e.g., Williams et al. 2007).

In 2011, the ASMFC HSC management board voted to base their annual harvest decision on the adaptive management plan. This marked the first time the board had formally adopted a plan with explicit management objectives for both the harvested species and a non-target species. The management board has used the adaptive management plan for decision making since 2013.

A significant challenge for the implementation of the adaptive management plan was maintaining a commitment to monitoring the abundance of the species. Funds for monitoring have been under threat of withdrawal or reallocation to other priorities. Funding for the HSC trawl survey was withdrawn in 2013–2015. As an alternative, we employed data from other HSC surveys, which sampled a smaller area than the offshore trawl survey, and fishery-independent surveys targeting benthic species other than HSC. Monitoring to support adaptive management requires a consistent institutional commitment.

The Double Loop Process

When the adaptive management plan was initially adopted by the ASMFC, the management board and the stakeholders in the technical committees called for a review of the plan (i.e., the double-loop process) within 4 years of initiation and at regular intervals thereafter. The double-loop process is detailed by Williams et al. (2007) as a way for decision makers and stakeholders to occasionally revisit the setup phase so that essential components of the decision analysis (e.g. models, utility functions, etc.) can be reviewed and revised (fig. 24.1). Establishing specified regular intervals for the double-loop evaluation was beneficial to our process because stakeholders could see that controversial aspects of the management plan could be revisited in the future, while at the same time preventing calls for a double-loop evaluation for advocacy reasons. The double-loop reevaluation is intended to be an objective reassessment of the management plan to update decision components with new information and perspectives. However, the process could be used to manipulate the decision analysis if done prematurely and for the wrong reasons, e.g., stakeholders prematurely calling for a reassessment of candidate models because they don't like the way model weights are evolving from monitoring data. Double-loop learning is described in general principles in Williams et al. (2007) but has seldom been implemented; to our knowledge, there is no other guidance on the process.

Setting up the decision framework required the stakeholders to work through controversial issues, and revisiting components of the HSC harvest decision framework had the potential to reopen those controversies and conflicts. Thus, before beginning the double-loop process we established working agreements among all participants that if consensus could not be reached on revisions to any component of the decision analysis, we would default to the existing component agreed to during the setup phase. We employed a deliberative, and where possible analytical, process to select options for changes to the decision analysis framework, and we again used the consensus-building approach, allowing any participant in the AMWG to propose changes or revisions as we went through each component. Parsimony, process transparency, accuracy of predictions, value of information, and alignment with changes in values/objectives/problem framing were the principles and primary modes of assessment for proposed changes to the framework. We reviewed the predictive models, harvest packages, and objective functions. We also evaluated inclusion of the biomedical take in the adaptive management plan.

Model Set Assessment

We began by considering revisions to the set of alternative models, based on work by Smith et al. (2013) showing that adaptive learning would proceed slowly under the current set of models. Further, McGowan (2015) used a Bayesian model weight updating analysis to compare competing models, including one linking knot productivity to arctic precipitation cycles and one linking knot survival to HSC spawning. The competing models were fit to observed population count data from the Delaware Bay stopover site. The analysis found that the arctic-linked productivity model weight was approximately equal to the Delaware Bay–linked survival model. The AMWG believed an evaluation of the model set, perhaps including building new models and executing new data analysis, would take up to 1.5 years to complete. The ASMFC management board desired a faster timeline for the double-loop review, so we moved forward with evaluation of other components of the decision analysis. Thereafter, our double-loop assessment focused on 3 components: (1) the objectives and the utility functions, (2) the harvest packages, and (3) the inclusion of the biomedical take into the plan (table 24.1).

Objectives and Utility Functions

We reviewed 4 possible revisions to the objectives and utility functions: (1) elimination of the sex ratio constraint on male harvest, (2) elimination of the doubling of female utility in the reward function, (3) change from a knife-edged (step) function to a sloped function for the HSC female abundance constraint on female harvest, and (4) decrease of the female utility threshold from 80% to 50% of carrying capacity. Eliminating the sex ratio constraints on male HSC harvest utility was proposed because the constraint was redundant with the sex ratio effect on fertility in the HSC projection model (McGowan et al. 2011b). We compared optimization of harvest packages with and without the sex ratio constraint on male harvest. If the frequency of harvest package selection had been similar regardless of the sex ratio constraint, then the utility constraint would be redundant with other model components. The comparison found important differences. For example, in the absence of the sex ratio constraint, a 500,000 male-only harvest (policy 3) was recommended; otherwise, a full moratorium (policy 1) was recommended (table 24.2). Generally speaking, without the sex ratio constraint, male HSC harvest was more liberal but female HSC harvest remained unchanged (table 24.2). Because the removal of the sex ratio constraint resulted in more liberal male harvest, the utility function from the setup phase was retained.

We used the same analytical approach to assess the impact of doubling the female harvest utility in the reward function on the optimal harvest strategy. In the setup phase, the value for female utility was set as twice the male utility to reflect market conditions. However, there was concern that the increased value for females override conservation considerations. When comparing the optimal harvest policy with and without the doubling, there was almost no change in the optimal harvest actions (table 24.3). The default utility function was retained because removal of the multiplier did not appreciably affect harvest policy, and higher female utility is logically consistent with market pricing.

Assessments for the other 2 proposed changes to the utility functions were based on discussions and deliberations of the AMWG. We considered converting the knife-edge utility function on female harvest (i.e., the threshold utility function where $u=0$ below

Table 24.1. **Components of the horseshoe crab and red knot adaptive management plan evaluated during the double-loop process**

Component	Setup phase default	Concern	Double-loop resolution
Utility function	1. Included a sex ratio constraint on male utility to guard against overharvesting males.	1. Sex ratio constraint is redundant with fertility function in the system models.	1. Removal of sex ratio constraint caused more liberal male harvest, thus the default constraint in the utility function was retained.
	2. Female utility twice that of males to reflect market pricing.	2. Higher female utility overrides conservation considerations.	2. Higher female utility logically consistent with market pricing and removal does not have appreciable effect on harvest package selection, thus the default utility was retained.
	3. Included a knife-edged constraint at 80% of carrying capacity on female utility to ensure high abundance associated with egg availability.	3. Sloped constraint recommended by peer review panel.	3. Simulations indicated no effect of change to sloped constraint, thus the default knife-edged constraint was retained. The 80% of carrying capacity threshold was consistent with the mechanism causing egg availability, thus the default threshold was retained.
Harvest packages	1. 5 harvest packages	1. Existing packages too restrictive.	1. Given current population levels, more liberal harvest was unlikely to be recommended in the near future, thus the default harvest packages were retained.
System models	1. Model set includes no effect, moderate effects, and strong effects between HSC and red knots.	1. The no-effect model does not realistically represent how red knot dynamics could be decoupled from HSC dynamics.	1. Although candidate models show support from observations, insufficient time prevented a full evaluation, thus the default model set was retained.
	2. Biomedical harvest not included.	2. Absence of biomedical harvest ignores important source of mortality.	2. Considered 2 approaches to include biomedical harvest: one to make biomedical mortality implicit within the projection models and the other to make biomedical harvest explicit to the harvest quota. Because of the principle of explicitness, adding biomedical harvest to the quota was recommended and is under consideration by the ASMFC board.

the threshold and $u=1$ above it) to a sloped function that gradually increased utility from 0 to 1. The peer review on the original adaptive management plan (Jones et al. 2009) recommended this change to the utility functions, and Smith et al. (2013) explored the consequences of using a sloped utility function in a simulation study. Smith et al. (2013) concluded that the sloped utility function would not increase female harvest rates and would not benefit the rate of learning in the adaptive management program. The AMWG concluded that changing the utility function would not benefit the plan and decided to retain the knife-edge (step) function.

The AMWG also discussed lowering the HSC abundance threshold from 80% to 50% of carrying capacity on the grounds that fisheries management models typically set harvest quotas to maintain populations at 50% of carrying capacity to maximize

Table 24.2. Changes in optimal harvest policy under the existing utility functions (status quo) and a modified utility function that does not include a sex-ratio constraint on the utility of male HSC harvest

		Harvest policy without sex ratio constraint (%)					
		1	2	3	4	5	Total
Harvest policy (status quo)	1	31.3	0.5	56.7	6.4	5.0	12,392
	2		8.6	91.4			397
	3			100.0			40,301
	4	0.9		42.5	50.4	6.2	3,622
	5			2.3	0.1	97.5	113,172

Note: Each cell in the table shows the percentage of outcomes that transitioned from the status quo harvest policy (on the left) to each of the harvest policies (columns) after removing the sex-ratio constraint. Total number of harvest policies is shown in the last column. Harvest policy numbers refer to specific packages: 1 = 0 males, 0 females; 2 = 250K males, 0 females; 3 = 500K males, 0 females; 4 = 280K males, 140K females; 5 = 420K males, 210K females.

Table 24.3. Assessing the changes in optimal actions under the existing utility functions (status quo) and a utility function without a doubling of utility for female harvest

		Harvest policy after removing $2 \times H^F$ from utility function (%)					
		1	2	3	4	5	Total
Harvest policy (status quo)	1	100					12,392
	2		91.4	8.6			397
	3			100			40,301
	4			5.3	94.7		3,622
	5	<0.1		3.2		96.8	113,172

Note: Each cell in the table shows the percentage of outcomes that transitioned from the status quo harvest policy (on the left) to each of the possible harvest policies (columns) when using the modified utility function. Total number of harvest policies is shown in the last column. Harvest policy numbers refer to specific packages: 1 = 0 males, 0 females; 2 = 250K males, 0 females; 3 = 500K males, 0 females; 4 = 280K males, 140K females; 5 = 420K males, 210K females.

population growth potential and therefore harvest yields (e.g., Francis 1974). We discussed this proposal but reminded the AMWG that the original purpose of setting the threshold at 80% was because this was a 2-species decision process, so traditional single-species fisheries models and approaches to managing the stock may not be effective.

Alternative Actions

Participants proposed revising the set of alternative harvest packages, primarily to add opportunities for harvesting female HSC. After 3 seasons under the adaptive management plan, female harvest was prohibited in the Delaware Bay region, and some stakeholders and ASMFC board members were dissatisfied. Alternative harvest packages with female harvest were considered but deemed to have a virtually 0 chance of selection in the near term given current population estimates (~4–7 million female HSC and ~45,000 knots) in relation to utility thresholds (11.2 million female HSC and 89,000 knots), and therefore AMWG recommended no change.

Biomedical Take

During the setup phase, the possible impacts of biomedical bleeding activities on HSC stocks in the Delaware Bay were largely excluded from our process because the ASMFC does not regulate biomedical collection of HSC blood (hemolymph), and because confidentiality agreements between the biomedical companies and ASMFC made the appropriate data unavailable. The confidentiality agreements state that when there are less than 4 companies operating in a region, company-specific harvest data will not be publicly released. We devised 2 basic approaches to incorporate biomedical take into the decision analysis: (1) implicitly insert biomedical take as an additional mortality in the horseshoe crab projection model and (2) explicitly incorporate biomedical take into harvest packages.

The projection model for HSC does not explicitly account for biomedical-related mortality in any of the projection equations (McGowan et al. 2011b; Sweka et al. 2007). A parameter for biomedical mortality could be added the model. In the original models, abundance projections were calculated for each sex separately as follows:

$$N_{t+1}^{AF} = N_t^J \times T_t^{J,AF} + N_t^P \times S_t^{PF} + ((N_t^{AF} - H_t^F) \times S_t^{AF})$$

$$N_{t+1}^{AM} = N_t^J \times T_t^{J,AM} + N_t^P \times S_t^{PM} + ((N_t^{AM} - H_t^M) \times S_t^{AM})$$

where N is abundance for adult (A), juveniles (J) or prebreeding (P) females (F) or males (M), $T_t^{J,AF}$ is the transition probability from juvenile to the adult female (or male) age class, S is the annual survival rate for prebreeders or adults, and H is the harvest amount. To incorporate biomedical take (bleeding-induced mortality) we would add an additional mortality term to the N_t^{AF} and the N_t^{AM} components of the equation to account for the bled crabs that die:

$$N_{t+1}^{AF} = N_t^J \times T_t^{J,AF} + N_t^P \times S_t^{PF} + ((N_t^{AF} - H_t^F - (N_t^{BF} \times m_t^B)) \times S_t^{AF})$$

$$N_{t+1}^{AM} = N_t^J \times T_t^{J,AM} + N_t^P \times S_t^{PM} + ((N_t^{AM} - H_t^M - (N_t^{BM} \times m_t^B)) \times S_t^{AM})$$

where N_t^{B-} is the number of bled female (BF) or male (BM) HSC and m_t^B is the mortality rate for bled HSC. This approach would require access to biomedical industry data to estimate the proportion of the population that are bled each year, and literature values would be used to estimate the mortality rates (m) due to bleeding (Leschen and Correia 2010).

The second approach sought to account for the bleeding mortality along with any harvest into the harvest quota. This required us to estimate the average biomedical collection rates for recent years with available data, multiply that by estimated mortality due to bleeding (same as m_t^B above and estimated from literature), and add these take rates to the existing packages. Acknowledging that ASMFC did not have the authority to regulate biomedical take, the total quota was maintained by reducing the allocation to bait harvest; under a bait harvest moratorium, biomedical take would still occur. For example, harvest package 3—which currently allows no female HSC harvest and 500,000 male HSC harvest—would allot female HSC take to the biomedical industry and cap male harvest at 500,000; that is, bait harvest would be restricted by the allocation to the biomedical industry. With this approach, the H_t^- terms in the model above would be adjusted to account for biomedical take allowances.

After much debate and consideration of these 2 approaches, the AMWG decided that either approach effectively accomplished the goal of incorporating biomedical take into the decision analysis, but most core group members preferred to account for biomedical take explicitly in the harvest quota. In general, this option was believed to be more easily understood and transparent.

Discussion

Decision analysis helped ASMFC untangle a persistently intractable problem. Previous management efforts had been marred by issue advocacy and mired in political maneuvering and litigation. Decision analysis shifted the discussion to objectives, data, science, and models and did not force any stakeholder to concede their uncertainties, give up their preferred management action, or drop their goals. On the contrary, decision analysis only works effectively if all sides bring their perspective to the table and actively participate in the process (Gregory and Keeney 2002). Our embrace of ecological uncertainty and competing objectives, and our commitment to transparency of the process through stakeholder meeting and external peer review, greatly enhanced trust in the outcome of the decision analysis.

To our knowledge, this is the first published account of an attempt to implement the double-loop process. The double-loop process offered an opportunity to reconsider past decision framing and process. In fact, during the initial setup phase, we often assuaged stakeholder concerns and moved past challenging decision points by reminding participants of the prospect for revisiting those components during the double-loop process. It is possible that the double-loop process also opens the window for process manipulation to achieve a preferred management action. Key to our success at avoiding this potential trap was the prior agreement among participants that if consensus on revisions could not be reached, we would retain the existing components agreed upon during the initial setup phase. A commitment to transparency, parsimony, and value-focused thinking helped ensure that the process was not manipulated.

Challenges remain with maintaining sufficient resources and commitment to the adaptive management plan. For example, maintaining a commitment to effective and high-quality monitoring is needed, either through funding for the offshore HSC trawl survey or a commitment to monitoring protocols by the field crews in lieu of other research studies in the field. The relatively stress-free decision process in place after the completion of our decision analysis and adaptive management plan has possibly led to complacency among stakeholders. The ASMFC must keep the AMWG engaged and the stakeholder groups apprised of the ongoing issues. The ASMFC must also maintain a commitment to transparency when making any changes or updates to the adaptive management plan to prevent erosion of stakeholder support.

ACKNOWLEDGMENTS

The authors thank the many participants in the development and implementation of the adaptive management plan, in particular, the members of the Adaptive Resource Management Workgroup: Jim Nichols, John Sweka, Greg Breese, Kevin Kalasz, Larry Niles, Rich Wong and Jeff Brust. Atlantic States Marine Fisheries Commission provided unwavering support for the adaptive management effort through their management board and committees. The authors thank Mike Millard and Ken Williams for providing helpful comments on an earlier draft of this chapter. Any use of trade, firm, or product names is for descriptive purposes only and does not imply endorsement by the US government.

LITERATURE CITED

Baker AJ, Gonzalez PM, Piersma T, Niles LJ, do Nascimento I, Atkinson PW, Clark NA, Minton CD, Peck MK, Aarts G. 2004. Rapid population decline in red knots: fitness consequences of decreased refueling rates and late arrival in Delaware Bay. *Proceedings of Royal Society London B* 271:875–882.

Francis RC. 1974. Relationship of fishing mortality to natural mortality at the level of maximum sustainable yield under the logistic stock production model. *Journal of the Fisheries Board of Canada* 31:1539–42.

Gregory RS, Keeney RL. 2002. Making smarter environmental management decisions. *Journal of the American Water Resources Association* 38:1601–1612.

Hata D, Hallerman E. 2009. Evaluation of the coastal horseshoe crab trawl survey for estimating juvenile recruitment and mortality. Supplemental report to the Atlantic States Marine Fisheries Commission Horseshoe Crab Technical Committee. Blacksburg, VA: Virginia Tech.

Jones M, Chen Y, Nol E, Tremblay J. 2009. Terms of reference and advisory report to the Horseshoe Crab Stock Assessment Peer Review. Atlantic States Marine Fisheries Commission Stock Assessment Report No. 09-02. www.asmfc.org/uploads/file /2009HorseshoeCrabAdvisoryReport.pdf.

Leschen AS, Correia SJ. 2010. Mortality in female horseshoecrabs (*Limulus polyphemus*) from biomedical bleeding and handling: implications for fisheries management. *Marine and Freshwater Behaviour and Physiology* 43:135–147.

Lubow B. 2001. *Adaptive Stochastic Dynamic Programming (ASDP): Version 3.2*. Fort Collins, CO: Colorado State University.

Lyons JE, Kendall WL, Royle JA, Converse SJ, Andres BA, Buchanan JB. 2016. Population size and stopover duration estimation using mark-resight data and Bayesian analysis of a superpopulation model. *Biometrics* 72:262–271.

Lyons JE, Runge MC, Laskowski HP, Kendall WL. 2008. Monitoring in the context of structured decision-making and adaptive management. *Journal of Wildlife Management* 72:1683–1692.

Martin J, Runge MC, Nichols JD, Lubow BC, Kendall WL. 2009. Structured decision making as a conceptual framework to identify thresholds for conservation and management. *Ecological Applications* 19:1079–1090.

McGowan CP. 2015. Comparing models of red knot population dynamics. *The Condor* 117:494–502.

McGowan CP, Hines JE, Nichols JD, Lyons JE, Smith DR, Kalasz KS, Niles LJ, Dey AD, Clark NA, Atkinson PW, Minton CDT, Kendall W. 2011a. Demographic consequences of migratory stopover: linking red knot survival to horseshoe crab spawning abundance. *Ecosphere* 2:art 69.

McGowan CP, Lyons JE, Smith DR. 2015a. Developing objectives with multiple stakeholders: adaptive management of horseshoe crabs and red knots in the Delaware Bay. *Environmental Management* 55:972–982.

McGowan CP, Smith DR, Nichols JD, Lyons JE, Sweka J, Kalasz K, Niles LJ, Wong R, Brust J, Davis M, Spear B. 2015b. Implementation of a framework for multi-species, multi-objective adaptive management in Delaware Bay. *Biological Conservation* 191:759–69.

McGowan CP, Smith DR, Sweka JA, Martin J, Nichols JD, Wong R, Lyons JE, Niles LJ, Kalasz K, Brust J, Klopfer M, Spear B. 2011b. Multispecies model for adaptive management of horseshoe crabs and red knots in Delaware Bay. *Natural Resource Modeling* 24:117–156.

Merritt EM. 2015. Evaluating methods for estimating Delaware Bay *Limulus polyphemus* abundance. MS thesis. Auburn, AL: Auburn University.

Millard MJ, Sweka JA, McGowan CP, Smith DR. 2015. Assessment and management of North American horseshoe crab populations, with emphasis on a multispecies framework for Delaware Bay, USA populations. Pages 407–431 in Carmichael RH, Botton ML, Shin PKS, Cheung SG, eds. *Changing Global Perspectives on Horseshoe Crab Biology, Conservation and Management*. New York: Springer.

Niles LJ, Bart J, Sitters HP, Dey AD, Clark KE, Atkinson PW, Baker AJ, Bennett KA, Kalasz KS, Clark NA, Clark J, Gillings S, Gates AS, Gonzalez PM, Hernandez DE, Minton CDT, Morrison RIG, Porter RR, Ross RK, Veitch CR. 2009. Effects of horseshoe crab harvest in Delaware Bay on red knots: are harvest restrictions working? *BioScience* 59:153–164.

Niles LJ, Sitters HP, Dey AD, Atkinson PW, Baker AJ, Bennett KA, Carmona R, Clark KE, Clark NA, Espoza C, Gonzalez PM, Harrington BA, Hernandez DE, Kalasz KS, Lathrop RG, Matus RN, Minton CDT, Morrison RIG, Peck MK, Pitts W, Robertson RA, Serrano IL. 2008. Status of the red knot (*Calidris canutus rufa*) in the Western Hemisphere. *Studies in Avian Biology* 36.

Smith DR, Brockmann HJ, Beekey MA, King TL, Millard MJ, Zaldívar-Rae J. 2017. Conservation status of the American horseshoe crab (*Limulus polyphemus*): a regional assessment. *Reviews in Fish Biology and Fisheries* 27:135–175.

Smith DR, Mandt MT, Macdonald PDM. 2009. Proximate causes of sexual size dimorphism in horseshoe crabs (*Limulus polyphemus*) of the Delaware Bay. *Journal of Shellfish Research* 28:405–417.

Smith DR, McGowan CP, Daily JP, Nichols JD, Sweka JA, Lyons JE. 2013. Evaluating a multispecies adaptive management framework: must uncertainty impede effective decision-making? *Journal of Applied Ecology* 50:1431–1440.

Sweka JA, Smith DR, Millard MJ. 2007. An age-structured population model for horseshoe crabs in the Delaware Bay area to assess harvest and egg availability for shorebirds. *Estuaries and Coasts* 30:277–286.

Williams BK, Nichols JD, Conroy MJ. 2002. *Analysis and Management of Animal Populations*. New York: Academic Press.

Williams BK, Szaro RC, Shapiro CD. 2007. *Adaptive Management: The US Department of the Interior Technical Guide*. Washington, DC: Adaptive Management Working Group, US Department of the Interior.

Index

Page numbers in bold indicate full case studies.